KU-536-535

The Environmental Impact of Railways

T.G. CARPENTER

JOHN WILEY & SONS
Chichester · New York · Brisbane · Toronto · Singapore

Copyright © 1994 T.G. Carpenter

Published 1994 by John Wiley & Sons Ltd,
Baffins Lane, Chichester,
West Sussex PO19 1UD, England
Telephone National Chichester (01243) 779777
International +44 1243 779777

Reprinted June 1995

Reprinted October 1996

All rights reserved.

No part of this book may be reproduced by any means, or transmitted, or translated into a machine language without the written permission of the publisher.

Other Wiley Editorial Offices

John Wiley & Sons, Inc., 605 Third Avenue,
New York, NY 10158-0012, USA

Jacaranda Wiley Ltd, 33 Park Road, Milton,
Queensland 4064, Australia

John Wiley & Sons (Canada) Ltd, 22 Worcester Road,
Rexdale, Ontario M9W 1L1, Canada

John Wiley & Sons (SEA) Pte Ltd, 37 Jalan Pemimpin #05-04,
Block B, Union Industrial Building, Singapore 2057

Library of Congress Cataloging-in-Publication Data
Carpenter, T. G.
    The environmental impact of railways/T.G. Carpenter.
        p. cm.
    Includes bibliographical references and index.
    ISBN 0-471-94828-4
    1. Railroads—Environmental aspects. 2. Railroads—Planning.
I. Title.
TD 195.R33C37 1994   93–46722
363.73′1—dc20   CIP

British Library Cataloguing in Publication Data
A catalogue record for this book is available from the British Library

ISBN 0-471-94828-4

Typeset in 10/12pt Times by Acorn Bookwork, Salisbury, Wiltshire
Printed and bound in Great Britain by Antony Rowe Ltd, Chippenham, Wiltshire

Contents

Preface   vii

Part 1   RAILWAYS AND PLANNING   1

1   Introduction   3

2   Environmental Planning   11

3   Railway Planning   28

4   Passenger Traffic   58

5   Freight   92

Part II   IMPACTS ON PEOPLE   111

6   Social Impacts and Public Perception   113

7   Noise and Vibration   129

8   Pollution   165

9   Visual Impacts   184

10   Construction   199

Part III   IMPACTS ON RESOURCES   215

11   Resource Use and Route Selection   217

12   Residential, Commercial and Productive Land   238

13   Nature Conservation   247

14   Heritage and Amenity   270

15   Railways in Scenic Landscape   289

16   Environmental Evaluation of Land Resources   323

Part IV   PLANNING FOR THE TWENTY-FIRST CENTURY   341

17   Environmental Rail Transport Solutions   343

References   363

Index   369

Preface

Two developments affecting transport have characterised the last decades of the twentieth century. One is a revival of train travel, particularly through the introduction of very high speed inter-city services. The other is awareness of the fragility of the earth's remaining natural resources; this has led to requirements for environmental assessment of all types of project including those providing transport infrastructure.

Recent advances in rail travel have been a comparatively minor feature among the sweeping changes which have taken place in transport as a whole. Aeroplanes have revolutionised long distance passenger travel. Road transport of people and freight has expanded immensely. New motorways or airport runway extensions compete for land space with industrial and housing demands and against pleas for sustainment of ecological reserves and scenic amenities. However, railways have been with us for 150 years and have already faced most of the environmental problems. Competition for land has continued to stiffen and higher speeds have raised some new issues; but the basic structure of railways and operating characteristics of trains remain. They provide a clear model of the impacts today of works which were implemented a century or more ago.

The first purpose of this book is therefore to inform environmental specialists and land use planners about railways.

The second purpose is to inform railway planners and designers about the environmental impacts which they may cause and to suggest how they can design optimal solutions.

In choosing between alternative routes and track alignments, the high speeds and loads which fast, commercially attractive train services require have to be balanced against construction and environmental costs. Evaluation of the latter in transport projects is an immature science and often leads to controversial decisions.

Therefore a third purpose of this book is to suggest ideas to economists and decision-makers concerning valuation of the land and other resources affected by transport infrastructure developments. This may help to devise means of choosing between solutions solving short-term problems, solutions which may be cost effective for the entrepreneur, and solutions which best sustain fragile natural resources.

No attempt to describe railway planning and the full range of environmental issues in one volume would have been possible without a great deal of help and advice. Martin Hyndman's keen interest launched the project. Elizabeth Carpenter spent many hours checking the clarity and composition of each section, often through several stages of adaptation. Stephen Ellis initiated me into train noise, undertook the measurements in all weathers and put me in contact with other enthusiasts.

I am greatly indebted also to those experts who reviewed particular chapters including:

- Jo Hughes for detailed comments on nature conservation in Chapter 13;
- the late Chris Stevens who reviewed Chapter 15 and provided ideas and encouragement on the fascinating subject of scenery;
- Graham Parry of DNV Technica who gave me detailed advice on noise prediction and other aspects for Chapter 7;
- G.J. Boschloo of The Netherlands Ministry of Housing, Physical Planning and the Environment who provided copious information and a clear explanation of Dutch environmental impact assessment procedures;
- Richard Hope who reviewed the sections on public perception;
- Dennis Drake who commented on several chapters both as a railwayman and as a general reader;
- Andrew Jarvis who threw light on modern approaches to environmental evaluation (Chapter 16);
- Brian Green who scrutinised the railway planning and engineering sections;
- Ian Spiers who advised me on parallel issues in road planning;
- Lawrence Gardiner who reviewed my draft dealing with air pollution in Chapter 8;
- Christopher Savage, who kindly contributed the annex to Chapter 6 describing railways in the Japanese environment.

Tom Carpenter
1994

Part I

RAILWAYS AND PLANNING

1 Introduction

1.1 HISTORICAL BACKGROUND

Most railways were built and many of their features were developed in the nineteenth century. Throughout the twentieth century new lines have been constructed to serve mines and other developments in the more remote parts of the world whilst expansion of networks in developing countries has continued. However, in Western Europe long distance passenger and short distance freight suffered a severe decline up to at least the 1960s. Thereafter much faster inter-city services started to reverse part of the decline.

Mechanically-operated railways were a development of the latter part of the industrial revolution. Taking over from horse-drawn wagons and canal barges, the invention of the steam locomotive led to railway domination in the first half of the nineteenth century. Soon after 1800 the principles developed by Newcomen and Watt for stationary engines were applied to locomotives; these were first used for haulage by rail from Northumbrian collieries and then achieved a major breakthrough in the 'Rocket' for the Liverpool and Manchester Railway in 1829. This engine proved the value of multi-tube boilers through which the hot gasses from the fire were drawn by the exhaust steam from the cylinders into the chimney. The principle was applied to locomotives throughout the world thereafter until the end of steam.

By 1850 locomotives with a power equivalent of up to 100 horsepower were hauling trains of about 100 tons and achieving 50 mph on test runs. Within the next 80 years such speeds were regularly exceeded by trains five times as heavy.

By 1900 the rail networks of industrial countries were substantially complete. The growth of national economies had become dependent upon transport. Ships continued to trade internationally but inland transport was performed predominantly by railways.

The operation of trains, their significance to people and the impact of railway infrastructure on land use were therefore well evident 100 years ago.

The twentieth century saw a continuing expansion in city commuter rail services as well as the introduction of electric traction on all suburban and many main lines and of diesel locomotives elsewhere; but medium and long distance passenger business was attracted by more flexible road vehicles and faster aeroplanes. Rail freight transport declined where it was no longer economic over the shorter hauls common on an island such as Britain. The pattern of freight movement also changed, particularly when coal lost its position as the dominant industrial and domestic fuel and when unit loads of mixed or high value cargoes became common.

By the end of the 1960s the British railway network extended for little more than half the 1914 peak of 20 000 miles.

1.2 CURRENT RENAISSANCE

Recent strains on the capacity of road and air transport as well as environmental objections have led to greater use of certain rail services.

On routes between cities provision of motorways and bypasses has accommodated much of the increased road traffic. But this same traffic causes congestion in the cities where there is no space for more roads or parking. Hence a need arose both for convenient fast rail transport between city centres and for improved or extended city metro rail services. Eventually there must be a limit also on further main road capacity, if only to reflect the finite vehicle capacity of the towns they serve. Together with the wider introduction of incentives to use rail transport, this could stimulate further transfer of medium distance traffic from the roads.

Now high speed trains increasingly compete with airways. Rapid passenger transport systems are being provided for provincial cities and to international airports. Some lines are being newly built; elsewhere there is still scope for intensifying or accelerating train services on existing routes.

The opening of the Channel Tunnel and the Common European Market have provided opportunities for a recovery of rail freight in Britain and across Europe. A European Summit meeting in December 1993 committed a 'task group' to implementation of £60 billion of investment in priority transport infrastructure projects. Less than 15 per cent of this amount is for road links but 65 per cent is for 'rail links including combined transport' (Parker 1993). Further east, countries like Poland do not intend to be left behind in speeding up long distance rail journeys.

In China, India, Russia and southern Africa new lines primarily for freight are still being constructed and planned; elsewhere existing lines are being upgraded for modern rolling stock or to carry new commodities.

1.3 ENVIRONMENTAL AWARENESS

Some of the public, their politicians and planners are aware of the railway revival. But they are also aware of the various impacts of industry and transport on the quality of life and on resources. New development of any type has to take account of a range of popular reactions as well as to satisfy statutory evaluation and approval processes.

Although the concept of 'environment' was not then identified, people in

1850 were already aware and writing about the noise and emissions of trains and the disruption of railway construction. Land acquisition and intrusion by railways had been a source of contention since before 1830. The word 'ecology' was coined by the German zoologist Ernst Haekel in 1878 by which time painters recognised railways as part of the landscape.

Public environmental awareness as such is a late-twentieth-century phenomenon. It has arisen through such factors as:

- publicity about levels of pollution and the unsustainable use of resources in energy, transport and industrial processes;
- a sometimes presumed right to a quality of life involving freedom from noise, inconvenience or even visual intrusion.

As we have seen, railways have been around for a long time. There has been some change in the pattern of traffic as motive power has changed from steam to diesel or electric, some of it now very powerful. But the basic form of trains, track and infrastructure has not changed during the twentieth century. The nature of railway-related environmental impacts is also well established. Thus railways provide an ideal model for environmental appraisal and planning, by modern techniques, of well-understood operations. Such appraisal may even be valuable in analysis of more recent industrial or transport processes to which the environment may be more sensitive.

Accordingly Part I of this book describes:

- the process of environmentally-sensitive planning which has to be applied to all forms of development (in Chapter 2);
- the particular features and functions of railways and trains, and planning for them in modern environments (Chapters 3 to 5).

1.4 ENVIRONMENTAL IMPACTS

Environment is surroundings. A railway is built to move passengers and goods but also has effects on its surroundings and on lineside activity. *Effects* are the physical changes that trains and railways cause. *Impacts* are the noticeable consequences of these changes.

Impacts can be classified as those affecting people or animals, those using up resources and those relating to people and resources.

People can be affected by transport systems which pass through or near the towns or villages where they live. Residents are often resentful of an intrusion into their relatively undisturbed surroundings, whether they live in pleasant suburbs or in country retreats. The intrusion is difficult to define but is not far different from that felt by nineteenth-century landowers who resisted the first railways. In those days they were particularly concerned about possible physical effects of the fire-breathing locomotives; but they

Table 1.1. Environmental impacts of railways

Economic and transport benefits/impacts	Direct impacts: faster, convenient rail services for passengers and freight	Secondary impacts: effects on other transport systems	*Part I* Chapter 3 Chapter 4 Chapter 5
Impacts on people			*Part II*
Social impacts	Jobs, housing, facilities	Equity/inequity Public perception Public participation	Chapter 6
Noise and vibration	Disturbance at lineside and near terminals	Property values Visual impacts of noise barriers	Chapter 7
Air and water pollution	Diesel engines Accident risks	Power stations Changes to drainage	Chapter 8
Visual impacts	Obstruction Intrusion	Views from trains	Chapter 9
Construction impacts	Disturbance by dust, noise and traffic	Disposal of spoil Transport of materials	Chapter 10

Impacts on resources			Part III
Energy use and climatic change	Depends on efficient use of fuels	Depends on sources of electric power	(Chapter 8)
Material assets	Manufacture of rolling stock and equipment	Disposal of old equipment Land reclamation	Section 11.1
Land resources General use	Land-take: in long strips of undervalued resources	Partition or severance of:	
Residential	Property loss	– communities, roads	Chapter 11
		– factory complexes	Chapter 16
Commercial	Production loss	– farms	Chapter 12
Agriculture	Production loss		
Nature conservation	Loss/disturbance of habitat	– wildlife corridors	Chapter 13
Cultural heritage	Loss of historic features	– historic units or related groups	Chapter 14
Amenity	Land-take	– paths, golf links, playing fields, etc.	Chapter 15
Scenic landscape	Intrusion; modifications to features	Effects on distant active land-forms	

were indignant also at the intrusion by construction workers and then of passengers viewing their private property. Nowadays noise is the most definable disturbance and this can be described, measured and predicted. Interference with access can also be examined methodically but assessment of visual intrusion is bound to be more subjective.

Animals are unable to offer opinions on the sight and sound of trains although farm stock appear unconcerned when seen from any high speed train. Wildlife thrives on enclosed railway land. Kestrels intently hunt, rabbits burrow and wildflowers spread alongside the track.

Global sustainability is threatened by transport systems as much as by industrial processes; but the relatively efficient use of fuels or electrical energy by trains makes them compare favourably with road and air transport in use of finite resources and in influencing climatic change.

The main *resource* affected by railways is land. Near towns, transport's need for space must compete with that of industrial, commercial, residential or recreational uses. In rural areas, large tracts may be devoted to agriculture or they may be of high mineral resource, recreational or scenic value. Natural habitat is often the scarcest land and its elimination the greatest threat to wild fauna and flora.

Impacts on people and resources relate to all man's uses of land, fuels and natural materials. Scenic landscape or historic features are worth preserving for aesthetic rather than utilitarian purposes. Those that survive come to be valued more highly than land which is drained, paved or built on for economic purposes such as transport or commercial development.

Table 1.1 summarises the environmental impacts which are relevant to railway operation and construction. Part II of this book deals with the impacts *on people* caused by railways, Part III with those *on resources* and on man's use of resources.

No single environmental scientist can comprehensively appraise all the external impacts of a railway. Whoever coordinates either railway design or its environmental assessment must have a broad grasp both of railway operations and construction and of environmental issues. Detailed assessment must be undertaken by specialists in noise, ecology, historic buildings, etc., in close liaison with railway operators, civil engineers and transportation and environmental economists. Then the engineering will be environmentally sensitive and subsequent evaluation of railway development may be equitable.

1.5 WORLDWIDE ASPECTS OF RAILWAYS IN THE ENVIRONMENT

Much of the evidence and analysis in the following chapters refers to railways in Britain. This is thought appropriate for the following reasons:

- Britain was an early developer of railways and has a wide and varied network; British engineers had a strong influence on many early overseas railway systems.
- Britain is an island with a great variety of topography, scenery, population density and competing commercial activities.
- Some of Britain's main lines have been upgraded for high speeds without the great cost of providing completely new alignments.
- Particular interest and controversy have been generated by the various issues and routes examined since 1988 for a new railway from London to the Channel Tunnel; this was known *inter alia* as the Channel Tunnel Rail Link (CTRL) until 1993 when the proposed East Thames Corridor route became the Union Railway.
- The whole raison-d'être of the British Railway System is being brought into sharp focus by its privatisation; the institution generally referred to in this book as British Rail or BR no longer exists and now comprises a rail track owner and commercial operating companies.

Meanwhile, new high speed railways facing similar environmental problems have already been built in Japan, France, Germany, Spain and Italy and are being constructed or planned in several other countries. New mixed traffic routes are still being built in Asia and mineral railways in the southern hemisphere.

It would be a formidable task to gather comprehensively all the planning experience and evidence concerning environmental aspects of these new railways. Nevertheless comments and published data have been incorporated concerning a range of relevant issues in different parts of the world.

Environmental appraisal itself was initiated in the USA; it was brought formally into planning agendas in Europe through the European Commission's directive and in developing countries by the international lending agencies. Environmental impact assessment is undertaken by a range of methods; the well-defined procedure in The Netherlands is described as an example. Public consultation approaches are mentioned for France, Germany and Switzerland. These approaches have to relate to public attitudes which differ regionally—for instance between car-oriented communities in the USA and Britain, and organised urban societies in The Netherlands or Japan; an informal view of Japanese perception and policy is included.

Japan was also the country which pioneered modern high speed trains. Reference is therefore made to the still expanding *Shinkansen* ('bullet train') network as well as to publicised noise and operational information about the different generations of French TGVs and other European passenger trains. Very long trains have for many years been a feature of passenger traffic in countries like Russia and for peak time suburban services in large cities everywhere where metro stations have the necessary capacity. Efficient use of track capacity and hence of land space in mass transportation worldwide is one of the major factors to which attention must be drawn.

Freight trains can be long enough to carry ship-size tonnages, especially in those countries like Russia, China and the USA where the distances involved usually exceed the economic thresholds for road transport. Relevant information about overseas freight requirements, combined road/rail transport, and freight train operations is therefore incorporated.

In assessing the prospects for railway development in the twenty-first century, special account has to be taken of the burgeoning, increasingly transport-conscious population of the Third World. Reference is therefore made to the particular circumstances and environmental issues of Asian and African countries.

Nearly everywhere, the impact of new railway developments can be predicted by looking at the effects of existing railways in similar environments as well as by examining the particular characteristics of trains.

2 Environmental Planning

2.1 INCLUDING THE ENVIRONMENT IN TRANSPORT PLANNING

In Chapter 1 we identified the types of environmental impacts caused by railways. This chapter describes how these impacts can be taken into account in the planning of development projects such as for new, intensified or accelerated rail services.

In order properly to include environmental factors in the planning of any transport project, the following stipulations are usually necessary:

- To introduce these factors at project *conception* as explained in Section 2.2.
- To adopt a comprehensive environmental *appraisal* programme to meet all the planning objectives and to satisfy statutory requirements; this is explained in Section 2.3, whilst technical *assessment* procedures are proposed in Section 2.4.
- To fit the appraisal activities into the appropriate *democratic* and *legal* framework (Section 2.5).
- To determine how *evaluation* of a railway development, and hence its justification and approval, may include environmental factors (Section 2.6).
- To ensure that the appraisal processes are undertaken in a practical and satisfactory sequence (Section 2.7).

Environmental impact appraisal as such is a relatively new element of planning. In the past many of the impacts were already recognised although the terms used and some of the assessment techniques had not been invented. Countless problems regarding property, land, scenery, air pollution and construction disturbance had to be resolved in the 1830s and throughout the subsequent railway building boom.

There were few major rail developments in Britain in the years immediately before the introduction of environmental appraisal; but a motorway system *was* initiated. The M1 was opened in 1959, 24 years before the issue of the Department of Transport's *Manual of Environmental Appraisal*. However, visual features at least were not ignored; the M1 motorway was designed in consultation with a 'very influential' Landscape Advisory Committee (Williams 1991). Whilst that committee was suddenly abolished in

1994, forms of consultation continue to play an important part in environmentally-sensitive design of new transport projects including the Channel
Tunnel Rail Link.

Increased recognition of environmental factors has been achieved through
'green' publicity and awareness of past mistakes, through environmental legislation, such as regulations setting limits on emissions or noise, and through
environmental appraisal procedures, which bring specific issues to the attention of planners.

Environmental awareness has accelerated the introduction of public involvement in planning. Legislation is not yet effective in all areas affected by
railways; this is just as well as many of the impacts are too complex to be
constrained within arbitrary limits. But environmental appraisal has promoted planning improvements through

- study of particular causes and effects and by methodical application of
  the results in measurement and prediction of impacts;
- formulation of alternative technical solutions and methods of minimising
  adverse impacts.

Meanwhile environmental appraisal does not always play an automatic or
significant role in project evaluation. It can certainly affect route selection in
drawing attention to the most jealously-guarded land resources; it may also
establish the need for specific measures such as noise barriers or landscaping.
Environmental enhancement measures can be costed, but the costs or 'disbenefits' of environmental damage are seldom quantified. Environmental
Statements, which present the results of appraisals, are still used mainly as
supplementary descriptive information.

2.2 CONCEPTUAL JUSTIFICATION

A railway project must be justified in the first place by a particular transport
need. For example, a greatly accelerated rail service could be needed where
the present route is ill-suited to fast traffic. The proposed solution could be a
new two-track railway link along a broadly defined route corridor at an
alignment to suit the envisaged type, speed and intensity of traffic. This basic
proposal would be conceived to meet the need. But before proceeding to the
expense and time needed to plan a detailed alignment, the concept must also
be justified or modified by checking

- whether a new railway solution accords with any imposed planning *policies* and, if so,
- what *particular solution* is the most suitable.

Environmental policies and environmental suitability are fundamental to
these checks. Engineers and planners charged with preliminary or pre-

feasibility studies of projects should commence with a thorough economic, socio-economic and environmental review of the basic need and project concept.

Policies for development of railways exist in Germany, France and Switzerland but in Britain apply only to special cases, e.g. encouraging certain freight or local public transport facilities. The Channel Tunnel Rail Link was constrained initially by policy, i.e. the Channel Tunnel Act which forbids direct government subsidy.

However, the need for environmental policy assessment at the concept stage of a project was recognised by The Standing Advisory Committee on Trunk Road Assessment (SACTRA), the British Government's environmental watchdog for *roads*, whose report (1992: 93, 16.18) states that

> No scheme should be admitted into the Roads Programme until its performance against strategic policies has been assessed and reported, at least in outline.

SACTRA's findings are equally applicable to railway development or to integrated transport planning.

The acceptability of *particular solutions* should be the early concern of the promoters. Financiers and entrepreneurs sponsoring transport schemes should insist on conceptual review as a means of avoiding expensive setbacks later. An environmental appraisal could reveal any need for design changes that might otherwise delay project implementation; or an environmental audit of railway operations could foresee unacceptable performance of expensive equipment.

Many modern railway developments are devised to relieve rather than to strain the environment. But such has been the concern about many recent road, airport, housing, commercial and industrial developments that any project is a potential target for criticism at every stage.

Serious attention to transport and environmental issues in justifying project concepts may in due course lead to clearer policies. The planners and the public alike can check whether or not a proposed transport plan accords with these policies. If it does, then the concept becomes less controversial; the area of contention moves to issues related to particular routes which is where full environmental appraisal and consultation come into play.

2.3 REQUIREMENTS FOR ENVIRONMENTAL APPRAISAL

The Commission of the European Community (EC) is the highest bureaucracy engaged in devising environmental policies and in ensuring their widespread implementation. The Single European Act confirmed EC's fundamental policy objectives—to preserve, protect and improve the quality of the environment; to contribute towards protecting human health; and to ensure a prudent and rational use of natural resources. These objectives are being attained through implementation of EC's 1985 Directive 85/335/EC which has to be applied to various types of projects including long distance railway lines. It is left to EC member states to decide whether tramways or underground railways should be similarly treated. In The Netherlands, for instance, environmental impact assessment is only compulsory for construction of railways with a length of more than 5 km outside built-up areas (Netherlands Government 1988:9). In cities, traffic is evidently a matter of urban rather than transport planning. In Britain, although the EC directive was not officially implemented until 1988, a substantial environmental impact assessment meeting the draft requirements was produced in 1985 for the Channel Tunnel.

Most railway developments worldwide are likely to be affected by environmental planning requirements; if these are not national regulations then they could be the policies and procedures of international financing agencies like the World Bank or the European Bank for Reconstruction and Development. The planning activities which are required are those which constitute what is commonly known as environmental appraisal.

Environmental appraisal is the whole sequence of collecting and processing information to reach decisions of environmental consequence. It comprises a number of closely related elements:

- environmental *assessment*;
- presentation of *information* and *consultation* with interested parties;
- environmental *evaluation* to help decision-making and to establish project feasibility.

Environmental assessment encompasses the collection, analysis and production of detailed data about the predicted impact of a proposed project. The following paragraphs clarify which impacts have to be appraised and what information has to be presented, whilst the next section sets out the steps to be taken in an environmental assessment. Consultation and evaluation are introduced in Sections 2.5 and 2.6.

The *impacts* concerned are broadly defined in EC's 1985 Directive. Article 3 requires that environmental impact assessment shall cover:

the direct and indirect effects of a project on the following factors:
  - human beings, fauna and flora,
  - soil, water, air, climate and the landscape,
  - the inter-action between the factors mentioned in the first and second indents,
  - material assets and the cultural heritage.

In Figure 2.1 these effects are related to the actual impacts of railways which were identified in Chapter 1.

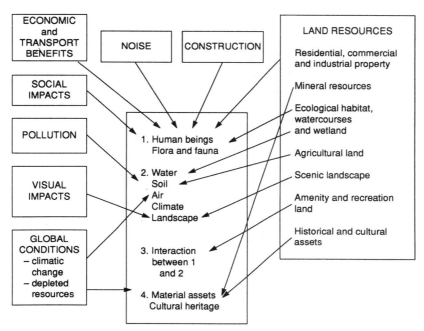

Figure 2.1. Railway impacts in EC categories. The numbered points are effects listed in EC directive 85/335/EC Article 3

Resources affected should include 'material assets' such as buildings and industrial, mining or agricultural land as well as habitat for 'flora and fauna'. Effects on land resources include those on 'soil, water, the landscape and the cultural heritage'. The latter, as well as human recreational activities, are affected by 'inter-actions' between humans and the landscape.

Impacts on 'human beings' are loss of property or of its utility, improvements or deterioration of transport services and access, pollution risks and disturbance such as noise.

The *information* required has been defined by a number of agencies. Four examples are given in Table 2.1, viz.:

1. US requirements—the classic model; note the requirement for a full description of 'the activity' (often resulting in thick or multi-volume reports) as well as specific mention of scarce resources (aimed at sustainable development).
2. EC's 1985 directive (Article 5) requirements—similar but more recent; allows much latitude in choice of format and detail but requires a nontechnical summary (enabling particular points of emphasis to be passed on to decision-makers).

Table 2.1. Statutory information about environmental impacts of projects

US environmental impact statements	EC information (to be provided by the developer)	UK Dept of Transport environmental framework	Netherlands environmental impact reports
Comprehensive technical description of activity proposed	Description of the project, comprising information on the site, design and size of the project	—	Description of activity and alternatives
Likely positive or negative impacts on natural and human environments	The data required to identify and assess the main effects which the project is to have on the environment	Effects on travellers Effects on occupiers of property Effects on users of facilities Effects on policies for conserving and enhancing the area	Decisions which will have to be made by government bodies
Description of alternative solutions and variations including alternative localities	Description of the measures envisaged in order to avoid, reduce and, if possible, remedy significant adverse effects	Effects on policies for development and transport	Existing environment conditions
			Consequences on environment
Description of all adverse impacts which cannot be avoided or minimised	—	—	Comparison of consequences of proposed activity, alternatives and no activity
Examination of scarce resources which would be irreversibly committed	—	Financial effects	Review of gaps in information
—	Non-technical summary of the information mentioned above	—	Summary for the general public
US National Environmental Policy Act 1969	Council Directive 85/337/EEC, Article 5	Manual of Environmental Appraisal, Part A (superseded in 1993)	Environmental Protection (General Provisions) Act 1989: 19

3. UK Department of Transport (1983) environmental framework for *roads*; this was superseded by requirements in line with EC's directive (Department of Transport 1993a); however, the earlier headings illustrated an approach which concentrated on how transport development affected *people* and *policies*.
4. Netherlands Government requirements for an environmental impact report; the emphasis is on assessment of serious *alternatives*.

Information can be presented in various forms and in appropriate stages; it is usually compiled eventually as an 'Environmental Statement'.

Excessive publication of paperwork may be thought an inevitable consequence of environmental appraisal. But it can be minimised by suitably structured documents:

- to provide only what is essential at *each* stage, e.g. at basic conception, at public presentation/consultation or in a formal environmental statement;
- so that essential but detailed material can be included in limited edition appendices.

2.4 THE ASSESSMENT PROCESS

The objective of environmental assessment is to allow judgements to be reached concerning the consequences of development proposals.

Judgement requires explanation as to how the environment is affected, a basis for measuring the significance of each effect and guidance as to how to evaluate the overall impact of the whole development.

Assessment can be undertaken as a separate exercise to appraise the impact of an already-defined project, for instance if there is a requirement for a completely *independent* environmental appraisal. Alternatively, it can be carried out as an *integral part* of both the conceptual planning and the detailed design stages. In either case the following steps are proposed.

1. Study the features of the project concerned and determine the nature of the existing environment ('baseline conditions'); identify the *types of impact* which are likely to be relevant and what sorts of *decisions* about them are going to have to be made ultimately; determine the *scope* of the environmental assessment.

Then undertake the next steps for *each* type of impact.

2. Determine the main issues or *implications* affecting planning and design. Which people or which resources are affected? What sort of physical or perceived impacts will result from the changes?

3. Identify or develop any specific *scientific correlations* relating cause and effect. Identify the available sources of information or methods of measurement.
4. Collect, analyse and produce *objective data* about each impact.
5. Identify or develop *standards* or thresholds of acceptable impact, degrees of severity or feasible methods of ranking effects of alternative options.
6. Predict the *magnitude* of impacts and thus determine their significance.
7. Determine the type, costs and benefits of available amelioration or *mitigation* measures.

Finally, for the whole project and for the combined economic and environmental impacts:

8. Determine how 6 and 7 can be incorporated in an *evaluation* of the project or of alternative schemes.

In bureaucratic procedures *monitoring* may mean checking that environmental assessment follows directives. In scientific terms monitoring can be systematic measurements to establish correlations or determine whether changes occur naturally or due to applied adverse or remedial measures.

2.5 THE DEMOCRATIC AND LEGAL PLANNING PROCESS

Democracy requires that the information from environmental assessments should be made available to all parties who will be affected or who can contribute to planning an acceptable outcome. EC's 1985 Directive Article 6 requires that, for any request for development consent:

- the concerned authorities should be given an opportunity to express their opinion;
- the request and Article 5 information should be made available to the public;
- the public concerned is given opportunity to express an opinion before the project is initiated.

In the context of planning new railways, democracy works through three groups of people, viz.:

1. Democratically elected government which
   - formulates policies, e.g. to encourage rail freight or to subsidise certain transport infrastructure;
   - makes laws and regulations, e.g. Channel Tunnel Act, Transport Act or specific laws for environmental protection;
   - approves or rejects major items of proposed expenditure, e.g. London's new underground lines.

2. Official (government controlled, publicly funded) or non-governmental (privately supported) technical bodies who require opportunities to comment on proposals and to participate in consultation, e.g. environmental protection agencies, heritage preservation bodies and national or local ecological groups.
3. People who will be directly affected by the proposed works.

The first of these groups acts mainly nationally, the last is purely local and the second can be either. Chapter 6 describes public consultation with particular relevance to impacts on people. It also examines the difference between local and national perception of railways. Promoters have to cope with both to gain approval for any otherwise viable project. In federal Switzerland both levels of democracy are taken so seriously that

- there have been *national referendums* on transport and environmental matters concerning, for instance, a new high speed railway line and new tunnels under the Alps;
- there is a comprehensive staged process of consensus-finding followed by *local authority negotiation* and action on property owners' objections.

In France four or sometimes even five levels of elected bodies are involved in planning decisions. French decentralisation since 1982 means that decisions are taken primarily at the bottom level, i.e. by tiny 'communes'. These tend to make their decisions for hard economic reasons, through chambers of commerce or similar associations combining the power of all communes where several are involved, rather than giving priority to the feelings of residents generally. Thus consultation at the Calais end of the Channel Tunnel mainly concerned jobs for the local work force and development of new enterprises. In Kent, at the other end, local concerns 'were primarily about the environment, quality of life and transport congestion' (Eurotunnel 1993).

In The Netherlands, public consultation takes place at two stages of environmental impact assessment (EIA). The first is at the 'scoping' stage after which guidelines for preparation of the Environmental Impact Statement (EIS) are issued. The second is public review of the EIS when it is published. Notable features of Dutch procedures are the firm scheduling (allowing about a month for each stage of public consultation), the wide range of people who may be involved and the emphasis on assessment as an aid to decision-making.

The public in The Netherlands is defined as 'every person or organisation interested in an EIA in some other way and wishing to join in the thinking during the decision-making process' (Netherlands Government 1990: 8). Besides the involvement of the general public and advisory services, a permanent EIA Commission assists in preparation of EIS guidelines and checks whether the EIS contents meet statutory requirements and follow the guide-

lines. For any particular project this is done by an EIA Commission working party composed of appropriate experts not connected with the project.
The ultimate decision-making process is also taken into account in preparing
the scope of the EIS. The latter has to be prepared in the light of earlier
government decisions and of those which have been made for the project.
The decision when presented has to give full reasons and to show how full
consideration has been given to alternatives, comments and recommendations described in the EIS.

In Europe generally, EC's Article 6 requirements are currently met by various consultation procedures reviewed in Chapter 6. The people involved
may be local government, environmental groups or private individuals.

Thus planning procedures deal with both national and local interests.
However, the force of possibly excessive local demands may inflate the cost
beyond what national interests perceive as necessary. According to whatever
law applies, approval of national projects is ultimately given or refused,
either by a parliamentary decision or by that of a member of the government or his or her department.

In Britain railways have in the past been promoted through parliamentary
bills. The main objective of such bills has been power to purchase land com-
pulsorily.

Parliamentary procedures require promoters to consult affected local communities before depositing their bills. Objections to schemes are considered
by parliamentary Select Committees, calling expert witnesses where necessary. Then, after initial local and detailed consultation, the final decision
becomes formal parliamentary business. After review by a Standing Committee they are formally debated on the floor of each house and passed or
rejected. The debates deal in detail only with the initial clauses (before the
guillotine falls).

The time needed by Members of Parliament in the committee stages,
where opposition has to be heard, may be onerous and possibly inappropriate to the technical task. Parliament is *not* directly concerned in the quite
different process adopted for new road projects (on which a minister decides,
based on his department's recommendations and with consideration of
environmental statements and reports on public consultation). Legislation
has been enacted which brings railways into line. The new rules include the
following:

- A Draft Order application to the Secretary of State for Transport; this
  requires 'research', i.e. planning and preliminary design, to the same
  extent as was needed for parliamentary bills.
- An Environmental Statement which must accompany the Draft Order
  (this statement was not compulsory with parliamentary bills).
- Notification of objections to published proposals, negotiation between the
  promoter and the objectors and consideration of outstanding objections

by the Secretary of State before he makes a decision and an Order (Statutory Instrument).

In addition the Secretary of State can call for:

- a Public Inquiry before his Inspector (usual for any large or contentious project);
- a formal Parliamentary Debate if the project is of national significance; this partial duplication of the old procedure seems likely for any major new railway.

The whole environmental appraisal process has to be fitted into whatever long-established or newly-conceived national decision-making procedure is in force. The Environmental (or Environmental Impact) Statement, with any associated explanatory material, is the vehicle for presenting the environmental assessment. It is provided as a technical aid to both planners and decision-makers. It may be used or challenged during the public consultation process.

2.6 ENVIRONMENTAL EVALUATION

2.6.1 Evaluation Problems

Step 8 of Environmental Assessment is to determine how the net impacts of a project can be incorporated into its evaluation. Projects are justified or chosen:

- by financial performance forecasts (particularly for commercial enterprises);
- by economic cost-benefit analysis (for socio-economic and government-sponsored schemes).

Financial performance takes the environment into account only if prices reflect environmental preferences. Market prices seldom do this; one exception is marginally higher tax on leaded petrol; another rarer one is a lower price for diesel fuel sold to railways than to less efficient or more polluting forms of transport. If the financial return of a project is attractive, then only strong government policy or legally established social or environmental standards are likely to prevent its implementation.

Economic analyses may be even less sympathetic since they often exclude environmental costs or benefits. For transportation, these analyses place values on some unpriced 'benefits', such as time saving or accident reduction, but rarely on loss of environmental assets. The Institution of Civil Engineers (1990: 30, 14.19) states

> The methods employed by the Department of Transport . . . suffer the impor-
> tant disadvantage that an economic appraisal of performance for individual
> schemes is evaluated against an assessment of environmental inputs which are
> not economically appraised.

Environmental impact assessment is supposed to ensure a more balanced evaluation than cost–benefit analysis alone provides. It is intended to take into account all a project's effects on the social or physical environment.

Total project evaluation can be undertaken by two alternative approaches:

1. By separate evaluation and ranking of the environmental performance of project options—to be compared then with each option's financial or economic performance.
2. By monetary valuation of environmental costs and benefits and by including them in cost–benefit analysis.

The first approach still requires 'balanced' judgement by decision-makers as to the relative merits of quite different features; it provides no means for accountants or economists to recognise these merits. Pricing environmental costs and benefits in the second approach is fraught with contentious valuation problems.

Both approaches can benefit from comprehensive environmental assessment during project formulation. That assessment draws attention to environmental issues; sensitive design can then produce solutions for some issues *before* the final evaluation stage. Probably a typical conservation view is that of the Royal Society for the Protection of Birds (described by Richards et al. 1992: 173) that the primary process should be environmental appraisal and that the cost-benefit analysis should then be applied to the best ecological options (e.g. for a railway route).

2.6.2 Separate Environmental Evaluation

Each environmental impact can be assessed

- in *quantitative* terms, e.g. the level of train noise predicted at certain points, or
- by *subjective evaluation* of qualitative data, e.g. concerning the significance of interrupting a view by a new railway embankment.

Quantitative information, particularly concerning the magnitude of an impact, can be compared

- with standards or thresholds which may establish acceptability or degree of severity,

- with corresponding information about alternative schemes or options for achieving the same objective.

The Institution of Civil Engineers (1990: 27, 4.6) notes that

> Apart from noise levels, few *thresholds* of acceptability are defined for environmental disbenefits of transportation, either in construction or in use.

Comparison with similar established cases could help to establish some realistic thresholds.

Even **subjective evaluation** takes numeric or objective data into account to establish significance, i.e. the duration of an effect, the number of people affected or the scarcity of the resource that is depleted.

Meanwhile in many, perhaps most, situations there are too many variables in an environmental issue for any clear threshold, quality measurement or indication of scarcity to be established, for instance in a complex ecosystem. There are also cases where different types of impact (affecting productive use, habitat conservation, scenery, heritage or recreation) each has its own optimum solution.

These complex but common situations are particularly difficult to reconcile in economic terms. *Matrices* are one way for specialists to present qualitative judgements about complex situations. The squares in a matrix can be used to indicate the significance, scarcity or other measure of various environmental criteria, listed in one direction; in the other direction are listed different types of occurrence, circumstances or land use. Various types of matrix have been devised for environmental appraisal, for instance in coast protection schemes (Coker 1992a: 133–134, 155–157). The data can be prepared, discussed and amended by different planners, specialists and other interested parties and classified or ranked numerically or by short descriptions. Comparable matrices can be presented for the present situation and in the future for a 'without project' case and for various development scenarios. For railway planning, it would be appropriate to prepare these matrices for sections of each route option.

The very complexity of the situations represented by matrices militates against a numerical valuation being put on a situation, for instance by adding up the total effect of any row or column. The entries in each box, as much as the overall conclusions, have to be agreed on whatever evidence, objective or subjective, the experts and interested parties agree to use; inevitably environmental 'trade-offs' are involved. Coker (1992a: 135) states that

> Ultimately, it is for politicians to decide the public acceptability of the trade-offs preferred by professional advisers in the final decision-making process.

Judgement on the combined environmental impacts of a railway project or any element of it must be presented with full justification; this may be in a formal environmental statement or in more specific aids to decision-making like matrices. Recommendations may comprise any of the following:

- Absolute embargoes on unacceptable solutions (such as serious intrusion into a national park or destruction of a Site of Special Scientific Interest).
- 'Best solutions' negotiated and agreed among environmental specialists, engineers and operators.
- Acceptance of certain levels of negative impacts.
- Recommendation of means of enhancing positive environmental benefits.

2.6.3 Monetary Valuation

Valuation of irreplaceable resources or 'Accounting for the Environment' is inherent in an economic philosophy for sustainable development. This philosophy was prominently expounded by Pearce et al. (1989). A few years later O'Riordan and O'Riordan (1993: 24) commented

> There has been rich academic literature on environmental valuation ... but ... the idea of granting prices and weights to environmental quality is still slow to reach effective fruition.

Whitelegg (1993) comments that

> there is still very little sign of progress towards a translation of sustainability into tangible targets and values relevant to consumption and pollution.

This is particularly true of transport projects and the use of land resources.

There are a number of different valuation approaches, some more relevant than others to particular types of environmental impact. Literature is building up on the subject as is controversy about the validity of some of the proposed methods. In the USA acknowledged experts (NOAA 1993) have given qualified agreement to the usefulness of some methods and there are signs that these might eventually become statutory. Nevertheless further research and proposals are needed at a time when all land and energy-consuming development needs examination in terms of diminishing resources and increasing global pollution.

In Chapter 7 the price of *noise disturbance*, an impact on people rather than resources, is related directly to the cost of reducing that noise to an acceptable level or of paying compensation. Most of the cost of *construction impacts*, described in Chapter 10, can be absorbed in the price tendered for work which has to take account of specified precautions and statutory constraints on pollution. Attempts have been made to value landscape, historic structures or other recreational land in ways representative of people's willingness to pay for its use or its existence. Such attempts are difficult to validate because of uncertainties about future scarcity and pre-ferences.

The market price of productive land (discussed in Chapter 12) can be readily assessed and is included in conventional cost–benefit analyses. How-ever, no price representative of the long-term value of the sort of lower uti-lity land described in Chapters 13 to 15 is readily available. Chapter 16 therefore proposes approaches which could be considered for the more equi-table evaluation of land resources.

2.7 TIMING OF ENVIRONMENTAL APPRAISAL

The different phases of environmental appraisal can be timed so as to pro-vide feedback mechanisms which deal effectively with many impacts and thus *eliminate* them from the difficult full evaluation process.

A long period is usually needed to establish the feasibility of a major transport project, complete the design, prepare a financing plan and obtain final approval. The various stages of environmental appraisal should be able to fit into this period.

Even before a development programme is conceived, environmental issues can play a significant part in the formation of national or regional transport *policies*. SACTRA (1992: 91, 16.03) recommends that

> in order to ensure that proper account is taken of all effects, the appropriate environmental assessments must underlie every stage in the hierarchy of deci-sions, from the making of national and regional policies downwards.

Principles to be borne in mind in scheduling the planning of an actual railway development arise from comments already made in this chapter.

1. Environmental issues should be defined and considered first at the project concept stage (see Section 2.2 above). A preliminary public statement at this stage could refer to these issues.
2. Environmental appraisal (Section 2.3) should involve continuous, inte-grated involvement of environmental specialists, engineers and operators throughout the planning and design activities. Documents explaining environmental issues should be provided for information and consulta-tion. These can be limited in scope as appropriate at each stage but can pave the way for an Environmental Statement whose findings will be nei-ther a surprise nor a set-back.
3. The stages of environmental assessment (Section 2.4) need not necessarily be followed in strict sequence. There are a number of iterative steps whereby a preliminary design is modified because detailed design shows flaws in earlier technical or environmental assumptions.

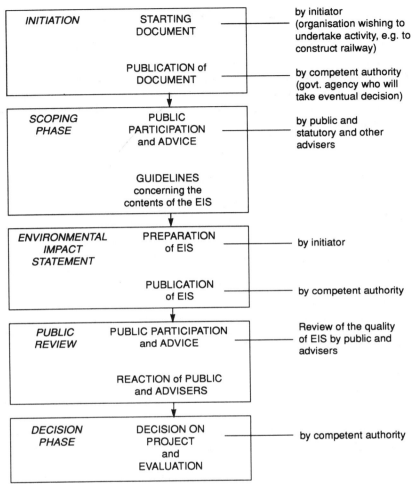

Figure 2.2. Typical sequence of environmental impact assessment (based on EIA procedure in The Netherlands from information supplied by the Ministry of Housing, Physical Planning and the Environment)

4. Public involvement should take place at well-defined stages, the amount of information disseminated and the degree of consultation taking place depending on the depth of detail to which planning or investigation has proceeded (Section 2.5).
5. Preliminary environmental evaluation (Section 2.6) should take place first so as to settle the simpler issues early in the process; the eventual method of full project evaluation should be determined in the light of a number of

preliminary evaluations—of different issues and with alternative approaches—starting at the early stages of planning.

In any particular legal or planning situation, there are likely to be statutory requirements regarding the timing and subject of various environmental statements or consultation processes. It is less likely that the detailed content and format of such statements will be strictly defined. So there may be considerable scope for the planner to include only the subject matter which is relevant at each stage.

Figure 2.2 shows an example, from The Netherlands, of the sequence of EIA procedures which are undertaken for projects.

The key to successful railway project planning is to combine environmental studies and investagations with the engineering design and operational planning. This should result in acceptable solutions to particular problems being derived as the total planning process proceeds. For any major issues which remain to be resolved, the aim should be to devise a satisfactory means of including environmental factors in full economic analysis.

3 Railway Planning

RAILWAY INFRASTRUCTURE

3.1 BASIC FEATURES

The function of any transport facility is the conveyance of people or goods. Modes of transport differ according to the *type of vehicle* used, the *space* it occupies and the *source of energy* which propels it. These characteristics in turn determine the *viability* of each mode.

3.1.1 Type of Vehicle

In *environmental terms*, vehicle design is related to carrying efficiency, space requirements and source of energy. In *commercial terms*, a vehicle's achievable payload, the speed and flexibility of the movement mode and the energy consumed are all factors which determine cost effectiveness.

The type of vehicle has to suit the people or cargo carried. Many passengers may be accommodated in a fixed route public conveyance but few in more adaptable private cars. Small vans can deliver parcels locally while bulk carriers or container lorries carry larger consignments over longer distances. Vehicles on rails are limited in width and height by lineside structures; units can be only as long and large as wheeled suspension, track spacing and clearances on curves permit; but they may be coupled together to form *trains* in a variety of combinations.

3.1.2 Use of Space and Mode of Movement

An aeroplane can move in *three dimensions*. It uses speed and considerable energy to overcome the force of gravity. Speed is also an important commercial benefit although the high energy use involved is environmentally damaging. Operational and safety requirements dictate the need for extensive ground space and support facilities; airports are therefore too costly to be numerous.

Steered conveyances on land or water have freedom to move in *two dimensions*. For mechanical and safety reasons the majority of land vehicles have to proceed upon a prepared smooth and strong pavement. All-weather road surfaces, adequate to carry the weight of animal-hauled carts, were prepared in Roman times. Metalled roads became a general necessity to carry more sophisticated and heavier vehicles propelled by their own engines.

Railway tracks constrain movement effectively to *one dimension*. The operator has only to control a train's speed to avoid colliding with other trains or obstacles. Rails also provide a low friction surface. Speeds are limited by the stability of the trains on the rails, their capability to stop within a safe distance and the motive power provided. Well designed track and rolling stock and long radius curves ensure stability; modern signalling and train control systems allow long stopping distances; with high performance traction these make railways a fast and high capacity form of land transport. Environmental disadvantages lie in the comparative inflexibility of railway alignment and related difficulties in avoiding obstacles or creating barriers.

3.1.3 Energy Source

Early-nineteenth-century problems in the mechanisation of previously horse-drawn coaches or barges were focused on the size of engines. Traffic either had to be hauled by cable from a static source of energy, such as steam or water power, or a mobile source had to be incorporated on the chassis of the load-carrying vehicle itself. A self-propelled *steam* 'locomotive' could not work effectively on too small a scale, nor could it carry a very large load on its own wheels. But the considerable earlier invention of coupled trucks running on twin rails provided the means of guiding trainloads long enough to make efficient use of a locomotive's capability.

At the end of the nineteenth century two further developments had important consequences. The invention of the *internal combustion engine*, using easily transportable petroleum, made compact engine units available. These could be used to propel 'horseless carriages' which eventually replaced all the small animal and hand-drawn vehicles which travelled on roads and tracks where railways had not penetrated.

At about the same time the use of *electricity* became feasible through the use of relatively light electric traction motors in vehicles which could pick up current along the route.

During the twentieth century there was a huge increase in petroleum-fuelled road vehicles, each carrying 1 to 50 people or up to 40 tonnes of goods. Diesel became the eventual fuel for railway locomotives which could haul hundreds of passengers or thousands of tonnes of freight.

But it was the transmission of electricity through track level ('third') rails or overhead wires which revived the opportunity for rail transport to compete against roads. Electricity offers advantages for environmentally-sensitive transport systems in:

- economies of scale and efficient use of energy resources when using a central generating source rather than individual power units;

- a wider choice among available means of energy production, some of which are less pollutive of the atmosphere or demanding on finite resources.

3.1.4 Railway Viability

A guided transport system such as a railway, conveyor belt or pipeline offers high capacity and speed. It may also hold environmental advantages over less constrained transport. But the capital and fixed operation costs can only be justified if the services provided earn sufficient revenue.

Table 3.1 summarises the basic characteristics of railway systems compared with roads and air transport. More specific comparisons are used in assessing the operational, commercial and environmental viability of particular railway schemes.

The fundamental characteristic of any railway is its track. Section 3.2 describes track elements and requirements for its alignment. Alignment in turn calls for civil engineering works, described in Section 3.3. Some completely new railways are being built but the track and civil engineering works which are already in place today are likely to remain a very substantial part of any future rail network.

Characteristics of trains and traction are described in Section 3.4 and other operational features in Section 3.5, whilst issues relating specifically to passenger and goods traffic are examined further in Chapters 4 and 5 respectively.

Section 3.6 deals with the capability of railways to meet demand for services. The severity of environmental impacts will depend both on the nature and intensity of rail services and on any need to provide extra infrastructure. Future rail traffic has to be forecast in the light of developments which will occur in *all* forms of transport as well as in industry. Railway planning should examine impacts on other transport (see Section 3.7) and on economic activity (Section 3.8).

3.2 TRACK AND ALIGNMENT

Track materials are mined resources; the track's appearance affects the visual environment; its underlying formation is a physical barrier; and its alignment determines how the line disturbs the land resources which it crosses.

Permanent way, as the track is generally known by railwaymen, comprises the following components:

- Steel rails, on the inner edges of which run the train's flanged wheels. Rails are replaced every 15 to 20 years and reground more often according to how much wear they undergo; old rails can be recycled.
- Rails are supported by and fixed to concrete, wooden or occasionally steel

Table 3.1. Comparative transport characteristics

Characteristic	Transport mode		
	Rail	Road	Air
Mode of movement			
Dimensions of movement	One (rail-guided)	Two (steered)	Three (steered)
Traffic signals	Fully applicable	Partial (at some junctions)	Internal (radio)
Automatic control	Widely practised	Seldom practicable	Landing only
Accidents	Rarely serious	Frequent, some fatal	Rare but always serious
Operational performance			
Speed	High between stations	Moderate; slow in congestion	Very high between airports
Direct access for people or freight at all points	Poor	Very good	Poor
Unit carrying capacity	High	Low for private, high for public	Medium (people), low (freight)
Commercial performance			
Inter-city	Fast, direct	Variable / Difficult access for city centres	Very fast between airports
Rural	Few services available	Very suitable	No service
Commuters	Efficient	Public transport effective	not applicable
Freight	Bulk and long distance	Medium distance distribution	Light, usually high value items
Environmental performance			
Energy efficiency	High if well loaded	High for buses; low for cars	Low
Use of land resources	Narrow but not very flexible	Wider but more flexible	High but only at airports
Noise	Loud but nearby only; intermittent	Moderate but common and continuous	Very loud near airports
Air pollution	Low	Medium/high	High

sleepers. Wooden sleepers are required on curves but concrete ones last longer and are often standard. Sleepers support and maintain the distance (gauge) between rails.

- Sleepers are founded on ballast, usually of broken stone, of sufficient depth to spread the load so that it can be supported without differential settlement by the placed (formation) or natural (sub-grade) earth beneath. Ballast also provides top surface drainage and is a medium for maintaining the precise level and gradient of the line and for providing superelevation on curves. 1000 to 3000 tonnes are laid per kilometre according to the conditions. Periodic cleaning is required and replacement about every 40 years according to how long the stones retain their shape and strength. Some old ballast is recycled and reused.

The strength of track is determined by the cross-sectional area of the rails, the type and spacing of the sleepers and the depth of ballast. Track strength requirement is dictated by the axle loads which it must carry, the speed of traffic and the annual tonnage carried.

The rails, their fixings to the sleepers or to bridge structures and the manner in which they are worn by train wheels will affect the smooth riding of rolling stock and the noise and vibration which is created.

Requirements for ballast are determined by its necessary depth and the availability of stone of adequate quality. It is sometimes dispensed with altogether on low traffic desert railways. But without ballast, there must be stable ground, frequent maintenance and severe speed restrictions.

Concrete 'slab' track support is an expensive but largely maintenance free alternative to ballast. The structure of slab track may also be designed to reduce vibration.

Permanent way is not in fact 'permanent', the title being originally applied to differentiate the finished track from the temporary rails laid by contractors during construction. Longevity of the track structure as a whole can only be assured by regular maintenance and periodic replacement of rails, sleepers and ballast.

Material resources are expended in permanent way in the following forms:

- Steel in the rails, which last for long periods if correctly laid, regularly maintained and not subjected to heavy wear at curves.
- Aggregates and cement in *concrete* sleepers—a relatively small call on natural resources compared with the material in road structures or airport surfaces; a rather more significant proportion of available resources in *timber* sleepers if hardwoods are specified.
- Hard, crushed rock in ballast.
- Energy output in manufacture of rails and sleepers, crushing of rock, and transport and maintenance of materials.

Track gauge (distance between rails) was a subject of early controversy. Almost by chance a standard gauge of 1435 mm was adopted in Europe and North America and now accounts for 60 per cent of the world's route mileage. Half the remainder is of wider gauge, half is narrower. On some islands like Ireland (1600 mm) or in a comparatively isolated sub-continent like India/Pakistan (1676 mm) broad gauges were adopted for the major routes; but these countries have few needs for transfers to international connections. This was not the case in Australia, where each state chose its own gauge and standardisation had to be achieved over several decades. Meanwhile major continental anomalies still exist:

- In Spain and Portugal, at the west end of Europe, and Finland and the former Soviet Union to the East, railways still operate on a wider than standard gauge necessitating changing the bogies (wheel sets) of trains that cross borders.
- In East Africa the 1000 mm gauge is incompatible with standard 1435 mm gauge in Egypt and the 1067 mm gauge of central and southern Africa.

Commercial and operational considerations could lead to conversion to standard gauge in the Iberian peninsular. Indeed the Madrid to Seville high speed line has been built to standard gauge as will be the planned new route from Madrid to Barcelona and France. Environmental implications of standardisation are significant where a change of gauge is taken as an opportunity to adopt a completely new alignment.

Narrow gauge railways, such as those in Africa, were built originally to save costs of track space and to use available materials. The choice of track gauge has no effect on the size of trains nor on their capability to traverse curves in difficult terrain. In fact all but the tightest curves (less than 150 m radius) are practicable on standard gauge which can be widened on curves, by up to an inch, to accommodate longer rolling stock. It is the *stability of trains* which determines at what speed curves may be safely negotiated.

The ideal *horizontal alignment* of track is as direct as can be attained in flat open country such as the 478 km straight crossing of the Nullabor Plain in south-western Australia. Curves have to be introduced to overcome topography on steeper ground or to evade substantial obstacles. A minimum radius of curve has to be provided to achieve a certain speed at a required degree of safety or comfort. Centrifugal force is proportional to the square of the velocity and inversely proportional to the curve radius. The effects of curvature can be reduced by cant (superelevation) of the track up to a practicable maximum height of the outer rail above the inner of about 150 mm and by a smooth transition from straight track into the curved canted section. There is also a safe limit of 'cant deficiency' wherein the train can travel *faster* than is ideal at the speed and cant concerned but at the expense of some passenger awareness. Trains travelling more *slowly* than the design speed for a curve can cause harder or unequal wear on the rails.

Railway routes can be aligned to avoid any curve at lesser radius than the design speed requires. Alternatively the curve radius which topography and avoidance of obstacles dictate, together with the practicable cant and cant deficiency of the rails, will set a limit on allowable speed.

Vertical alignment concerns the track gradients which are necessary to climb up and down sloping terrain. In practice, in hilly country, railways follow valleys as far as curves and gradients permit. Except on very steep mountain 'rack' railways, limits on gradients are imposed by the motive power of the locomotives and the weight of trains. When gradients become too steep or curves too tight then railways must cut through the heads of the valleys and any intervening high spots or side valleys.

Route selection, introduced in Chapter 11, is a key element in determining the impacts of railway construction on land resources. The faster the train service provided, the straighter and more inflexible is the alignment and the more significant are the environmental impacts; whilst, in steep or densely occupied country, achievement of an acceptable alignment can require substantial civil engineering works.

3.3 CIVIL ENGINEERING

3.3.1 Engineering Features

The cost of civil engineering works is a key element in choice of a route for a new railway. Final design solutions and precise track alignment then involve more detailed considerations of the impacts which the railway infrastructure and its trains will have on the surroundings. The main features of railway civil engineering are:

- the permanent way itself;
- the underlying formation of earthwork or depth of cutting that is needed to attain the grade across undulating land;
- drainage structures, permitting water to flow under the formation; these must protect the railway from damage in floods but preserve the resource value of the waterways;
- other bridges, retaining walls and tunnels needed to attain the required alignment across the topography.

Figure 3.1 represents the features in a typical cross-section of permanent way and formation. Figure 3.2 illustrates how railway civil engineering relates to local topography and environment.

Some principles of railway engineering were already established in the design of canals. These had required excavation, earthworks and construction of major bridges and tunnels. The main differences were in the straighter course of railways and their ability to tackle modest gradients where canals gained height in a series of locks.

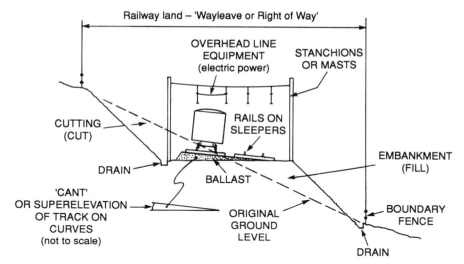

Figure 3.1. Permanent way and formation

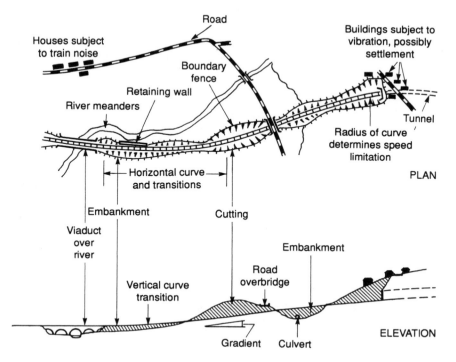

Figure 3.2. Civil engineering for railways

3.3.2 Railway Land Requirements

The overall width of track formation depends on the width of rolling stock and the necessary distance between the trains. Additional width may also be needed for ancillary equipment such as supports for overhead electrification. Typically, a British two-track line will be built on an 8 metre width of ballast within a total of 10 or 11 metres on level ground between drains ('cess strip'). The corresponding figure for the latter on French *Lignes à Grand Vitesse* is 13 to 14 metres. The width of railway 'wayleave', represented by the distance between fences, varies from 9 metres on older lines to as much as 17 metres on level ground on some new high speed lines. Slopes of cuttings and embankments can add a great deal more. In undulating to mountainous country the average width could be 20 to 30 metres. Four-track configuration will add about 7 metres to any two-track formation width.

3.3.3 Drainage Structures and Bridges

Any very long structure inevitably interferes with the pattern of surface and ground water movement across the land it traverses.

Interference may be least where a railway or road crosses gently sloping country or along a wide valley parallel to a river; only at tributary crossings is it necessary to provide a bridge or culvert. More serious interference may occur in a narrower valley where the transport route has to take a straighter route than a meandering stream, crossing it frequently. In these circumstances the engineering problems are:

- to provide adequate passage for flood water at an acceptable level of risk—this is primarily an exercise in hydrology and hydraulics;
- to minimise damage to the resource value of the stream beds, banks and any associated wetland—this is essentially nature conservation, dealt with for railways in Chapter 13.

High or long bridges across valleys, commonly called viaducts, have been important visual features since Roman times; but, besides crossing rivers and valleys, railways frequently have to pass over or under roads. In such cases the structure itself and the railway and road approaches have to be planned as a single unit. The location, layout, appearance and method of construction all determine the ultimate effects on the environment.

3.3.4 Embankments and Cuttings (Earthworks)

Engineering and cost principles in earthworks design include the following:

1. The most economic solution usually balances cut and fill and thus avoids either a shortfall or an excess of material.

2. Slopes of embankments and cuttings, and hence the volume and cost of their construction and the area of land-take, are dependent upon the vertical height and nature of shearing planes for cohesive soils like clay but are constant for each type of granular soil and can be very steep for sound rock.
3. Where the nature of the soils requires shallow slopes but where land space is very costly, the space required for earthworks can be reduced by structures. Retaining walls can reduce the width of cuttings and viaducts can shorten the length of wide embankments.

Trains on embankments cause most noise in the surrounding area; in cuttings train noise is muffled but passengers are denied a view.

3.3.5 Embankments versus Viaducts

The engineering choice between an embankment and a bridge or viaduct is determined by the volume and nature of available fill material for the embankment and by the width and number of spans required for a viaduct. Circumstances obviously differ widely but the breakeven height above which a viaduct becomes more economical is often 20 to 30 metres.

Environmentally, a viaduct may be preferable at even lower heights because it causes less disturbance to the land area beneath and may provide an aesthetically more pleasing structure. Additionally, a solid barrier causes an obstruction across a valley which affects both the view and the microclimate in its vicinity. The Great Central Railway, completed as a high speed route in 1899, enhanced the aspect of most of its valley crossings by incorporating only short approach embankments to its long and graceful viaducts. Much earlier, the Thames flood plain outside Windsor had been crossed by the rail branch from Slough on a 2 km long multi-arch viaduct; the height of the track above ground level is only about 5 m but the resulting structure is arguably more attractive than an embankment of the same length. There are numerous examples of low, space-saving viaducts in more densely inhabited urban situations, notably the 6 km long 878 arch structure built for the London to Greenwich Railway in 1836. Some spaces 'underneath the arches' became useful for light industry so that only the space taken by the piers constitutes railway land-take.

3.3.6 Tunnels versus Cuttings

Tunnels remain the most viable solutions for major deep crossings—under the English Channel, for instance, or beneath the Alps. Other tunnels were built in the past to obviate gradients which modern trains could tackle without difficulty.

Where new lines still have to cut through hills at comparatively shallow depths there are geological, design, cost and environmental factors which determine whether a tunnel or a cutting is appropriate.

Geological factors include the stability of the material in cutting slopes and within or above a tunnel. Some railway tunnels proved dangerous and costly to build or reconstruct through unstable material and sections have had to be converted to open cutting. This happened on coastal lines through Folkestone Warren and in Scotland at Cockburnspath.

Design issues which affect the costs of tunnel construction are the speed at which trains are to traverse them, the diameter of the tunnel and whether single or twin bores will accommodate the up and down tracks. A train requires more tractive energy to overcome air resistance in a tunnel than in the open air or to run in a narrow one-track bore than in a wider one for both tracks. On the other hand disturbing air pressures, which may require speed limitations, can occur as trains pass each other in opposite directions in a single bore.

Construction costs depend not only on excavation quantities and conditions but also on ease of access into tunnels or cuttings and on methods for movement and disposal of excavated material. Particularly in urban surroundings, surface space is a critical factor. Cut-and-cover tunnels (shallow cuttings contained within retaining walls and roofed over) were a common feature of many early metro lines.

Environmentally, tunnels hide most of the perceived impacts of a railway; they are therefore solutions often recommended by local objectors—although not by rail travellers. Possible adverse impacts to property owners which may arise from tunnelling are settlement (which is controllable during construction) and vibration (which is not a common problem but, at least for existing buildings, is unavoidable). Tunnel excavation could also disturb ground water. Long tunnels may require supplementary ventilation or drainage facilities which can involve substantial surface structures. Access to tunnels for maintenance purposes or in emergencies is more difficult than to cuttings.

3.3.7 Railway Infrastructure in Urban Environments

Even when some of the early railways were constructed, formidable obstacles were posed where they approached city centres. Often the solution was to follow road routes. But, unlike trams, trains cannot share space with road vehicles. Common solutions were to build the track either on elevated steel frameworks or brick arches above streets or submerged in cut-and-cover beneath them.

New railway lines into cities make use first of any existing but redundant railway routes, then of derelict or underutilised industrial areas and last of new, deeper underground routes.

3.3.8 Widening Railways

Considerations concerning the choice and design of earthworks and structures for new railways apply equally when widening routes to accommodate more tracks. But additional constraints on space may be imposed, for instance by buildings or conservation land bordering the original wayleave area. In these situations structural solutions such as bridges or retaining walls may well be more practicable than additional earthworks.

Widening of tunnels is practicable where the rock is strong enough to need no structural lining. Widening of existing bores in softer material can be fraught with difficulty where the existing walls are lined with bricks or concrete; a new and separate bore is usually the chosen solution.

The lengthening of over-bridges is a particular problem, involving new abutments and longer spans to accommodate extra rail tracks. Various ingenious methods have been devised for lengthening bridges over motorways and railway engineers are familiar with very rapid replacement of bridge span units.

3.4 TRAINS AND TRACTION

The environmental significance of trains relates to their size, power and speed; if their dimensions and requirements for vertical and horizontal alignment are not suited to existing lines, then new routes—with all their implications—may have to be planned. Particular environmental features of passenger and goods trains are reviewed in Chapters 4 and 5.

Trains traditionally consist of 'locomotive' engines hauling 'rolling stock' along the track. All freight trains are still made up of wagons hauled by locomotives. But many modern passenger trains are semi-permanently assembled diesel or electric multiple units (dmus or emus). In these the motors are incorporated within or underneath one or more of the coaches. The whole set of coaches or a combination of sets is controlled by a single driver at whichever end is in front.

3.4.1 Train Characteristics

The size and weight of trains and the power of their locomotives increased steadily throughout the nineteenth century and well into the twentieth. These increases took place in spite of width and height limits which could rarely be changed. Typical 'loading gauge' dimensions, generally applicable in the countries concerned, are shown in Table 3.2.

Adherence to loading gauge limits ensures adequate clearance for structures beside or above the line, at platforms and between passing trains. The heights of overhead structures have to be greater where overhead electrification is installed. This has resulted in special techniques to raise bridges and

Table 3.2. Typical 'loading gauge' limits

	Max width (m)	Max height (m)	
British railways	2.75	3.95	(generally)
	3.25	4.55	(for structures originally built for 2140 mm track gauge)
Western Europe	3.10	4.50	
Former USSR	3.40	5.30	
USA	3.30	4.90	

Source: Hollingsworth 1977: 208, 209.

in ingenious designs to minimise the space taken up by the overhead equipment.

The gross weight ('tare' or dead load plus 'pay' or live load) of rolling stock determines both the axle loads, which the track and bridges must support, and the total weight of trains for which motive power has to be provided.

Increase in the tonnage of trains can be achieved:

• by strengthening the bridges and permanent way to carry heavier axle loads;
• by lengthening trains.

The length of trains has to be related in turn to the available space at stations or depots or, on single track routes, to the length of crossing loops. Two-set *Trains à Grand Vitesse* (TGVs), 400 m long, produced a need to return to the long platforms which had been adopted at some European city stations in earlier days. Platforms 800 m long are being provided for loading road vehicles on and off Channel Tunnel shuttle trains.

The ability of the trains to travel at high speeds depends upon track control and signalling capability, braking capacity, traction power available and limitations imposed by the radius of curves. For standard (1435 mm) gauge track, trains running at a modest 80 km/h require a minimum radius of curves of about 500 m. Even on canted track, the fastest freight and ordinary passenger trains travelling at about 125 km/h do not normally negotiate curves of radius less than 1000 m, whilst 200 km/h 'InterCity' HST 125s require 2000 m and 270 km/h TGVs need 4000 m. Top speed sections of later TGV routes have been built with curves at a minimum of 6000 m radius or more. Where the terrain makes such almost straight alignment possible, cant deficiency can be reduced to provide additional passenger comfort.

3.4.2 Traction

Steam locomotives are still used in some parts of the world where plentiful supplies of coal and labour make them economically preferable. They are obsolete in modern practice in most of Europe and North America where they have become heritage machinery.

British steam locomotives achieved about 2500 horsepower (hp), the largest French compounds claimed 5000 hp and the most powerful US steam units 7000 hp. The largest single-unit diesel locomotives can provide about 4000 hp. Single electric locomotives can produce up to twice as much power. But the great advantage of all modern traction units is their capability for working together as linked units under the control of one driver.

Increased power capacity is needed:

- to overcome air resistance which at higher speeds is effectively proportional to the square of the velocity;
- to overcome steep gradients.

There is no tendency to use more powerful units for urban rapid transit systems. Essentially they remain similar to the multiple unit electric trains invented as long ago as 1897 except that more compact engines now occupy no more than underfloor space. Modern suburban diesel and electric multiple units continue to provide about 250 hp per passenger coach. But separate power units in British HST 125 sets provide about 560 hp per coach and German ICE, Spanish AVE and French TGV sets 1000 hp or more.

Environmentally there are two issues which arise from the use of high power traction. One is the capability of trains to climb short, very steep gradients on alignments using less land than would be taken for more circuitous lines. This is likely to offer more freedom in route selection, discussed in Chapter 11. The other issue is whether more powerful engines use more energy to perform the same task. In freight transport high power achieves economies of scale in hauling large tonnages. On passenger trains, power is used mainly to accelerate and maintain high speeds but it will be seen that this is not necessarily wasteful of energy compared with slower trains.

The environmental advantages of electric traction have been mentioned. The operational advantages are as follows:

- Rapid acceleration, particularly useful for passenger stopping trains.
- Increased adhesion if power is applied directly to several axles.
- Ability to take energy from a remote source, continuously up to the rated capacity or above that rating for short periods.

The disadvantages of electrification include the following:

LIVERPOOL JOHN MOORES UNIVERSITY
LEARNING SERVICES

- High installation cost which has to be justified by heavy traffic demand; where this demand is insufficient diesel traction is more economic.
- Accidents to railway workers in contact with live rails or wires.
- Possible failure of the power supply or transmission system.

The last two are evidently rare occurrences. Incidental adverse effects cited occasionally are electrical disturbance by transmission lines and power pick-up equipment or by signalling circuits at nearby sound recording studios (Fox 1992). However, such effects on outside electrical systems or human activities are also rare.

Other minor disadvantages of some modern traction or rolling stock occasionally become evident, for example:

- Some diesel multiple units, designed for medium distance passenger services, have trouble negotiating steep gradients, for instance on certain branch lines; in the future further standardisation of stock to suit the services in highest demand might make such difficulties a reason to discontinue operations on steep but only marginally viable routes.
- Disc-braked and small-wheeled stock are less able to deal with autumn leaves on the track, increasing the need to call on special leaf clearing machinery; the associated delays to trains, which occur in continental Europe as well as Britain, become more serious in the face of demand for faster, more intensive and reliable services.

Modern suspensions, small wheels and disc brakes are designed to meet environmental objectives, particularly noise reduction, as well as to ensure safety and smooth riding.

3.5 OPERATIONAL FEATURES

Track, suitably supported by civil engineering works, provides the rail 'way' for trains. The characteristics of trains determine how they can make use of existing lines and define design parameters for new lines. Actual train services have to be planned:

- to make best use of the line capacity so as to accommodate the traffic demand;
- to ensure safe operations;
- to maximise reliability of service.

Environmentally, the first of these is of most significance, it being desirable to minimise any need for *new* infrastructure. Safe and reliable services are necessary incentives to persuade passengers or freight agencies to use what may be seen as an acceptable and sustainable mode of transport.

3.5.1 Line Capacity

The number of trains which can be accommodated is determined by the number of tracks which are provided and the number of 'train paths', per hour or per day, which the speed of trains and signalling/control systems will permit on each track.

Single track lines have much the least capacity as one line has to carry trains in both directions; an 'up' train cannot enter a single track section until a 'down' train has cleared it. Suitable lengths of double track have to be provided as crossing loops between the long single track sections. It is the length of these crossing loops which limits the length of freight trains. In remote areas 'stations' at loops are often among the few centres of human activity in the locality.

Single track railways are normal across the wide, sparsely inhabited spaces of Asia, Africa, America or Australia. In Europe single track is common on rural lines and is utilised for some main line sections even in northern Germany.

In environmental terms the difference between railway formation for single or for double track is significant only in the most scenic landscape.

The capacity of *two-track* lines is that of each track, one in each direction, 'up' and 'down'. One-direction track capacity is the number of possible train paths multiplied by train capacity (in freight or passengers). The number of train paths is determined by the speed of the *slowest* trains and the headway, i.e. the interval (permitted by the signalling system) between successive trains on the same line—as low as 3 minutes on many main lines, in the future possibly 2 minutes (already common on metro systems). Uniform speed is desirable not only to maximise train-carrying capacity; track wear is minimised if all trains travel at the speed for which the cant of the track on curves has been set.

It is the need to accommodate some slower trains and to provide extra capacity on busy routes which led to the adoption of *four-track lines—*fast and slow, up and down—along key main lines and in metropolitan areas.

Two-directional tracks (extra tracks for either-way working) permit capacity to be increased in the direction of any *peak* traffic or up steep gradients. Some sections of US long distance freight routes have even recently been increased to three-track. Faster trains can be observed *overtaking* slower ones even on two-track sections.

A means of increasing the volume of both fast and slow traffic may be the widening of existing railway formation to take additional tracks or provision of a new pair of tracks on an alternative, different speed alignment.

However, any type of new construction may have environmental implications.

Another factor in line operating capability is the suitability of routes for the optimal and timely running of trains. Passenger services over medium distances have for many years been provided at regular intervals, i.e. hourly or more frequently at the same minutes past every hour. Convenience in journey planning may attract travellers as much as speed.

Regular interval services have been a main feature of timetables for over 50 years on heavily used electrified networks. In improving such services for the optimum use of stock, Swiss Railways (SBB) aim to bring the running time between all defined 'hubs' of their system to just *less than one hour*. SBB's *'Bahn 2000'* plan, if it can eventually be financed, includes 130 km of *new* 200 km/h route, particularly to achieve this objective, for instance to cut 22 minutes off the existing, operationally awkward, 71 minute schedule between Berne and Basle. The new works would make available four tracks between these cities, two to accommodate the faster direct passenger trains and the other two to accept the anticipated extra volume of freight traffic via the Berne area following duplication of the Lotschberg Tunnel route.

3.5.2 Safety

Safety is primarily avoidance of accidents. Causes of train accidents are as follows:

- Derailment due to instability when travelling at excessive speed as a result of driver misinformation or inattention, or due to wheel/suspension or track defects.
- Collision with obstacles fallen on to the track or placed by vandals: high fencing, preferably not visually obstructive, should discourage illegal entry on to the tracks but the effectiveness and necessity for fencing is arguable; it has not been provided on some German *Neubaustrecke* lines.
- Displacement of track by landslips or collapse of bridges; the Tay Bridge disaster of 1879 was not unique and passengers have drowned in similar more recent accidents in the USA and Wales.
- Collision with vehicles on crossings or obstacles protuding from bridges damaged in road accidents.
- Collision with other trains due to failure to respond to signals, to faults in the signalling system or to brake failure.

Safety in respect of the last and most common type of accident ultimately depends on braking systems, secure couplings and signalling and control. Considerable improvements have been made to allow for the safe stopping of increasingly fast trains. Signalling equipment is itself an important part of the physical infrastructure of a railway. Old-fashioned manually-operated semaphore signals and points provided a robust system; whilst occasionally subject to mechanical faults, they were inter-locked to provide clearly seen fail-safe devices. Where less visible, equally fail-safe but complex electronic or computer circuits are in use it is essential that the designer understands the operation of trains. It is equally essential that train crews and operators maintain the same vigilance that drivers and signalmen did before—such rare train accidents as do occur still result from human as much as electronic frailty.

Meanwhile, there is a possibility that safety constraints, perhaps of unnecessary severity, may be placed on train operation in new long tunnels intended to promote more sustainable transport routes. Speaking about a proposed 54 km long Brenner 'base' tunnel, Dr Peter Steiniger (1991) warned that 'safety concerns can exercise a disproportionately high influence on construction and operating costs—for example the requirements of the inter-governmental safety commission for the Channel Tunnel mean that practically all existing European rolling stock will be banned from using it'. By the end of 1993, Dr Steiniger's fears were realised in escalating costs of the Brenner tunnel.

3.5.3 Reliability

Dependable services suffer a minimum of unplanned delays. Mechanical faults affect traction units or sometimes operating equipment like points or gates. Electrical faults can causes train stoppages due to lack of power for traction or through fail-safe action in signalling circuits. Human deficiencies include the non-appearance of train operating or signalling staff. Delays to trains *en route* for any reason can delay all following trains where there are no overtaking tracks available.

Routine maintenance activities sometimes require closures of tracks. But if this occurs on double track sections, 'single line working' can be introduced on the one line which remains open.

Curtailment of services, or even derailing of trains, can arise as a result of exceptional weather conditions—snow or sand may drift on to the track or floods may undermine the track or even bridge foundations. In the more uninhabited parts of the world like Sudan—where traffic is not dense enough to justify infrastructure to withstand every eventuality—'wash-outs' from sudden storms can occur such that all traffic has to be suspended until the line has been rebuilt. Windblown sand can also disrupt operations of desert lines unless it is laboriously removed.

The cost and delays of emergency repairs or maintenance measures have to be compared with those of providing expensive capital works; so works such as protective barriers or extra culverts may be forgone where the occurrence of storms or floods is known to be rare and their location unpredictable.

TRANSPORT PLANNING ISSUES

3.6 CAPABILITY AND DEMAND

Railway development schemes can only be justified if there is an assured demand for transportation. The degree of *environmental impact* will then depend on the number of trains provided to meet that *demand* and on whether existing rail infrastructure has the *capability* to carry those trains or whether new lines are required.

Capability and demand are inextricably linked. The cost of providing an improved inter-city rail service can only be justified by a sufficient number of travellers being attracted from road or air services; whereas low demand for rural train services may not support even the operating cost of existing trains. One level of demand is needed merely to continue services on an existing line. A higher level must be attracted to justify increasing railway capability.

Railway planers must therefore first assess how railways are suited to the demands that may arise and then make traffic forecasts for the rail services to be proposed.

3.6.1 Suitability of Railway Services

Planning for rail services involves:

- assessment of the *capability* provided by railway infrastructure;
- recognition of *opportunities* to provide competitive railway services;
- assessment of potential *demand* and competition;

3.6.1.1 Capability

The capability of a railway system to carry traffic comprises:

- facilities for picking up or delivering passengers or freight at conveniently located stations or depots;
- railway infrastructure and rolling stock capable of carrying heavy or bulky loads within the limitations of permissible axle loads and loading gauge;
- speed permitted by track quality, alignment and signalling, and achievable by trains;
- train services offered on each line, their frequency, capacity, convenience and cost;
- line operating capacity.

Stations are reviewed in Chapter 4 and freight depots in Chapter 5. Infrastructure, rolling stock, speed and operations have already been discussed in this chapter.

Line capacity was introduced in the last section. A single railway track or a road lane dedicated to bus or tram traffic can each carry about 20 000 passengers per hour in one direction. A road lane can only carry about one-tenth as many passengers if they are travelling two to a vehicle in private cars. In particular circumstances rail can provide even higher peak capacity, e.g. in long commuter trains with wide multi-door access and high 'crush' capacity.

Track capacity for freight varies widely according to the type of cargo, the composition of full and empty train loads and the pattern of passenger traffic sharing the same line. But, by way of illustration, trains carrying 1000 tonne loads at six minute intervals could transport 10 000 tonnes per hour on one track. If freight services were allocated six hours per day for 300 days per year then a total of 18 million tonnes could be carried annually in one direction on a track available for other trains during the rest of each day.

This capacity is high where intensive traffic can be maintained over most of a day. In fact most capacity constraints or line congestion arise out of very high peak time transport demand. Overall capacity can be increased by *directing* freight or *attracting* passengers to use less intensive periods in the middle of the day, at night or at weekends.

However, where railway capacity is ample, the inflexibility and cost of rail transport is unlikely to attract a large proportion of new traffic except by focusing on specific opportunities.

3.6.1.2 Opportunities

Opportunities for increased rail services lie in:

- fast inter-city passenger traffic, mainly on existing routes in Britain but extensively on new lines on the Continent;
- commuter trains to cities, including on lines serving new residential or employment areas;
- longer distance freight traffic, e.g. to Europe, and as 'combined' (part road/part rail) transport; mainly on available existing routes but with new terminal facilities where necessary.

It is therefore in these areas that transport planning should recognise the capability of railways and for which traffic demand forecasts should be prepared.

3.6.1.3 Demand and Competition

Demand for extra rail traffic arises where it becomes more attractive than road or air alternatives. Much of this extra traffic can be accommodated within the capacity of existing rail systems. But, where new transport links must be built, it has to be shown how railway development can be both a commercial solution and one which best suits the environment.

If railways are to be more competitive with roads they must reduce the gap in costs per passenger-kilometre or per tonne-kilometre of freight. Automation of train operations and reversible multiple unit rolling stock have reduced labour charges. Maximum use of existing tracks can minimise capital expenditure. But it is in the efficient use of energy (electricity or fuel) that advantages of rail lie. Higher taxes on fuels to make their prices more nearly reflect their ultimate scarcity might benefit both railway accounting and global sustainable development.

In practice many railway services only remain commercially viable because they are subsidised—generally for social or environmental reasons. However, in the long term, benefits arising from general but environmentally-oriented taxation could be politically preferable to subsidies. Whilst taxes, in one form or another, 'are always with us', subsidies for particular causes are controversial and subject to periodic public spending cuts.

There are marked inequalities between different European countries in the grants, subsidies and taxation applied to transport. As a result there is no 'free market' in EC international rail services.

Meanwhile, to plan particular improvements in railway services, the proportion of total transport demand suitable for rail transport should first be estimated. These estimates should then be refined in the light of actual capability that can be offered under one or more railway solutions.

3.6.2 Traffic Forecasts

Forecasting future rail traffic involves:

- estimating *total transport requirements*, i.e. by any mode for all types of traffic which railway services could provide;
- *allocating* those total requirements among each transport mode, with an estimate of a range (highest and lowest) of proportions which railways could expect to carry;
- *refining rail traffic forecasts* within that range and for each possible railway development solution.

Figure 3.3 shows how these stages in forecasting future railway traffic relate to proposed developments and their environmental consequences.

Forecasts of total transport requirements are usually made through techniques based on transportation trends, macro-economic predictions, local development and land use plans, and travel surveys.

In most modern situations, a great deal of transport is carried by road. Therefore, even if directed ultimately at railway development, any transport

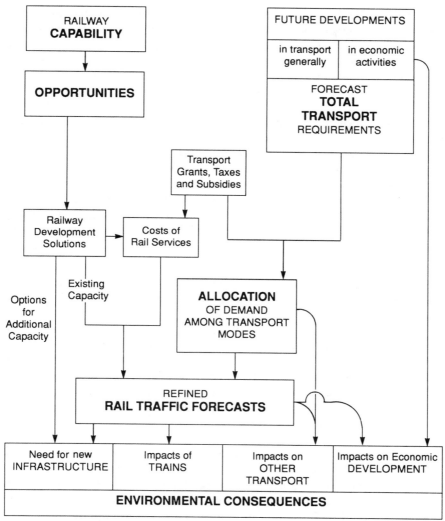

Figure 3.3. Forecasting future rail traffic

study must consider the traffic pattern of the *road* networks. Current or historic road traffic data in Britain is available from the Department of Transport's National 'Trip End' Model and other more recently developed data bases. Such data is derived from traffic counts and sample origin/destination surveys. Models of flows on road systems can be prepared; these models can be used to simulate changes in traffic flows which would occur if road networks are altered or new traffic is introduced.

Railway traffic information is simpler. Networks are closely defined, trains are scheduled in the timetables and ticket sales and freight transport contracts can be analysed. Rail traffic can therefore be incorporated with more complex road (and air or water transport) data to simulate total transport patterns. But a total transport model can be simpler than that of a complete road network if its ultimate purpose is only to study railway traffic. Those minor roads or types of traffic which do not concern the railway route can be omitted.

Forecasting techniques and inputs to computer models are valid only if the assumptions about the future are realistic. Recent historic trends are perhaps the easiest but least reliable guide for forecasting. Periods of boom in any activity usually encourage unjustified optimism about how long they will continue, whilst during recessions there may be misjudgement about how quickly recovery will follow. Recent trends for rural and freight rail use to decrease, and for south-east England commuting and French TGV traffic to increase, may not be typical of what will happen in these sectors in the next decade.

Allocating transport requirements to rail, to public and private road vehicles and to other transport modes is also known as 'modal split' or 'disaggregation'.

Transport services have to be provided by the means, or combination of different means, which strike the optimum balance between costs of provision and operation, benefits to users and environmental impact. Integrated solutions may have to be found for each type of traffic, e.g. commuters, recreation, bulk freight, local goods distribution. In planning and justifying various transport solutions a number of issues have to be investigated and questions answered, viz.:

1. Anticipated commercial performance   Will demand and revenue be sufficient that a railway or bus company can operate profitably?
2. Environmental performance   Which are the preferable modes on a particular route in terms of noise, pollution and land use? What are the global consequences of each solution in terms of energy consumption?
3. Effects of change in one mode of transport on the others (see Section 3.7)   Will improvements to rail services remove bottlenecks on roads?
4. Socio-economic factors   What are the relationships between transport and other aspects of development and social planning? (Section 3.8)
5. Economic and political justification for transport and travel   What part of the anticipated traffic should take priority and what should be discouraged? Are economic or fiscal measures (grants, subsidies, taxes) likely to be applied to effect social or environmental policies?

Rail traffic forecasts can be prepared taking into account:

- high and low forecasts allocated to rail in the estimates of future total transport requirements;
- the capability and comprehensiveness of the railway improvements proposed;
- the changes in demand which have arisen where similar improvements have been made in rail services elsewhere.

3.7 IMPACTS OF RAIL DEVELOPMENTS ON OTHER TRANSPORT

New railway lines are provided to serve specific demands, for example:

- a need for increased capacity, e.g. through or under London;
- new traffic sources like the Channel Tunnel or the planned Denmark–Sweden link
- requirements for higher speed services between cities.

Meeting these demands by rail solutions is the direct objective of such schemes. The indirect impacts on other traffic systems may include reductions of traffic or easing of congestion which might be described as environmental. Figure 3.4 indicates the sort of traffic effects which new railway development could cause on other traffic systems.

3.7.1 Impacts on Other Rail Traffic

New railway lines should capture some of the traffic using existing but less well-suited routes. Very little passenger traffic between Paris and Lyon has been carried on the original SNCF main lines since the TGV routes were introduced; and, whilst the Union Railway will be primarily for cross-channel traffic, special fast commuter services from Kent or Essex may use it to attract a proportion of higher fare-paying commuters from the slower services on existing lines. In Italy extensions of the *direttissima* routes north of Florence and south of Rome are planned to provide high speed, high frequency services particularly to relieve congestion on the busiest parts of the existing network.

These three examples are for passenger-only or predominantly passenger services. Rail freight traffic has decreased over several decades in the face of more flexible and cheaper road competition. But recovery of rail freight could arise if more competitive services could be introduced with new international connections such as the Channel Tunnel. It is anticipated that eventually a substantial proportion of freight which formerly crossed the Channel on lorries will now be carried on to the existing railway networks connecting with the Tunnel in south-east England and northern France. Eventually these networks may have to be further upgraded or even supplemented to

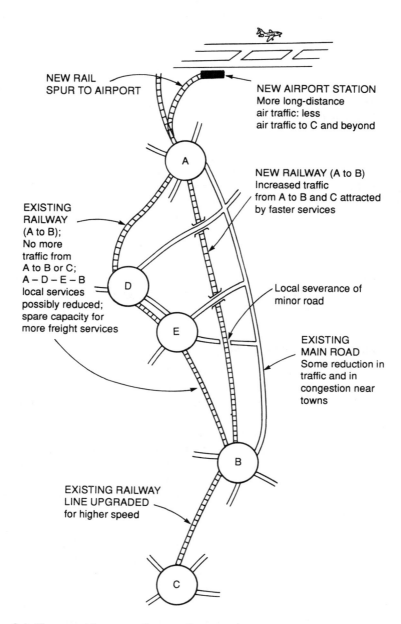

Figure 3.4. Transport impacts of new railway works

accommodate heavier traffic. Alternatively, new rail links to the Tunnel may yet be designed to accommodate mixed traffic.

3.7.2 Impacts of Railway Development on Road Traffic

3.7.2.1 Regional impacts

An important environmental benefit of new railways is removal of traffic from crowded roads. This benefit is being stressed in Switzerland and Germany where announced intentions are to build new railways instead of more roads.

In rare cases, railway construction may be undertaken specifically to reduce or avoid heavier road traffic. Switzerland does not wish to accept lorries weighing over 28 tonnes on its roads. The very long Gotthard and Lotschberg 'Base Tunnels' are planned new railways which, among other things, will provide 'piggy-back' transport of 4.2 m high lorries under the Alps and across Switzerland. Swiss federal approval has been given to the Lotschberg project in principle subject to EC withdrawal of pressure for a road freight corridor through the country for 40-tonne lorries. Exploratory tunnel investigations have commenced.

In Britain provision of rail/road freight interchange depots is intended to reduce the volume of goods vehicles on the roads. The introduction of 44-tonne 6-axle lorries may paradoxically *improve* the situation on roads if the use of larger lorries is restricted, as intended, to journeys to and from *railheads* and reduces the total *number* of vehicles on the roads and their net contribution to air pollution.

Department of Transport (1989) statistics show that, in the previous 20 years, *total* passenger movement by all modes in Britain increased nearly three-fold and freight tonnage doubled. But rail passenger traffic had remained at the same proportion and rail freight had reduced to less than half its 1950s level. Rail passenger-kilometre and freight tonne-kilometre in 1988 were only 7 and 12 per cent respectively of the road and rail combined total. Proportions of rail traffic are higher if short distance movements are ignored. Nevertheless, even a dramatic increase in railway traffic is likely to make only a marginal difference nationally to the much larger volume of road traffic. However, along particular transport corridors such as Kentish motorways the impacts may be much more significant.

3.7.2.2 Local impacts

Local road movements are affected when new stations are provided and when streets are cut by railway development or have to be remodelled. The changes concern:

- vehicles delivering passengers or freight to stations;
- local transport of railway personnel and equipment;
- short journeys altered or forgone by pedestrians or vehicles because their former route is cut off.

For road networks near to stations there are well-established traffic forecasting and management techiques. These have been used in planning new London railway terminal developments and for parkway and airport stations in more rural or motorway-connected environments. Measures to mitigate any adverse effects of changed road traffic patterns resulting from station development can include highway and parking improvements or better integration into other urban transport systems. If neither of these is feasible, the station may be in the wrong place.

Local railway-related road transport will seldom be a problem. Any plan for local road modifications can take into account the access requirements for routine or emergency work on the railway.

The effects of new obstacles to pedestrians and short distance movements are more complex. In Britain, the Department of Transport's practice where communities are severed by new trunk roads is to assess the degree of severance on a four-point scale. Such an assessment could be satisfactory for comparative ranking of the effects of alternative railway schemes. The variables which determine severity include increases in journey length, any partial obstacles introduced and the number of people affected.

3.7.3 Impacts on Seaborne Traffic

A decrease in demand for *ferry crossings* across the English Channel is expected as a result of faster Eurostar, trans-European rail freight and Eurotunnel shuttle services.

Ports served by *intercontinental ships* were a focus for early railway services and continue to be so for some long distance bulk and container traffic. But in certain instances railways can also offer alternative 'land bridge' transport. An example is movement of freight from Japan or China to Europe on the Trans-Siberian Railway. Such services could become more competitive as political, procedural and operational barriers are removed.

Inland waterways remain significant for shipment of bulk commodities on certain routes in northern and central Europe, particularly related to the Rhine and Danube rivers and connecting canals. However, the well-established waterways systems complement rather than compete with railways for freight unsuited to road transport.

3.7.4 Impacts on Air Traffic

Main line railway services to airports can *increase* the long distance air traffic through those airports at the expense of others less well served by surface transport.

High speed inter-city rail services can cause a *reduction* in short or medium distance air travel, considerably at around 400 km, more marginally up to 800 km. Amtrak's calculations for extension of high speed rail services to the New York–Boston line (about 350 km) claim that congested air traffic will be reduced by 37.8 per cent (Railway Age 1993).

However, real reductions in total traffic between two points by *all modes* may be offset when more travellers are encouraged by *any* improved service. Up to a point, supply generates demand.

3.8 IMPACT ON ECONOMIC DEVELOPMENT

3.8.1 Transport and Development

It became almost a truism that transport begets development. In Europe, the industrial revolution arose out of technology but burgeoned because of railways. Later, for instance in Africa, the development of Kenya as a country arose out of construction of a railway from the Indian Ocean to Uganda. The need for improved transport services remains overwhelming in many countries. However, in the wealthier states increased mobility has encouraged the migration of people, light industries and services to semi-rural surroundings. Facilities like new roads may be more of a local convenience than a national economic necessity.

Forecasting industrial and commercial development is a function of economic planning. Environmental interest in development is chiefly in how its implementation will affect land use. Transport and land use are closely related and should be planned together.

Three types of economic development related to rail transport justify examination, viz.:

- pioneering development in less populated country;
- 'developing country' (Third World) situations;
- neo-industrial or post-industrial development.

3.8.2 Pioneering Development by Railways

Most of Canada, the United States, Argentina and the former Soviet Union were opened up by nineteenth-century railways. The process is still continuing in Siberia and parts of Africa. In the former the process is one of virgin land development; this is characterised by new opportunities for exploiting natural resources with all the associated environmental hazards.

In Africa transport-led development started from seaports and proceeded up rivers and then along railways into the interior. Today rail services continue to play a vital part in maintaining the economies of land-locked countries. The 'Tan-Zam' railway, constructed with technical and financial assistance from China, was a large railway construction project by any standard. The 1860 km long line, opened in 1975, runs on the 1067 mm Zambian/Southern African track gauge rather than the East African 1000 mm gauge otherwise in use in Tanzania. Thus it connects the port of Dar-es-Salaam with a distant but major inland freight market.

3.8.3 Developing Countries

Countries like India have long passed the stage of basic transport-related industrial development. Continued heavy use of railways by both passengers and freight has been one result of insufficient public and private financial resources to invest in roads and road vehicles. Economic benefits of rail use in the future may be best ensured through

- rationalisation and modernisation of freight trains and transfer facilities to give goods produced the best chance in free markets;
- provision of subsidised mass transit systems which will result in better industrial and living environments around cities;
- fast inter-city passenger services, as part of the pattern of improved communications which has become a common feature in wealthier, more developed economies.

3.8.4 Neo-Industrial and Post-Industrial Development

Britain is typical of countries where there has been a gradual decline in the volume of heavy railway-served industries involving commodities such as coal or steel. Railways do not serve the same dominant role in carrying freight for the variety of widely based light industries and services in which most workers are now employed. Major new railway construction for traditional industrial uses is therefore rare although there are occasional needs for short links to new factories, power stations or road/rail freight depots. However, better general communications including *passenger* railway services can result in socio-economic or generated benefits.

Socio-economic effects of railway development are impacts on people, such as job creation or housing demand, and are described in Chapter 6. *Generated development* is that which arises because new transport facilities have become available. For instance, high hopes of job creation at Lille are placed on the city's new station on the Paris–Channel Tunnel TGV route (see Chapter 6).

Transport or development economists consider regional and macro-economic factors in attempting to forecast generated economic growth or decline; but it is evidently very hard to quantify such effects. In post-industrial economies the relationship between provision of transport and economic growth is subtle, if it indeed exists. Demand for roads for cars and lorries is influenced by convenience and current profitability rather than economic argument.

However, peripherality—remoteness from centres of activity—can undoubtedly be a constraint on regional development. This constraint can be mitigated by good transport links. Typical peripheral European countries are:

- Portugal, separated as it is from the centre of Europe by the rest of the Iberian peninsular and its non-standard railway gauge;
- Scotland, whose industrial heart is about 800 km from the Channel Tunnel.

Central Scotland is in fact served by two main line railways by which passengers can reach the London area in five hours and freight in half a day. Further transport thence to Europe has to share the bottlenecks in southeast England which will exist until a Channel Tunnel rail link is built. However, Scottish and northern English industrialists have been very concerned lest that link and improvements to its main southern freight connections should be upgraded to accommodate a European *loading gauge*. This expensive action would make any areas where the lines were *not upgraded* more peripheral than they were before the Tunnel opened.

In summary, for Western European conditions, improvement of rail services

- may not in itself provide a significant boost to *regional* economic development, although any improvement in fast communication, distribution and freight forwarding services is encouraging, particularly in peripheral areas;
- is only one element in road-dominated transport systems; any relationship between transport and economic development should be examined in terms of combined (inter-modal) transport for freight and integrated transport and communications systems for people;
- does provide *local* stimulus to manufacturing industry and distribution services near freight depots, and to commercial activities near stations.

The *environmental significance* of rail-related economic development therefore lies in the local impacts of railway stations (see Section 4.6) and freight depots (Section 5.5).

4 Passenger Traffic

4.1 PASSENGER RAILWAYS AND THE ENVIRONMENT

Rail passenger traffic today falls predominantly into two key sectors—inter-city travel and city commuting. Both concern train services to concentrated areas of commercial activity. *Business* is the predominant motive for travel by train.

Based on figures for travel by each transport mode in Great Britain quoted by Turton (1992: 71), reasons for rail travel are:

- Business including education (6%)   57%
- Leisure including social/entertainment   18%
- Shopping (12%) and other purposes (13%)   25%

If the shopping/other category is divided in proportion among the other two categories then occupational journeys—those which *have* to be made—account for three-quarters of the total; the remaining quarter of rail journeys are made for less essential reasons, usually implying a choice, e.g of when to take leisure, where and how to get there.

The high proportion of railway journeys which people are obliged to make relates to that part of the population who find it *most* practicable to use public transport to get to work or school in major cities, or to make business journeys by train to city centres. The journeys arise where it is too time consuming or expensive to travel by car.

Leisure travel does make incidental use of inter-city services and off-peak suburban trains; it also contributes a substantial proportion to the more slender payload on the marginal lines remaining in less populated areas. More significant to total demand for rail services is the proportion of people who find enjoyment and convenience in travelling by train whether for business or anything else. Some businesspeople find rail travel pleasurable and preferable, some definitely dislike it and will avoid it where possible; passenger environment is therefore a significant factor to those travellers who have any choice.

Country dwellers seldom have any alternative to road transport in order to reach a town or railway station. Only in special geographical situations, as in parts of the Highlands of Scotland, do long distance railways still serve isolated areas. However, in the increasing area of well-populated country adjacent to the major conurbations, opportunities for rail access to nearby towns has in some cases led to revived or more frequent services, for instance in the Midlands, South Wales, northern England or Strathclyde.

British Rail (BR)'s passenger traffic rose from 39 billion km in 1952 to 42 billion km in 1957, then dropped to a low of 31 billion km in strike-prone 1982 before recovery to over 40 billion passenger kilometres in the late 1980s (Department of Transport 1989).

Reasons why rail's share of traffic has *decreased* in the past relate to:

- personal wealth, which rose steadily for many years in Europe up to the late 1980s, increasing people's propensity to travel but encouraging private transport and with it choice to live without railway services;
- the costs of private transport, particularly the price of fuel, which has fallen in real terms in recent decades;
- changes in employment and leisure opportunities and location.

Reasons why rail's share may *increase* in the future are faster, more convenient rail services for inter-city and international travel and, within urban areas, patterns of transport where railborne rapid transit systems and public bus services together take priority over private road vehicles.

Uncertainties concern the future structure and competitiveness of railway services and the attitudes of governments whose policies, legislation, taxes and subsidies can encourage or discourage particular modes of transport.

Inter-city trains are increasingly in the high speed category; only thus can they compete with the convenience of private cars or the rapidity of aircraft, neither of which has rail's easy access to city centres. For short and medium distances it is this access to commercial centres which promotes *commuter train* demand. New metro lines provide a higher capacity than new urban roads.

The main *environmental impacts* of new railways are land-take and severance. These may have serious consequences for property and communities in towns and for semi-natural resources in the country. Passenger trains require routes through residential and industrial areas for commuters and along almost straight high speed alignments for inter-city travellers.

Speed is a parameter affecting the amount of disturbance caused by trains. Possible *indirect* impacts of high speed trains related to long-term resources and sustainability which warrant consideration include:

- 'time pollution', by which high speed travel negates any value in the remoteness of distant places—a 'distance-gobbling' effect which 'pollutes space, time and the mind' (Whitelegg 1993: 76–80);
- greater use of energy resources than by slower trains.

Time pollution is a serious concept, whether it is nineteenth-century railways opening up the countryside to create suburbs or a modern tendency

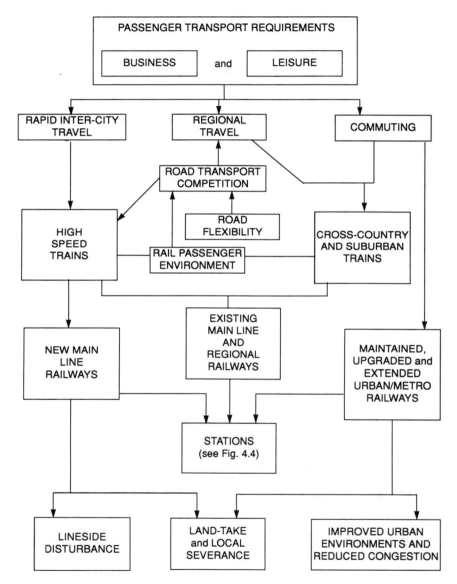

Figure 4.1. Passenger requirements, train services and environmental issues

to overestimate the importance of saving time. The concept is diametrically opposed to many theories of economic growth. However, the ultimate consequences are not addressed further in this book, if only because railways—and public transport generally—may not be the main offenders. Our present concern is to identify the tangible impacts of railways so that they can be compared with alternatives which may be more or less environmentally sensitive.

As evidence in Chapter 8 will show, it is not proven that a high speed train uses up more energy per passenger-kilometre than a slower regional or commuter train. Probably this is because of the high load factor, the maintained high speed and the long distance travelled without stops by the more powerful and well-patronised trains. Rail traffic as a whole is more economical in energy consumption than all air and most road traffic.

There is a strong correlation between national per capita energy consumption and the extent to which people use railways. The percentage of travel (in passenger-kilometres) which is undertaken by rail is 0.4 per cent in the USA, 7 per cent in Britain and 40 per cent in Japan. The energy consumed in tonne of oil equivalent per capita is 1.2, 0.7 and 0.5 respectively (Institution of Civil Engineers 1990: 10).

Figure 4.1 relates passenger requirements and services to the main environmental consequences of running high speed trains and of developing new railways.

The next section reviews passenger services which have been provided during the twentieth century in Britain and those which are being planned for Europe in the first half of the twenty-first century. Section 4.3 describes the characteristics of high speed trains, their impacts on their surroundings, their track alignment requirements and the environment of their passengers. Section 4.4 reviews the issues in providing improved or new lines to accommodate these trains. Section 4.5 deals with urban railways and Section 4.6 with all types of stations.

4.2 PASSENGER TRAIN SERVICES

It is a constant theme of this book that the best use of existing railway infrastructure should be a first planning goal.

Britain's railway network was substantially completed by 1910. Table 4.1 illustrates, in terms of typical train services, how that network has since been used. It shows the number of daily scheduled services and average speed of passenger trains on typical routes at three different dates spanning 83 years. Conclusions can be drawn about the capability of British networks to cope with inter-city, regional and suburban services over that period.

Meanwhile, undoubted and spectacular reductions in journey times have recently been achieved between major cities in continental Europe by means of completely new lines. Plans are in hand for more such lines as well as for extensive upgrading of existing routes. Comparisons of performance possible on new lines with that on existing track are therefore very relevant.

4.2.1 British Train Services since 1910

4.2.1.1 Inter-city services

What were formerly called express train services underwent little change during the first half of the century when steam traction predominated except for electrified services in the South-east. Some of the superior performances in 1910 reflect keen rivalry between different railways services whilst by 1950 the emphasis was on recovery after the war rather than on speed. Subsequent accelerations resulted from dieselisation, extended electrification and the introduction of the 'InterCity' (IC) high speed train (HST) concept. Figure 4.2 shows graphically the effect of some of these accelerations; but note that improved services occurred due to specific investments or operational changes and not gradually. The data in Table 4.1 hides short periods of faster prestige expresses; for instance, from 1935 to 1939 the Silver Jubilee covered the 432 km from London to Newcastle with one intermediate stop in 4 hours (130 km/h) whilst in the late 1980s the Tees–Tyne Pullman ran the same distance at a scheduled average speed of 175 km/h. The first run was ended by the Second World War, the second was slightly decelerated to allow for more stops.

There will be similar periods of sustained improvements during the next 80 years and similar phases of consolidation, rationalisation and perhaps even deterioration. If sustainable development is the goal then planning for railways, land use and social development for less than a century ahead could be considered short-term.

On certain routes the capability of trains to carry passengers nearly twice as fast as before has created a market for twice as many trains. As a result business trips or routine commuting over greater distances have become attractive. Times of 1 hour 40 minutes from London to Birmingham or Bristol, 2 hours to Cardiff or 2 hours 40 minutes to Manchester and soon to Brussels are unbeatable for travel between city centres.

But some routes are less adequate to cope with higher speeds or increased traffic. In southern England a new line is needed to connect London to the Channel Tunnel as well as main line realignments to remove speed restrictions elsewhere. Note that the improvements in speed achieved on main lines elsewhere in England have not been possible in the crowded networks south and east of London.

4.2.1.2 Regional, country or cross-country services

Table 4.1 includes a few typical major cross-country links. These lines have seldom maintained the high speeds of the main London-oriented lines, but

Table 4.1. Typical British train services in the twentieth century

Route	Distance (km)	Average Speed of 3 fastest trains (km/h)			No of trains in 8-hour period		
		1910	1950	1993	1910	1950	1993
1. London to Glasgow	646	75	73	125	3	4	13
2. London to Edinburgh	633	74	80	148	4	4	10
3. London to Newcastle	432	75	84	158	4	7	16
4. London to Truro	450	69	69	103	4	3	3
5. Manchester to Torquay	428	52	49	78	4	3	3
6. London to Leeds	299	78	80	136	12	9	10
7. London to York	304	78	82	169	5	5	16
8. London to Cardiff	266	85	85	141	3	4	9
9. Birmingham to Torquay	296	55	54	89	5	4	8
10. London to Bristol (TM)	191	95	82	121	7	6	10
(to Bristol Parkway)	180	–	–	137	–	–	8
11. London to Birmingham	182	91	77	110	15	9	13
12. London to Poole	183	72	68	101	6	7	24
13. Birmingham to Cardiff	190	64	66	88	5	8	12
14. London to Salisbury	135	81	83	104	5	6	9
15. Birmingham to Bristol	143	69	65	97	5	5	10
16. London to Dover	124	68	74	79	6	8	15
17. London to Ashford	90	64	70	86	7	5	15
18. London to Brighton	81	74	81	95	13	15 +	24
19. Manchester to Leeds	70	60	56	74	14	12	24
20. London to Reading	58	83	83	158	16	17	> 27
							(in 2 peak hours)
21. London to St Albans	32	60	58	96	7	9	20
22. London to Staines	31	53	49	60	5	5	6

Sources: David and Charles (1968), *Bradshaw's April 1910 Railway Guide.*
Bradshaw's Guide, Nov. 1950, Blacklock.
British Rail Passenger Timetable, May to Sept 1993.

have generally achieved modest accelerations and an increased frequency of service.

A large number of duplicate routes or those serving smaller towns were closed in the 1960s. Closures between the 1950 timetable and the 1993 one accounted for nearly half of the total railway route mileage in Britain. Figure 4.3 illustrates the changes which have occurred in two rural areas. In the area where the Great North of Scotland Railway (a relatively small company!) operated north and west of Aberdeen, only a quarter of the route mileage remains on only one route. The surviving service provides 10 trains each way taking 2 hours 20 minutes between Aberdeen and Inverness serving eight intermediate stations. In 1950 there were five through trains or connections between these cities taking from 3½ to 4½ hours with many more intermediate stations and a host of connections to lines now closed.

Lincoln was formerly served by trains in nine directions radiating from the city. Today two remaining lines cross at Lincoln so one can still travel in four directions. However, there are now *more* scheduled trains running out of Lincoln than there were in the past. No town the size of Louth or Fraserburgh can today justify its own branch line railway. Market Rasen and

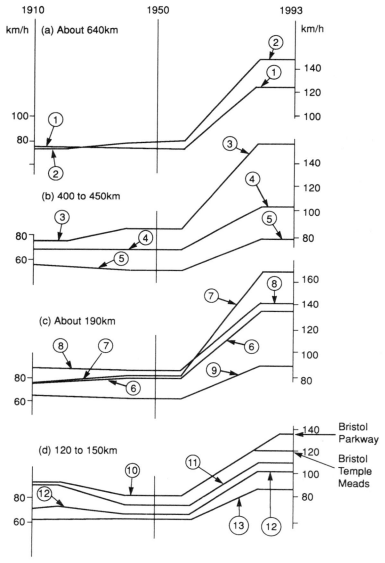

Figure 4.2. Twentieth-century changes in passenger train services (numbers refer to typical services listed in Table 4.1)

Elgin have a rail service only because they lie on routes connecting larger towns. Even the continuation of services to the city of Lincoln must be regarded as tenuous.

4.2.1.3 Suburban services

Table 4.1 shows two services at main stations from which commuters travel about 30 km into London. Services from Staines work into an intensive operating pattern which has been established for many years. There is no track space to permit radical improvements. The service from St Albans is able to use the underutilised and comparatively recently electrified main line out of St Pancras as well as the Thameslink Tunnel through London; the latter is a fine example of how a new service can be provided using old infrastructure (a line formerly used only for very limited goods traffic).

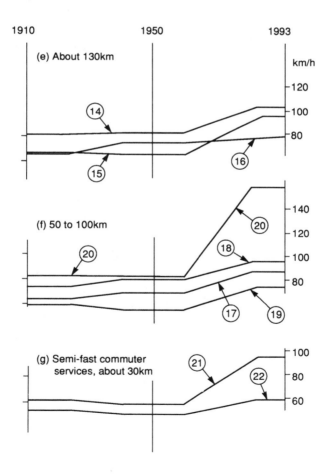

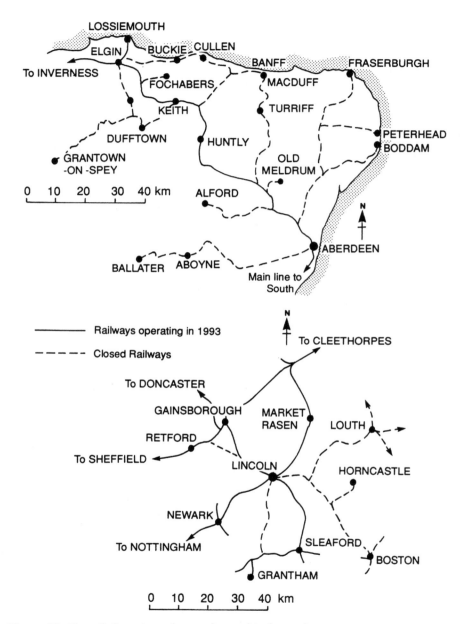

Figure 4.3. Twentieth-century changes in rural train services

4.2.2 Plans for European High Speed Services

The French TGV system and connecting lines are expected by 1995 to result in the following cuts in train times over those scheduled recently (Mathieu 1991):

Paris to Lille 1 h (instead of 2 h)
Paris to Brussels 1 h 20 min (2 h 25 min)
Paris to Amsterdam 3 h 21 min (5 h 16 min)
Paris to Cologne 3 h 9 min (5 h 5 min)
Paris to London 3 h (5 h 12 min)

France and Germany have government-supported commitments to new dedicated high speed railways. In France by the mid-1990s the majority of main line passenger traffic is expected to be handled by TGVs. Of 5700 km used by these high speed trains in France, 1260 km will be completely new line. For an expanded network encompassing Spain and Italy, French Railways (SNCF) plan a total of 4700 km of new line in a total high speed network of 12 000 km by the year 2010. For Western Europe as a whole the Community of European Railways in 1989 announced a plan for a 30 000 km network of which one-third would be new high speed lines (up to 300 km/h), one-third would be lines upgraded for 200 km/h and the remaining third would be existing links or extensions (Mathieu 1991; International Railway Journal 1991).

All this may prove over-ambitious, both in finding capital investment of about £100 billion and in the face of road and air competition. Other difficulties in formation of a pan-European rail network relate to technical differences and allocation of national responsibilities. Agreement has to be reached on electric traction voltage, loading gauge, axle loads, signalling systems and speeds. Operating and maintenance duties have to be allocated and an equitable division of revenue has to be devised. In particular areas, such as eastern Germany, there may be other more urgent transport priorities. But technical and transitional difficulties are unlikely to influence development later in the twenty-first century.

Britain waited for 100 years to be connected directly to the European railway system. The Channel Tunnel is itself a new railway. It is expected to carry all passengers who formerly travelled by train and transferred to ferries or jetfoil. The shuttle trains will transport that proportion of cars and buses whose occupants prefer the faster crossing. Many former airline customers will also take advantage of the dramatically reduced transit times by rail from London to Paris or Brussels and beyond.

Estimates for British Rail (BR Board 1988: 3) envisage Channel Tunnel passenger traffic of some 15 million passengers per year in the first full year of operation rising to about double that amount in 30 years. Even if the average related rail journey in Britain is only 200 km (cf. London to Paris

465 km), the latter prediction represents an addition of 15 per cent to pre-Channel Tunnel BR inter-city rail passenger-kilometres.

4.2.3 When are New Lines needed for Faster Services?

Where completely new rail services have been introduced in Europe there have been marked recoveries in passenger revenue. The number of passengers travelling on TGV *Sud-Est* (TGV-SE) rose from 12.2 million per year in 1980, before TGVs entered service, to 22.1 million nine years later when 85 per cent of the journeys were made by TGV (Whitelegg 1993: 93). Paris and Lyon are premier cities of France; businesspeople can travel from one to the other by TGV in two hours.

The parallel situation in Britain is to travel from London to Cardiff, capital of Wales, within two hours including stops or to Leeds, the major city of industrial South Yorkshire, in a few minutes longer. Although the distances are about a third less than that from Paris to Lyon, the relative commercial positions of these cities are similar in each country. The increases in passengers and profitability resulting from high speed inter-city services in Britain have possibly matched those which followed the introduction of TGVs in France.

Table 4.2 compares scheduled train services on existing and new lines. The services from Paris to Lille in 1992 (conventional electric locomotive-hauled trains) and in 1993 (TGVs) are compared with those to stations (Stoke and Doncaster) at similar distances from London reached via the West Coast and East Coast Main Lines. If the timings were adjusted at the same average speeds for a standard distance of 250 km, then the WCML and ECML trains take 108 and 95 minutes, the 1992 French express train 125 minutes and the TGV 84 minutes. Note that 30 per cent of the new TGV line to Lille

Table 4.2. Comparative performance on existing and new TGV railways

Route	Distance (km)	Scheduled time (h min)	Av. speed (km/h)	
200–270 km				
London to Stoke	241	1 44	139	1 stop (1993)
London to Doncaster	251	1 35	159	1 stop (1993)
Paris to Le Mans	211	1 51	114	Old route (1992)
Paris to Lille	268	2 14	120	Pre-TGV (1992)
Paris to Lille TGV	237	1 20*	178	TGV (1993)
Paris to Le Mans TGV	211	54	234	TGV (1992)
350–450 km				
London to Newcastle	432	2 47	155	2 stops (1993)
London to Newcastle	432	2 20	175	(1985)
Paris to Lyon TGV	427	2 00	213	TGV non-stop

*Line two-thirds open, to be accelerated to about 1 h later.

was not open in 1993 and that further accelerations, probably to achieve a one hour service from Paris, may be achieved. However, similar accelerations are practicable when the speed limits on the WCML and ECML are raised to 250 or 225 km/h as is planned (Prideaux 1991).

The distance from Paris to Lyon, on the first TGV route to be opened, is slightly shorter than the line from London to Newcastle (432 km). But the TGV timing from Paris to Lyon is only 20 minutes faster than that of the 1985 Pullman to Newcastle. Current trains to Newcastle take rather longer because of more intermediate stops—a critical issue in attracting passengers from the various cities along the line.

The TGV *Atlantique* services to Tours and Le Mans achieve average speeds of around double that of the conventional French expresses, although without the intermediate stops. The conventional trains are representative of all pre-TGV services—30 years ago they were among the fastest services in Europe, the new electric engines being capable of consistent running at the then maximum of 145 km/h.

Upgrading of existing lines would not have been sufficient for the accelerations that have been achieved by new routes in France or Spain which have halved some inter-city timings. There are undoubtedly many opportunities for profitable new routes as well as other cases where improved existing lines would be economic. One of the fundamental principles in the planning of TGVs remains that direct TGV services should be provided to 'towns and cities *far beyond* the new lines' (Berrin 1992). Similarly in Spain, where special new lines have been built or are planned, the possibilities of upgrading existing routes for 200 km/h running have been recognised.

4.3 HIGH SPEED TRAINS

4.3.1 Environmental Impact

The capability of modern trains to run at high speed on *existing* lines is a key factor in determining any necessity to construct *new* lines, with their consequences on land resources. The environmental impact of the trains themselves concerns noise and other effects of their operation on people in the vicinity. This section therefore reviews the features and operating requirements of modern high speed trains, describes how they cope with curves and gradients and introduces the subject of noise generation by passenger trains. The passenger's environment is also reviewed.

4.3.2 Train Features

The characteristics of high speed trains are multiple unit composition, aerodynamic shape (streamlining), high power capacity, acceptable vehicle dimensions and smooth riding, particularly on curves.

4.3.2.1 Multiple units

Multiple unit trains are powered reversible sets of rolling stock which are not normally remarshalled between journeys but can easily be added to other units to form longer trains. Individual power units or coaches can usually be detached or added without undue difficulty. A number of variations in the pattern of power traction location are in use. Many high speed trains incorporate power cars at either end, for example:

- diesel-powered Spanish (RENFE) *Talgos* or British (BR) InterCity HST 125s; these operate on existing routes, at maximum speeds of up to 200 km/h where alignment and operating conditions permit;
- electric French (SNCF) *Trains à Grand Vitesse* (TGVs), German (DB) Inter City Expresses (ICEs), Spanish *Alta Velocidad España* (AVE) or cross-Channel (BR/SNCF/SNCB) Eurostar trains; these can operate at their top speeds (270 to 300 km/h) only on specially built routes.

These trains are typically about 200 m long, with eight coaches and two power units, or 400 m long for the eighteen coach Eurostar sets or two TGV *Sud-Est* (TGV-SE) sets running in tandem.

BR's ECML electric InterCity (IC) 225s are push/pulled by single locomotives although operable when necessary from a control unit at the unpowered end.

Japanese Railways' first generation of 'bullet' trains, running on the specially built *Shinkansen* tracks, incorporated motors driving directly on to every axle of the entire 16-coach train. The latest Series 300 stock for the '*Nozomi*' super-expresses use asynchronous motors applied only to about two-thirds of the axles; they also incorporate a number of improvements including a 17 per cent weight reduction through use of alloy construction as well as wheel noise reduction measures and radically revised nose shape.

The layout and location of one or more engine units affects adhesion on the track, the flexibility to rearrange the rolling stock within sets and—with diesel engines—the pattern of noise generation along the train. The number and design of the bogies (pivoted-wheeled undercarriages) also affect noise as well as passenger comfort.

4.3.2.2 Aerodynamic shape

Air smoothing or 'streamlining' is increasingly important, as speed rises, in reducing air resistance as well as lateral disturbance between passing trains. The train nose and the 'skirt' or lower casing of the coaches are the main areas justifying attention in air-smooth design. BR studies (Prideaux 1991: 225) concluded that the pressure wave is as strong for flat-fronted locomotive-hauled trains at 176 km/h (110 mph, the current speed limit on the West

Coast Main Line) as it is for wedge-nosed IC 225s at 225 km/h (140 mph, the proposed speed for an upgraded East Coast Main Line). Similarly, for rolling stock, Semmens (1994) calculated that BR Mark III coaches cause only two-thirds as much resistance at 90 mph (144 km/h) as earlier Mark I stock and less than half that of pre-nationalisation LMS stock.

4.3.2.3 High power capacity

High power is required for rapid acceleration, ability to climb gradients and high top speed. More energy is also required to haul a train in a tunnel than in the open air. Tunnels and gradients are inevitable features of less costly and less environmentally damaging routes through hilly country.

Trains have to incorporate enough power to start (sometimes with only one of two power units operative) on the steepest gradient on which they are likely to be stopped. Signalled stopping points may be located so as not to stop trains intentionally on the severe gradients, such as 1 in 30, which are incorporated on some modern lines. Such gradients are short enough that they can be overcome by the momentum of a train if it is already running at high speed. The price of that high speed is the amount of energy expended. The power capacity of a TGV *Atlantique*, with less than 500 passenger seats, is 8.8 megawatts (11 800 hp). This is 2.4 times that of the electric locomotives that for many years have hauled Scottish expresses, carrying at least as many people, up the 4-mile 1 in 75 climb to Shap at rather more than half the top speed of a TGV. Steam or diesel-electric locomotives hauled the same loads rather less quickly, say 100 km/h point-to-point, with a quarter of the TGV's power. We have seen in Chapter 3 that TGV-type multiple units delivered four times as much power per coach as do most conventional dmus or emus; these last work mainly on regional and sub-145 km/h services. However, due to economies of full trains and the long distances covered, TGV high power characteristics relate to *capacity* and capital cost rather than to energy consumption.

4.3.2.4 Dimensions

Limits to the dimensions of trains (loading gauge) were also mentioned in Chapter 3. Again these are important criteria in determining whether new trains can use what may be environmentally-acceptable existing lines. For British trains travelling to Europe through the Channel Tunnel there is no difficulty fitting into the ample proportions of the European loading gauge. However, the rolling stock design has to cope with some physical differences such as lower station platform levels.

The cross-section of trains becomes a 'kinematic envelope' rather than a fixed shape if they tilt as they pass around curves. Therefore tilting trains have to be of reduced dimension compared with ordinary trains to fit into the same loading gauge limits.

4.3.2.5 Smooth riding

Smooth riding implies freedom from jolts or vibrations, obtained by suitable bogie and suspension design, as well as a comfortable ride around curves. The limitations imposed on the radius of curves by centrifugal force at speed can be partly overcome:

- by increasing the cant, cant deficiency and rate-of-change of cant within safety and passenger comfort requirements;
- by use of tilting trains.

Tilting is designed for the comfort of passengers, not train stability. It is claimed that a capability to tilt in response to centrifugal force can increase permissible, i.e. comfortable, speeds around curves by up to 20 per cent.

Two systems are adopted for tilting trains:

1. *Passive*, in which coach bodies are permitted to swing over to an angle in excess of that provided by the track cant, purely in response to centrifugal force.
2. *Active*, where body sensors or electronic detectors anticipate curve transitions; these bring hydraulic systems into operation which reduce passenger awareness by acting slightly quicker than any passive response.

Japanese railways have used both approaches. Spanish 'pendular' *Talgo* trains tilt passively whilst Italian *Pendolinos* or the Swedish X2000 system utilise active equipment.

Anything which can be done to improve the operational performance of trains on existing alignments and particularly on curves reduces the need for, and environmental implications of, constructing new straighter railways.

4.3.3 Noise

Noise is the most publicised impact of high speed trains and is investigated fully in Chapter 7. Passenger train noise increases as speed rises. Nevertheless there is scope in design of rolling stock to the extent that many new train sets are both faster and quieter than their predecessors. British electric and diesel trains are running at 200 km/h with a maximum noise level (slightly over 90 dB(A) at 25 m) which is similar to that of the previous generation of West Coast Main Line electric trains at 160 km/h. The first series of TGVs produce a measured noise level of over 97 dB(A) at 270 km/h but design criteria for even faster (300 km/h) Eurostar and TGV *Atlantique* trains require that they should not exceed 96 dB(A) even at the higher speed. Most locomotive-hauled trains on BR's Western line have been replaced by quieter Class 165/166 'turbo' trains.

For electric trains the main source of noise is wheel/rail contact. This can be reduced by development of quieter wheel mechanisms and disc brakes on rolling stock. It can be minimised by maintenance (regrinding) of continuous welded rails, by supporting them on resilient pads, and by suitable design of the interface between wheel flange and rail.

Wheel/rail contact can be avoided altogether by magnetic levitation (*maglev*) trains which 'float' at speed. Like aircraft they use wheels to take off and land and they can run at speeds almost comparable with aircraft (400–500 km/h). Their electric engines are very quiet and running sound is believed to be entirely aerodynamic. An environmental disadvantage could be the exceptionally sudden occurrence of this aerodynamic disturbance.

4.3.4 Passenger Environment—Comfort and Safety

Passenger satisfaction is relevant because it influences any preference for rail travel. The net impact on the wider environment is then dependent upon the location, speed and intensity of rail services which these passengers demand. If rail solutions prove the most environmentally friendly, then the use of passenger trains should be encouraged by providing convenient, comfortable, enjoyable and safe train travel.

Convenience of railway travel is related to the location of stations, frequency of services and journey time. High speed is a priority for the regular medium or long distance traveller; to a more occasional passenger the thrill of speed may itself be a significant attraction.

Comfort first requires that every passenger has a seat. This is usual if not obligatory in long distance trains. It is not practicable in many commuter trains in which capacity is measured in 'crush' terms rather than seating accommodation.

Requirements for smooth riding characteristics of trains have already been discussed. Measures taken in modern rolling stock to ensure comfort, such as smooth suspension and air-conditioning, reduce the sensation of speed. But passengers cannot be fully insulated from track irregularities, sudden braking or the passage of curves. The scope for improving conventional spring-and-damper suspensions on carriage bogies to continue to provide satisfactory ride quality on curves at higher speeds is approaching its limit.

The advantages of tilting trains in negotiating curves more rapidly would be offset if the tilting caused any discomfort to passengers. Italian experience on active *Pendolino* tilting trains (Semmens 1990) and Spanish on passively tilting *Talgos* are encouraging in this respect. The issue is important in so far as higher speeds can be attained on existing tracks.

The rush of air past a train should be no problem to passengers. But 'pressure pulse' discomfort, as trains enter tunnels or pass other trains travelling in the opposite direction, has been reported as considerable on the Rome–Florence *direttissima* line (Sullivan 1989). Reduced speeds could be one solution; aircraft-type sealing and pressurisation of the carriages are another. One might question the wisdom of sealing-in trains in view of the possible effects when the air conditioning is defective. Over-heating is a common cause of discomfort.

As will be seen in Chapter 7, noise generated by trains is not often of serious concern to passengers.

Enjoyment of travel *per se* depends upon pleasant station conditions (reviewed later in this chapter) and visual satisfaction. The latter calls for wide windows and well-aligned seating as well as interesting views from the train (discussed in Chapter 9).

Safety of trains has been reviewed in Chapter 3. As far as passengers are concerned, priorities are as follows:

- In serious accidents, capability for rapid rescue of people involved and avoidance of subsequent accidents on adjacent tracks.
- For more trivial breakdowns, rapid supply of information as to what action is needed to complete interrupted journeys.

4.4 HIGH SPEED RAILWAYS

Completely new lines carry *Shinkansen* and *Train à Grand Vitesse* (TGV) services in Japan and France. Others have been built in Germany and Spain and are planned for Belgium, Switzerland, the United States, Australia, Taiwan and Korea. In Italy new sections of *direttissima* line have been constructed from time to time over many years to replace more sinuous routes. We have seen that new TGV lines are planned for a total of 4,700 km by the year 2010 whilst the Community of European Railways is looking forward to an ultimate new track length of 10 000 km.

Environmental concern about new railway routes will no doubt continue to arise because they have to pass through and take land already fully devoted to established rural or urban activity. The stringent requirements for high speed trains on curves or gradients materially affect the choice of route alignment and make small deviations to save particular land resources difficult.

Environmentally-sensitive high speed route alignment has to take topography and obstacles into account simultaneously. How to deal with land resource obstacles—property or conservation features—is the subject of Part III of this book. Problems of railway alignment in undulating topography or across inhabited land are closely related to speed. Comparative freedom of space exists for new routes in much of France and also in Australia or Texas. But there is no such freedom in well-populated areas like south-east

England, northern Switzerland or the coastal strip of Portugal where high speed alignments are nevertheless being planned. In these circumstances the speeds attainable on existing railway routes must be examined before deciding where completely new sections are required.

The power of modern engines is such that it is track condition and alignment rather than motive power which determines attainable speeds. *Maximum speeds* on existing tracks in Britain are currently 177 to 200 km/h for several main lines—225 km/h could be permitted on the East Coast Main Line, as soon as upgraded signalling and automatic train protection is installed, and up to 250 km/h has been proposed as soon as funds are available for IC 250 electric trains to be introduced on an upgraded West Coast Main Line (Prideaux 1991). These improvements depend upon commercial justification rather than on technical feasibility. The only completely new sections of line justified on these routes will be where present alignments do not permit high speeds.

On completely new railways, French TGVs, German ICEs, Spanish AVEs and Japanese *Nozomis* work to maximum permitted speeds of between 270 and 300 km/h. Still higher speeds are planned for other new lines, such as 350 km/h for a new Sydney to Melbourne line in Australia. This speed would be practicable on some sections of TGV and AVE alignment.

The remoter future may see 500 km/h *maglev* trains for which a 43 km long test track is in use in Japan. However, any form of junction facility is difficult to provide on non-contact or monorail guidance systems and their operations offer less flexibility than conventional track systems. Such developments are therefore likely to be justified only for single, very intensively used routes. In such circumstances they could be both environmentally suitable and economic.

Limitations on achievable speeds are posed by:

- curves, since there is a minimum radius which permits smooth running of trains at speeds that are comfortable for passengers;
- gradients, in that trains must be powerful enough to climb them;
- the distance between tracks, especially in tunnels, because of aerodynamic effects for unsealed trains.

The *radius of curves* is the most common constraint on the speed and smooth running of passenger trains—for reasons already described.

Steeper gradients than before can be tackled by powerful modern motive power. TGVs can maintain top speed up short inclines as steep as 1 in 30. The new Cologne to Frankfurt-am-Main *Neubaustrecke* line in Germany has been constructed with gradients up to 1 in 25. In Britain less steep but 5 to 15 km long inclines like Stoke Bank on the East Coast Main Line or Shap or Beattock on the West (steepest sections at 1 in 178, 1 in 75 or 1 in 74 respectively) are no longer seen as obstacles. More than 20 years ago electric trains had topped Shap at the maximum speed then permitted of 160 km/h whilst on a more recent test run a Class 91 electric locomotive hauled IC 225 coaches *up* Stoke Bank at 225 km/h (140 mph), 22 km/h faster than *Mallard's* 1938 world record for steam achieved *coming down*.

Aerodynamic limitations may result from lateral accelerations (jolts) caused by a pressure pulse as high speed trains pass each other in opposite directions. Speeds can be accommodated for air-smoothed trains passing on BR track spaced at 3.4 m between centres. But on new lines a wider spacing is usually chosen—up to 4.5 m for later TGV routes (Roumeguere 1991). However, the pressure disturbance of passing trains is more serious *in tunnels* because of the more restricted space. The effect can be avoided if the expense of boring single tunnels for each track can be justified. Permissible speed will depend upon the size of twin-track tunnel. The incremental cost of a larger tunnel for each increase in speed is therefore relevant. For most new lines the tunnel dimensions adopted have usually been adequate for speeds only slightly slower than the designed open air maximum.

Completely new lines can be built to accommodate modern alignment and track spacing requirements, motive power, loading gauge and overhead electrical equipment—but only at considerable construction and land-take cost and with associated environmental consequences.

In Britain, the Channel Tunnel Rail Link (CTRL) is the most urgent candidate for a new high speed railway. Existing routes across Kent are unsuited to additional or faster traffic but the long distances between Britain and European destinations demand high speeds to compete with air transport. British Rail (BR)'s share of the revenue from trains between Paris or Brussels and London will be maximised if its part of the system can match continental speeds.

In fact the nature of the CTRL route and the need for tunnels and curves are likely to make speed restrictions, rather than maximum speed, the critical factor in determining journey times. At the time of writing, design speeds on the CTRL have not been finally announced. For the original 1989 route 225 km/h was being considered as a maximum on most of the surface sections, 200 km/h in most tunnels and 160 km/h in a long tunnel under London. It remains to be seen what will be practicable for a route entering London from the East, still involving tunnels and possibly accommodating freight trains as well.

As we have noted, there is a capability for relatively high speed running on several well-aligned British routes. *Upgrading of existing routes* is therefore an important option. The first step is usually to remove *local* speed restrictions at the sharper curves where this can be achieved by relatively minor realignments. The next is to raise the general speed limit. The practical difficulties of accommodating speeds in excess of 200 km/h have been identified in Britain as signalling and warning systems, possible impacts with outside objects, level crossings and fencing; most of these are safety features for which solutions are available or being developed.

4.5 URBAN RAILWAYS

The environment of city railways cannot be divorced from that of urban transport generally or from city planning. Therefore it is necessary to consider the urban environment as a whole and how its transport has developed to meet changing needs.

4.5.1 The Urban Environment

Adverse environmental impacts are associated both with the problems of urban transport and with their solutions. The problems concern congestion and the noise, vibration and fumes of traffic. The solutions are new or widened routes or restrictions on vehicles. Unless they are underground, these road or railway routes and their related stations and car parks require acquisition of land and demolition of buildings. Restrictions on traffic reduce the convenience of private cars and affect the economics of goods distribution. Nevertheless, restrictions are increasingly seen as an ultimate solution.

Century-old photographs of cities like London reveal that street congestion is nothing new. Horse-drawn vehicles moved slowly and street space was less organised but as equally densely packed as in a modern traffic jam. The nuisance of horse droppings is probably mild in comparison with motor vehicle emissions but the noise of hooves on cobbled surfaces was often in excess of that caused by modern traffic, especially since the use of horns has diminished.

Clearance of land space for early main line railways and their large city termini did involve widespread demolition of housing, much of it occupied by the poorer classes with few occupation rights. The poor living conditions meant that most housing areas needed eventual redevelopment anyway although some crowded districts, not affected by railway construction, remained unimproved for 100 years.

In Europe, most of this is history. But poor housing conditions still exist in many overcrowded Third World cities today. In such environments, traffic congestion, air pollution and noise are no more than secondary issues; but modern railway systems may be an essential part of necessary urban renewal.

By contrast, in some western cities out-migration has severely depleted the population of central and formerly industrial areas. Redevelopment in areas like the London Docklands aims at repopulation by a wide range of socio-economic groups. It is unlikely to succeed unless it can offer both an attractive living environment and a satisfactory public transport system.

4.5.2 Development of Urban Transport

Before railways, most people had to live within walking distance of their work. Factory development during the industrial revolution required high density housing close to the factories, a pattern relaxed later with the invention of bicycles. In the early days only owners and managers could afford carriages enabling them to live further away.

4.5.2.1 Urban railways

Passenger railways provided two opportunities for change. Firstly, trains from the new city termini could reach outlying areas very quickly. Suburbs therefore arose, from which professional people and managers could travel daily to their city work; artisans and clerks followed and, as average wealth increased and employment in service industries replaced factory work, more people moved to the expanding suburbs.

The second opportunity was the construction of railways for relatively rapid short distance transport within cities themselves. Electric trams, which replaced the horse-drawn variety, were the initial rail solution. But increasing road congestion could only be reduced by providing railways on their own reserved strip. To create more space by demolishing buildings or to tunnel beneath their foundations was much more expensive than constructing lines above the roads or beneath them.

Elevated railways were a common early feature above city streets but they constricted subsequent improvements to those streets. In American cities and elsewhere many were gradually replaced by underground 'subways', often on the same route. In other American and European cities partly overhead and partly underground systems continue to operate. The majority of early underground railways were built on the cut-and-cover principle, excavating far enough below the road to provide headroom for trains and then replacing the road structure on top. But routes along streets were too inflexible, since trains cannot turn corners; and acquisition of increasingly valuable land to construct more satisfactory surface alignments was prohibitively expensive. As a result, later underground routes were bored in tunnels at sufficient depth to avoid building foundations and other underground services and to take advantage of any favourable geological conditions.

4.5.2.2 Urban roads

Road vehicles can change direction easily. A bus can handle as many people as a tram without being constrained to a fixed route. Cars are even more flexible in that they carry people over the entire distance of their journey; but they require a disproportionate amount of street space per passenger.

This deficiency combined with the random nature of their journeys makes cars better suited to medium sized towns and rural areas.

Urban roads have proved inadequate to cope with the number of cars owned by a large proportion of the population. There is neither room for vehicles to move along nor space for them to park on. These difficulties have encouraged town councils to limit the use of cars by charging for their movement in congested areas, by restricting parking space and sometimes by providing an adequate public transport system.

4.5.3 Total Transport Planning

All but the smallest towns have some minimal provision for public road transport—buses and taxis—to supplement pedestrian movements. In medium towns, for instance up to the size of Oxford or Cambridge, buses prove adequate if they are afforded priority on otherwise congested sections of road.

However, for population thresholds exceeding 200 000 to 500 000, public demand justifies a more comprehensive transport system. In North America or Western Europe, where most of the population have cars, the threshold is likely to lie at the upper end of the range. At the lower end, in the less affluent world, few have such opportunity to drive themselves; in the former Soviet bloc the great majority of people rely on cheap public transport. As far away as Almaty, capital of Kazakhstan in Central Asia, transport systems comprise buses, trolley-buses and trams; and, now that the city's population exceeds a threshold figure of 1 million, a metro is being built.

The Soviet strategy was one classed as 'command and control'. Such an approach is generally derided in free market economies where there are major differences in conditions between one city and another and where patterns of employment, housing and travel are often changing. However, parameters such as population thresholds at which rail transport should be introduced were determined from long experience in many cities in the former Soviet Union where the activities of the population were for many years organised on standard models.

Non-communist countries followed very different strategies. One, which might be regarded as the norm in many western cities, is transport planning in response to demand, contrained only as far as current land use policies dictate. This approach is ill-suited to rapidly changing circumstances like those in post-1990 Russia. On the whole demand-responsive approaches favour private (and road) transport where 'command and control' in the less affluent societies favours public (including rail) systems.

Neither strategy is sustainable. More appropriate urban and suburban planning attempts to shape land use patterns, not only to ensure a pleasant and rewarding environment for several generations but also to *minimise* needs for routine travel and therefore for transport, especially private transport.

4.5.4 Land Use and Development Planning

Zones restricting or encouraging housing, commercial, industrial or recreational development should be defined to suit both the urban and suburban environment and a comprehensive transport system.

As in the past, use of common transport corridors into cities saves space and concentrates disturbance. Chicago originally had suburban railroads both above and beneath its streets. Today, outside the city, metro trains run along the central reservation of a motorway; Washington international airport could eventually be served the same way. In three Scandinavian capitals land use, transport planning and decentralisation of urban activity have been tightly coordinated over several decades. This has resulted in continued city accessibility using a limited number of suburban corridors (Turton and Knowles 1992: 89).

For 30 years Singapore's railways have been a key element of a physical planning regime for a city state which, in spite of the island's small size, has accommodated both population and economic growth. The system has facilitated promotion of high density development without overcrowding or congestion. In Edinburgh planning for both transportation and the environment started by looking at forecasts of the city's economic future.

4.5.5 Integrated Transport

Local passenger transport is too closely related to both urban development and total transport planning to justify deep analysis of railway systems in isolation. In a city with a successful integrated traffic policy, travel is more likely to be a pleasurable experience, involving swift and convenient interchanges and use of period or multiple tickets equally valid for use on metro trains, trams or buses.

Requirements for transport in urban environments include the following:

- A comprehensive, convenient and affordable transit system into and within city areas.
- Flexibility to cope with changes in locations of employment or residence.
- Interchange facilities between transport modes (including car parks in outer areas).
- A reasonable choice of mode (e.g. underground railways for speed, surface transit for accessibility and better passenger environment).

'Mass transit' became a catchphrase for city transport in mid-twentieth-century urban planning. The name was appropriate in that very large numbers of peak time travellers had to be moved. However, the term mass rapid transit or simply rapid transit was adopted for most of the systems which were introduced to solve the problems of the 'masses' in crowded trains and of delays and *slow* transit times suffered on congested city streets. Today rapid transit usually means railways, although it can include public road vehicles running in lanes dedicated solely to their use.

4.5.6 Railway Rapid Transit Systems

Urban railways can be above or at surface level, in shallow cut-and-cover or in deep bored tunnels. The first public underground line in the world was the Metropolitan Railway in London, constructed in cut-and-cover to take steam trains from Paddington to the City. Thereafter 'metro' became synonymous with city railways, usually underground.

Supplementing any main line services bringing commuters into cities, there is a range of lighter categories of city railway. The following descriptions are typical of a number of different definitions.

Rapid transit or *metro* usually refers to underground railways in central areas but they may be at surface level or elevated. Often automated, with an on-board attendant, they carry a maximum of 30 000–60 000 passengers per hour at average speeds of 30–40 km/h; stations 1–2 km apart, less in city centres.

Light rail transit implies driver-operated trams or light trains, mainly using surface networks on their own right-of-way with stops every 500 to 1000 m, carrying up to 20 000 people per hour at about 20 km/h.

Advanced light transit is similar, but fully automated without a driver and usually faster.

Light (and medium/low speed) surface rail systems do not require heavy foundations, long curves or elaborate safety precautions; nor do they have to be segregated like conventional railways. They are electrically powered so cause no direct air pollution, make little noise and utilise available rights-of-way such as former rail, highway and off-highway alignments. Light rail tracks can be crossed by vehicles and pedestrians and therefore do not sever communities (Institution of Civil Engineers 1990: 39).

4.5.7 Surface or Underground Systems

In city centres and under major waterways it is usually inevitable that metros should be underground. From the point of view of the passengers and, in less densely built-up areas, of capital cost, surface systems are usually preferable. However, there may be surprising opinions on the subject. When the Bay Area Rapid Transit (BART) was being planned for San Francisco the various proposals had to be approved by the voting public; the latter pay for public transport systems either through taxes or by subscribing to bonds. In Berkeley in 1966 the voter was given two alternative propositions—one for a surface system with only a short subway section in Berkeley city centre and one for complete (cut-and-cover) underground everywhere within city limits. Of those voting, 82 per cent favoured the much more expensive complete subway (Sarre 1975). The subway campaigners argued that the surface line would depress economic values, would be ugly and harm the environment and would be a symbolic cultural divide. Many voters at that time would not have intended to use the rapid transit system themselves but were convinced that they would be able to drive on streets that would no longer be congested once *other people* had taken to the trains.

Singapore's Mass Rapid Transit System consists of both underground lines in the city centre and elevated sections in the environs. The latter are generally in the centre of major highways. It is claimed that these lines have resulted in little disturbance to the urban fabric.

4.5.8 Underground Railways

As long as they were built in shallow depth cut-and-cover tunnels, the rectangular cross-section of metro railways could easily provide sufficient clearance for main line rolling stock or suitable new stock of the same loading gauge. The invention of electric trains, able to operate in more confined space than steam, and of improved soft ground tunnelling techniques led in the 1890s to the introduction of deep circular section 'tube' railways. Above a minimum size in which men and equipment can work, the construction cost of any bored tunnel increases with its cross-sectional area.

Many metro system planners therefore provided only such a tunnel size as would be sufficient for the envisaged trains, designed for short rather than luxurious journeys. Thus London tube lines were tunnelled to an equivalent diameter of less than 12 ft (3.66m) and Glasgow's metro tunnels to only 11 ft (3.35m). But Moscow's metro was originally planned in the light of London Transport's subsequent experience and runs in 15 ft (4.57m) diameter tunnels which allow much more standing room in the wide eight-car trains. For comparison, each tunnel for the much faster cross-Channel trains is excavated at nearly 25 ft (7.6m) diameter.

Thus the size of single tunnel bores is related to the size of trains but also, on longer faster sections, to ventilation and speed criteria. Plans for modern bored tunnels to extend *main lines* across cities are not uncommon; one main line, the Channel Tunnel Rail Link, and possibly Crossrail also are envisaged under London, as well as tube extensions. Construction of new railways under cities like London attracts local controversy of the sort associated with any commercial or infrastructure development. However, an underground solution is likely to be more practicable than any more circuitous surface route.

4.5.9 Environmental Consequences and Future Urban Railways

The primary benefit of metro railways is their much higher carrying capacity per track than that of most road traffic lanes. Railways provide an alternative to congestion on the streets. If rail systems carry a large proportion of the travelling public then the paved areas can be safer and less crowded. Streets can be restricted to pedestrians, cyclists, trams, buses, taxis and a limited number of priority category private vehicles. Thus an effective rapid transit system for city dwellers or workers can complement an improved environment.

Local air pollution, a serious hazard in some cities, cannot be ascribed to electric trains.

Underground railways can cause vibration and settlement problems. Geotechnical and structural analysis and design can provide solutions to the first which generally occurs only in certain geological strata and at shallow depths. Settlement can also be avoided by careful design aimed at controlled behaviour of the ground and other underground structures, particularly during construction. However, noticeable environmental impacts of underground railways often occur as disturbance in the streets above during their construction. Space must be found for surface working sites and access shafts. Equipment and materials must be brought in through the streets and excavated soil must be removed. The problems of allocating land space and managing surface traffic during construction will be major items in construction costs and many even affect the alignment of underground routes and the location and design of stations.

The impacts of new *surface* lines may relate to the visual appearance of the track and stations or to modifications or severance of the road system. Manchester's Metrolink and Sheffield's Supertram network are successful examples of new facilities which integrate with other transport and city layouts by making use of both streets and dedicated alignments. Noise, vibration and land-take present potential problems similar to those of new railways outside cities. Generated noise levels are lower and effects on buildings in close proximity can be attenuated by low barriers. Vibration calls for attention, especially at steel bridges. Land-take is a problem where tall or concentrated buildings have to be avoided or removed but alignment is less critical than for faster lines in that metro trains can accommodate tighter curves.

Experience with light rail lines in urban or suburban areas is that, unless house demolition is involved, their merits are soon recognised by the people they affect (Institution of Civil Engineers 1990: 38).

The basic discomfort of crowded underground trains has been mentioned. Various strategies are available to improve passenger environment, such as:

- provision, in the first place, of wider tunnels and longer platforms to allow for higher capacity trains;

Table 4.3. Metros and Light rail systems (number in operation in 1991)

	Europe except USSR	North America	China and India	World total	Country with most systems	
Metro	28	12	3	78	USSR (13)	USA (10)
Light rail	62	20	0	99	Germany (20)	USA (16)

Source: based on data quoted by Turton and Knowles (1992: 91).

- provision of spacious stations with wider platforms and passages;
- provision of quieter trains, for instance by increasing use of pneumatic-tyred wheels or improved permanent way design; reduction of station noise by applying sound-absorbing material to ceilings, walls and under-platform surfaces (see also Figure 7.9);
- extension of flexible working arrangements in employment to reduce peak hour crush.

The future of urban railway systems depends directly on the extent to which commuter services are required and the price at which they are provided. In Europe most metro systems are at least partially subsidised and most people can readily afford the fares. In the old Soviet bloc the metros were highly subsidised and fares nominal. In the Third World, on the other hand, many people are so poor that either they spend a major part of their income on bus or train fares or they have to walk long distances.

In the western world, rapid transit requires a range of related rail and road based systems, only partly requiring new basic infrastructure if systems which can be upgraded or converted are already in place. In the developing world, city populations are still rising rapidly and more adequate urban transport is urgently required. Table 4.3 shows the paucity of metro systems available for the 700 million people living in Chinese and Indian cities.

It is likely that as many new metro systems will be built in the first half of the twenty-first century as have been built in the second half of the twentieth. But the new systems will have to be planned as integral elements of each developing city or conurbation.

4.6 PASSENGER STATIONS

4.6.1 The Functions and Locations of Stations

Stations play a key role in any railway journey. Their first function is to safely accommodate the trains and to permit them to arrive and depart on time and in an appropriate sequence. This is a matter primarily of track layout and platform allocation.

An equally important function is to accommodate the passengers during the period in which they await their train, buy tickets, meet people, cross crowded concourses or change to other trains or different modes of transport.

Figure 4.4 shows how the needs of trains and passengers at existing or new stations are related to environmental issues.

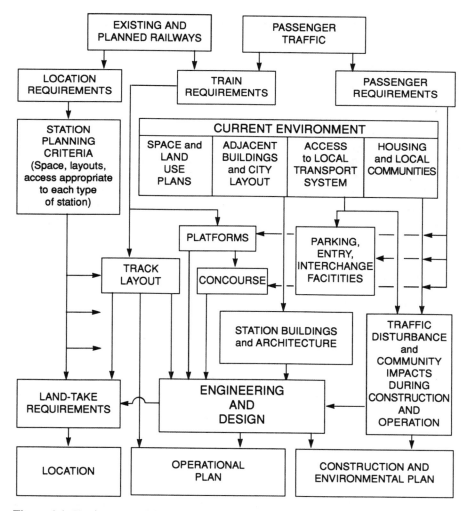

Figure 4.4. Environmental issues in station development

Appropriate location of stations is paramount in attracting custom. Siting of new stations has to take account of:

- access and convenience for passengers;
- available space including at existing stations or on spare railway land, often developed a century earlier;
- environmental impacts and effects on other local commercial activities and transport systems.

Use of obsolete or underused station sites or railway yards minimises expensive land-take. But city stations should also be suited to rapid transit connections. Available disused railway space is a main attraction for the otherwise surprising option of Stratford in north-east London that was at one time mooted as a main station for European traffic. However, such sites do not generally offer the connections to tram, bus and suburban rail networks associated with more central stations; whilst provision of new rapid transit services or connecting urban motorways may disrupt the physical and social structure of local communities.

The environmental relevance of stations lies in their use of land space, often in prime locations, and in their impact on surrounding rural or urban communities. From outside, major stations are also important public buildings whilst their interior merit depends on spacious accommodation of passengers and trains (Figures 4.5 and 4.6). The place of stations as architectural heritage is discussed in Chapter 14. Where a number of commercially viable alternative locations exist, environmental considerations should affect the choice.

4.6.2 Passenger Requirements

Typical situations where new or improved stations are required are:

- in cities, to deal with new high speed inter-city or international traffic;
- at airports;
- in new towns and in planned semi-rural or suburban developments to serve new or revived regional rail services;
- 'parkway' and city 'gateway' stations, for passengers transferring from their cars;
- on new underground lines.

4.6.3 Existing City Stations

Many older European city stations have long and numerous platforms. In steam days extra tracks were needed for engine and train manoeuvres which reversible or push-pull units do not need to perform. Long platforms were provided in anticipation of traffic demands which, except on certain routes

Figure 4.5. Main railway stations are important public buildings and features in urban scenery (Tournai, Belgium)

Figure 4.6. York—spacious platforms and natural light through a wide, lofty curving roof

and in the former Soviet Union, seldom materialised from the 1950s until the days of two-set TGVs or Eurostar trains.

Therefore older stations often have spare capacity. Sometimes excess platform lengths can be used to accommodate new ancillary facilities such as more passenger space; at London Paddington, platforms were shortened at the buffer end to provide more concourse space for passengers in transit. Elsewhere the spare platforms have been able to accommodate new or more frequent services. In a major expansion for more passenger traffic at Utrecht, where there is only lateral space for two more platforms, the plan is to *lengthen* existing platforms to nearly 1 km so that each platform takes two or more trains snaking past each other. Passengers will use airport-type travelling pavements along the platforms (Binney 1992).

The majority of medium speed passenger trains are short—often two or three coaches and seldom more than twelve. However, contemporary international trains of eighteen coaches or more have created a requirement for long platforms which exist at some continental stations, although seldom in Britain.

In the mid-nineteenth century numerous different companies built their own city termini in competition with each other. As a result a large number of competing stations were built, for instance in London, Glasgow or Paris. Some stations have become redundant and some have been redeveloped for other uses; some may still have space available for the new services unless the new traffic is too heavy or the trains are too long. If existing stations cannot accommodate trains then new facilities have to be provided—adjacent to an existing station or at an new site.

In provincial cities, use of a single *hauptbahnhof* (chief station) has proved an acceptable solution. Existing station areas may allow extensions to incorporate facilities for new trains.

4.6.4 Improved and New City Stations

In Madrid the redundant Atocha station was developed and re-opened as the terminus of the Madrid–Seville high speed line. In Paris, TGVs can be accommodated at long existing station platforms at Gare de Lyon and Gare du Nord, although Montparnasse station had to be extensively rebuilt; so many services are now operating or are planned that provision of new platform space for TGVs in Paris has to be considered.

Two alternatives were considered for the North London terminal of the high speed Union Railway from the Channel Tunnel. These were located at two closely adjacent existing stations. At St Pancras, the chosen option, there is already ample accommodation for a terminus provided some of the platforms are extended, involving additional land-take. At nearby King's Cross a more radical option was for a new low level station; the new station was seen as part of extensive commercial and recreational development. The total environmental impact of any new station has to be assessed together with any such development.

The platform extension option has already been implemented at London Waterloo for the continental passenger train terminal, constructed as a slightly elevated and considerably extended structure on one side of the existing station area.

The low level option was chosen beneath Zurich Hauptbahnhof station, also to accommodate new international traffic. The entire station has been transformed within the original shell to provide an environment for passengers which is spacious, well serviced and directly connected to the city shopping centre and bus station (Chicken 1993).

Design features of importance to passengers in *all* stations are those which permit free-flow at entrances, ticket offices, waiting areas and platforms. Mathematical modelling techniques, some developed originally for airline passenger terminals, are available to plan people's movements in both normal and exceptional circumstances.

Additional passenger handling space is required wherever existing stations have to cope with increased *international* traffic. For some years, the platform for TGV trains to Paris at Geneva station offered poor and crowded immigration facilities.

Constructing buildings in the 'air space' on top of surface level stations or siting stations completely underground maximises use of valuable land resources. But the impact on railway passengers and workers can be negative. Artificially lit platforms and subways were a dull replacement for lofty transparent roofs and spacious concourses. Fortunately a trend in the 1960s and 1970s towards utilitarian concrete structures and dark interiors is being reversed for what Binney (1992) describes as 'stupendous stations' in a new heroic age of European railway architecture.

4.6.5 Airport Stations

Many airports are located on existing main line railways or on diverted or extended routes. Main lines serve Zurich, Amsterdam, Frankfurt, Birmingham and London Gatwick airports. Spurs off existing railways were provided as part of major airport development schemes (Paris, Geneva and London Stanstead) and one is being constructed from the West of England Main Line to London Heathrow Airport.

The new TGV station at Paris Charles de Gaulle airport is an example of of what can be achieved in providing welcome facilities for rail transport. As is usual at airports, the station is constructed in shallow cut. But, instead of standing in a covered underground environment, passengers on the platforms look up through the slender steel and glass roof and see the aircraft.

Airports themselves require extensive land resources. Their railway stations therefore make no extra demands for land or new road connections.

Planning of stations at international airports has to be integrated with the movements of *long distance* air and train passengers and their luggage.

If motorists not travelling by air wish to use the airport stations as railway parkways, they can be accommodated in open or multi-storey parking already provided for air passengers. The stark appearance of car parks may be less intrusive in the environment of an airport complex than it would be in otherwise unspoilt countryside.

4.6.6 Parkway Stations

Parkway or park-and-ride stations are stops in rural or suburban locations, explicitly intended to attract car drivers. They provide ample parking space for the many people who normally travel from home by car but see advantage in using train services for the major part of journeys. These stations have been successfully developed in Britain, mainly because of the intensive business focus in London and the fast HST services to the capital. Stations like Bristol Parkway meet business trip demands for a wide rural and suburban car driving population.

Parkway stations have been considered for the various Channel Tunnel Rail Link routes with the intention of supplementing the line's international revenue with that from high speed London commuter traffic. Park-and-ride stations are not included on planned German and Swiss high speed lines and were apparently less than successful in south-eastern France—possibly because insufficient trains stopped at the stations. However, in northern France and the low countries conditions are more akin to those in densely populated southern England. In trying to suit the many cities which cannot be located on high speed routes, compromise solutions may prove attractive to park-and-ride travellers. On the TGV *Nord* route 'TGV-Picardy' station has been located at a motorway intersection between Amiens and St Quentin, neither of which town is on the new line (Freeman Allen 1990).

The main planning issue and environmental impact of new parkway stations is land-take for parking and road access. Rural land required may be of high natural resource value but, where new towns are strategically placed by main line railways, spacious design and appropriate road layouts can make their stations suitable. A search was made for alternative parkway station sites on the East Coast Main Line north of London and accessible from the M25 motorway; it concluded that the existing station at the new town of Stevenage was satisfactory for road access and the best site from the cost, environmental and railway operation points of view. Reading, Watford and Luton are similar main line town stations outside London doubling as park-and-ride stops.

4.6.7 New Towns and New Development

Existing railway stations are usually incidental to, rather than part of, the local transport system in British new towns. The upgrading or construction of stations at Stevenage or Milton Keynes followed later than the main new town development. In more recent European new towns, controlled road traffic and rapid transit rail systems have been included in the basic planning. Railways and stations fit into the basic layouts of new towns like Zoetermeer in The Netherlands.

Even old cities can be adapted to new railway concepts. Inner-city redevelopment was specially planned around a hub at Lyon Perrache, one of the two stations in the city served by the then new TGV services.

In suburbs and semi-rural areas the remaining smaller stations continue to serve commuter and regional trains. In some circumstances new and even reopened stations have been provided. Environmentally these improved facilities cannot be contentious. New introductions are a modest reversal of the extensive closures which have taken place in the past. Where stations previously had goods yards, the space may still be available for redevelopment or for such car parking as total transport planning wishes to encourage.

4.6.8 Underground Metro Stations

Underground railway stations are a considerable convenience for travellers, but they can be congested and claustrophobic unless they are spacious. For metro stations, wider platforms and higher ceilings are no doubt more expensive to construct but provide a more pleasant environment; compare the almost dangerously narrow space on the older London or Glasgow underground island platforms with the spacious areas of the heroically decorated Moscow Metro stations.

5 Freight

5.1 THE PATTERN OF RAIL FREIGHT TRANSPORT

5.1.1 Introduction

'Freight' first meant transport of goods *by water* but is now applied also to land transport. 'Goods' were commodities transported *by rail* at a time when railways dominated the business.

Much popular, environmental and political opinion in Britain favours transfer of heavy goods traffic from the roads 'back on the railway'. This chapter opens by examining the nature of freight movements and the areas in which rail transport is most appropriate. It is evident that most railways have capacity to carry more. So Section 5.2 of this chapter describes how the composition of freight trains, their speed and motive power are planned to make optimum use of that capacity.

Section 5.3 identifies the environmental impacts of freight train operations, mainly on *people*. The impact of rail freight on *land resources* depends on:

- whether new freight-only lines are required or whether existing track, some shared with passenger traffic, will suffice (Section 5.4);
- requirements for modern loading, unloading or transfer facilities which are today more necessary than the intermediate marshalling yards and small goods stations of the past (Section 5.5).

Section 5.6 returns to the apparent environmental preference for more rail rather than road transport. It identifies the financial obstacles and the means which are being suggested for attaining a more favourable balance, particularly in Britain and Europe. The future of freight operations in countries where rail is still dominant is also mentioned.

5.1.2 Competition with Road Freight

Why keep heavy lorries off the roads? Answers concerning environmental impacts on roadside buildings or communities are valid only in so far as these impacts may be avoided by railways. Deterioration of road surfaces and structures and the resultant costs of maintenance are relevant where a railway could accept the traffic with little or no extra track maintenance.

Research was undertaken for French motorway construction and operation companies concerned about the effect of increasing numbers of heavy lorries on their road maintenance costs. The result led to serious examination with French Railways of the possibility of a *new freight railway* as part of a 'rolling road' concept; this would avoid the construction of an environmentally damaging *second* motorway to duplicate the most heavily used part of the existing one (*Railway Gazette International* July 1991).

To enforce environmental/transport policies 'both Swiss and Austrian authorities have severe restrictions on the maximum permitted size and weight of heavy lorries, together with well advanced plans for the provision of piggy-back *trains* for carrying "Eurotrucks"' (Eurotunnel 1991a). In Sweden severe restrictions on long distance road vehicles have benefited rail freight since 1974. German policy has always favoured rail freight.

Two questions must be answered before speculating on any likely increase or change in rail freight traffic, viz.:

- Why are roads more attractive for freight transport?
- In what roles can railways provide a competitive service?

The *flexibility of road transport* arises from the capability of road vehicles, within obvious space and weight limitations, to collect or deliver consignments anywhere. Until the nineteenth century, movement of goods by road had been universal except where waterborne transport was feasible at sea, up navigable rivers or, latterly, on canals. The need to procure an adequate return on investment in railway construction precluded routes other than those serving ports, towns or large sources of raw materials. Roads connected all outlying places to the railway system and have always been used for local collection and distribution services. When motor vehicles replaced horse-drawn transport, railways became even better served by road feeders; indeed railways first provided their own bus and lorry services.

Any *advantage for railways* lay in their capability to carry frequent loads in large consignments. First they proved this over relatively short distances, such as between mines and quays. Later most railways came to thrive on long distance freight traffic. The greater the distance for which carriage was charged, the less handling was required and the more profitable was the business.

This commercial consideration still applies. But, as a result of increasing road competition, various aspects of rail (and coastal and canal) freight gradually became less profitable. There was also a significant change in the nature of goods traffic. The requirement for bulk point-to-point transport of raw materials fell whilst demand rose for movement of a greater variety of widely distributed manufactured or repackaged goods.

The invention and proliferation of containers, first for transport on ships, may have given an initial advantage to powerful railway haulage. But lorries were soon capable of hauling container-sized loads wherever road strengths and capacity allowed.

Competitive rail freight services continue to target those markets for which there is a high demand for frequent point-to-point train loads. The length of the rail journey must be sufficient to be economic compared with equivalent road transport. Not only must railway charges cover the cost of operating their own transport system but they must undercut road tariffs in respect of any extra costs which shippers incur in moving goods between rail depots and points not served by trains.

5.1.3 Breakeven Distance for Economic Rail Transport

According to *The Financial Times* (6 November 1990), BR stated that, with certain exceptions, rail freight transport over lengths of less than 200 miles (320 km) is uneconomic. No doubt the exceptions were regular bulk goods movements such as coal between mines and power stations. More specifically, BR Railfreight (Hansford 1990) have estimated that the breakeven distance at which rail transport of a *single unit* becomes economic is 250 to 300 miles (400 to 480 km).

Other planners (including Kilvington 1990) have asserted that 'the majority of goods' can be moved more economically by rail than road at over 300 km. But, according to a letter to *The Times* (Seymer 1992), even a continental railway operator has claimed that 600 km is the minimum distance over which rail could compete with road for container traffic.

Economic calculations (Runge 1989) for 'rolling road' piggy-back carriage of lorries by rail in Germany demonstrated that such a service might present a real alternative to direct road transport for distances of over 200 km. This very low figure may only apply in Germany where rail involvement is stimulated by the regulatory system and control on journeys by means of permits. The quoted 600 km threshold at the opposite end of the range might assume that the volume of traffic is modest or that the origins and destinations of the cargo are relatively remote.

The average threshold distance at which rail freight is viable could be reduced in the future if road transport costs rise faster than those of rail. Environmental pressures may eventually have this effect although the timing and extent are unpredictable. More practicably, containers or other unitised loads can be moved by both road and rail, each on appropriate sections of a single journey. The feasibility of inter-modal transport is more dependent upon the relative location of the transfer facilities to the points of final origin and destination than to the distances involved.

5.1.4 Bulk Freight

A large proportion of railway bulk goods movement was in coal in the days when that fuel was dominant in industry and homes. Coal production in Britain in 1950 was 230 million tonnes. By the end of the century it may have dropped to one-tenth of that amount. Energy sources are now mainly fuelled by natural gas (transported by pipeline) or by petroleum products (also moved by pipelines as well as by rail and road tankers). Bulk coal use is now mainly in electric power generation; even that market is threatened by use of liquid and gaseous carbon fuels which are cheaper and more readily exploitable whilst they last. The amount of coal carried by rail has accordingly been greatly reduced. In Eastern Europe, however, the pattern of energy and industrial fuel use, based on both 'hard' and 'brown' coal, did not change materially in 40 years from the 1950s. Subsequently, in changed economic circumstances, there has been a collapse in the demand, production and therefore transport of raw materials.

Nevertheless the total quantities of bulk goods carried by rail in countries like Russia are so great that railway freight is likely to remain dominant in world transport. Nor is the demand for mineral traffic in China or the southern hemisphere likely to be diminished in the long term. Indeed in China the current economic resurgence has produced greatly increased demand for freight transport, outstripping capacity on some lines. Goods concerned include coal, a commodity which, perhaps surprisingly, is also increasingly being carried on US railroads. In the latter case this is partly for environmental reasons; eastern power stations are seeking lower sulphur coal that can be supplied from western mines.

Iron and steel production has fallen off with the decline of heavy industry, first in Western and then in Eastern Europe. Building materials like cement and concrete aggregates continue to be carried in 'block' semi-permanently coupled point-to-point trains; so are fertilisers, some petroleum products and train loads of manufactured goods, including cars.

5.1.5 Small Consignments of Goods

Formerly most railways were obliged to provide a 'common carrier' service over the whole of their system, i.e. to transport all types of merchandise whatever the size and distance and including road delivery at either end. For the smaller consignments and over the shorter or least used routes this service was often provided at a loss; tariffs were too low, being imposed in the face of cheaper road competition.

Smaller items were packed in wagon or part-wagon loads at numerous goods stations. These wagons were shunted and collected by 'pick-up' goods trains which proceeded slowly from station to station fitting into the scheduled train paths as best they could. The wagons were then sorted at marshalling yards into train loads for their next—final or intermediate—destinations. Dr Beeching claimed that, by the 1960s when he became Chairman of British Railways, the average British railway wagon carrying general merchandise passed through four and a half marshalling yards per transit. As a result its average speed from loading to unloading was ½ mph (Davies

1982: 14). Considerable rationalisation followed, including a reduction in the number of resorting processes. But, in islands the size of Britain, comparatively few journeys are long enough to make rail wagon loads an economic form of transport and their use has been all but discontinued.

5.1.6 Containers and Intermodal traffic

The type of goods which travel in containers or truck loads is often of high value, produced in new, sometimes isolated, factories and distributed from out-of-town warehouses. These depots are frequently situated where there are no railway facilities. Therefore if railways are to continue to carry general merchandise widely it must be as a contribution to inter-modal or 'combined' transport.

Combined transport (CT) makes use of both road and rail vehicles to carry transferable loads. The units which are used to contain these loads include the following:

- Stackable, top-lift *containers*, also referred to as 'deep-sea containers'; these form the major proportion of current CT traffic; they are internationally standardised boxes, suitable for cellular storage, come in 20 ft (6.1 m) and 40 ft (12.2 m) lengths and are counted as TEUs (twenty foot equivalent units).
- Non-stackable, bottom-lift *swap-bodies*, sometimes called 'inter-modal containers'; these are usually lightweight, less standardised and more fragile than conventional containers;
- *Demountable wagon bodies*: a concept under development for use within a generous loading gauge; and tractor-less *semi-trailers*.

Whilst road/rail transfer is a relatively new concept, transfer of containers between ships and trains is long-established global practice.

5.1.7 Rail Freight in Britain and Europe

Transport statistics for Britain show that only 12 per cent of total road and rail freight transport (tonne-km) was moved by rail in 1988 and is probably still declining (Department of Transport 1989). BR's share of freight over distances exceeding 150 km was 15 per cent and over 500 km was 27 per cent (Eurotunnel 1991a) Corresponding figures quoted for Germany were 35 and 41 per cent at the same distances and recent French performance was similar. However, it appears that the proportion carried by rail has dropped sharply since the mid-1980s in both countries.

The average freight journey by *any mode* in Britain is less than 150 km. In view of the higher threshold figures for rail viability, this partly explains the low proportion of freight carried by rail. Somewhat surprisingly therefore, the average length of a *rail* freight journey in Britain in the 1980s was only 120 km; this may now be even less with the abandonment of most wagon load traffic. The distance for railborne containers was 300 to 400 km and for agricultural, food and drink products nearer 500 km; these are items for which most origins and destinations are not served by railways. The relatively low average figure arises from the predominant carriage of bulk commodities such as coal and coke (65 km), metals (140 km) and construction materials (150 km); these three classes of goods together made up over 80 per cent of the tonnage and nearly 60 per cent of the tonne-km in 1988. Oil and petroleum products are carried an average of 260 km but the railways carry this traffic only where there are particularly convenient railheads, most of the traffic being taken by coastal tanker, barges or pipelines.

Within Britain rail freight is likely to continue to find markets in:

- bulk cargoes like cement, construction materials or coal;
- combined transport;
- longer distance hauls of unit loads.

A considerable increase in rail freight is anticipated following the opening of the Channel Tunnel. This is largely on the grounds that direct rail connections to European destinations will greatly increase the length of journeys above the economic threshold. Railway wagons have crossed the English Channel on ferries for many years but the speed of the direct rail crossing through the tunnel should greatly encourage this traffic.

To carry freight through the Channel Tunnel BR planned 750 m long trains, each with a trailing weight of 1600 tonnes. The tonnage of rail freight across the Channel is expected to rise from the former (train ferry) figure of 2 million tonnes to 6–7 million tonnes very soon, increasing to between 8 and 16 million tonnes, according to different forecasts, within 30 years (BR Board 1988).

In western continental Europe we have seen that the proportions of freight traffic carried by rail have been as much as double those in Britain although the trend is still downward. The greater continental proportions may be due to the longer average distances over which direct rail services have been available; they may also be due to the superior organisation of road freight distribution in Britain. In Eastern Europe railway-borne freight proportions are much higher but the change from state-run to commercial operations will undoubtedly affect the issue. Wellner (1991) reported that in former West Germany long-haul freight had fallen to about 25 per cent of the total whereas on the eastern side of the country railways still carried 70 per cent.

Any scope for an increase in the proportion of freight carried by railways in Western Europe is very dependent upon fair competition as well as on railway operators' readiness to provide appropriate international and inter-modal wagons. It could also be encouraged by international marketing and control systems more akin to those operated successfully by private enterprise companies on the roads.

5.1.8 Very Long Distance Freight

In Eastern Europe and Russia both industry and transport continued until recently to follow the pattern of the 1950s. Centrally planned economies maintained coal and steel production by creating internal markets rather than by permitting private enterprise and Western style post-industrial economic changes. As a result the goods transported and the type of railway operations remained substantially as they were.

In the former Soviet Union this industrial background combined with the wide market and huge distances was well suited to rail freight. Recent volumes of traffic have approached 4000 billion tonne-km, about 10 times the equivalent figure for the rest of Europe and 27 times the tonnage for the EEC/EFTA countries.

Further electrification, even on some long distance routes, and increased containerisation of general merchandise have been common developments. But in Eastern Europe, India or China, or indeed in the western United States the formation and operation of long distance freight trains has not changed substantially in recent years. This is either because the pattern of industry has not radically changed or, in the more commercially competitive conditions of the USA, because rail traffic can still be commercially attractive over the distances involved—often exceeding 1000 km. In the USA the distances involved are so great compared with those within Europe that 47 per cent of freight was carried by rail (and 21 per cent by inland waterways) in 1986 (International Road Federation 1989). In the former Soviet Union even greater distances and a lack of road competition explain an even higher rail proportion of over 80 per cent, much more than in any Western European country.

The Trans-Siberian Railway, as a land bridge for traffic between the Far East and Europe, has been cited as a serious challenger to much longer sea routes. However, the problems of severe climatic conditions, unbalanced and unreliable services and inadequate freight handling services will first have to be overcome.

Modern freight trends such as containerisation and rail carriage of road lorries can be implemented without great difficulty *provided* no great change in the track infrastructure or loading gauge is required. A high volume of traffic and revenue has to be ensured to justify any improvement to the infrastructure. In countries like the USA, where railroads are no longer profitable for passenger traffic, freight has to be sufficient on its own to support existing operating and maintenance costs.

5.2 FREIGHT TRAINS

The nature of freight trains—their speed, power, composition and frequency of running—determines what disturbance they may cause in their surroundings. Their operational requirements determine the extent to which they can use existing rather than new railway infrastructure.

The following trends in modern railway freight services may be noted:

- Higher speed, already 120 km/h on BR and 140 km/h proposed for the next generation of rolling stock; maximum speeds of 160 km/h are being considered for Britain, planned for German freight and are already offered by SNCF's fastest freight choice 'Fret Chrono'.
- Increasingly powerful or multi-unit locomotive haulage to achieve these speeds.
- 'Block trains'—sets of rolling stock, semi-permanently coupled together and suitable for repeated 'merry-go-round' operations on the same route.
- Combined transport operations.
- 'Piggy-back' carriage of road vehicles.

Higher train speed achieves faster delivery to destination or transhipment terminals, better use of track capacity and improved compatibility with passenger service operations on mixed traffic lines. A capability for fast delivery is advantageous not only for urgent and perishable goods but to give flexibility in logistic planning; for instance, many retail and manufacturing companies with limited storage space seek 'just-in-time' delivery for which a freight company needs a range of available speeds and tariffs. Faster freight trains can provide the range capability if schedules and line and terminal operations can ensure the essential reliability.

The effect of speed in line capacity can be illustrated by BR's calculations for the Channel Tunnel (BR CTRL: 1988). These showed that a 120 km/h train through the Tunnel occupied 1.5 to 2 'standard paths' whereas a 100 km/h train required 3.5 to 4.5.

In noting these trends for faster running of freight trains in Western Europe it should not be overlooked that most of the world's rail freight (mainly in Russia but also in the USA) consists of very long train loads on very long journeys. The weight of these trains requires greater motive power but running speed, whilst important, is less critical in optimising track usage.

More *powerful haulage* is not necessarily achieved by design of yet more powerful locomotives. With steam the necessity to raise steam and to drive each engine individually led to the use of huge units such as the Union Pacific 4–8–8–4 'Big Boys'; these had a rated tractive effort of 135 000 lb (61 000 kg) and were capable of developing 7000 horsepower (hp). With the push-button starting and multiple operation capability of modern engines, the necessary power capacity can be made up of suitable numbers of units developing the optimum power output for each type of traction—up to

4000 hp for diesel units, about twice as much for electric. Adequate combined braking capacity must be provided for the weight of trains hauled and the gradients to be descended.

In Britain single or double units, providing 2000 to 4000 hp, haul heavy freight trains up to 4000 tonne gross, although average loads are less.

Typical of North American operations are Rocky Mountain coal trains of 11 000 tonnes headed by four 3000 hp units in front with two more in the middle; another four units are attached at the rear up the steepest gradients; thus total power availability is 18 000 to 30 000 hp.

In the former USSR 9000 tonne trains are standard on some routes whilst lines specially dedicated to iron ore or other minerals utilise still heavier trains; LAMCo in Liberia run bauxite trains of up to 12 000 tonnes on a 274 km route whilst Hammersley Iron's 384 km long line in Western Australia uses 23 000 tonne trains, each 1750 m long. A new coal line in South Africa carries 21 500 tonne trains and a similar iron ore line 20 800 tonnes. On the latter the ultimate carrying capacity was demonstrated by movement of a world record 70 800 tonne train. This required 50 000 hp from 16 locomotives at the front, centre and tail of the 7 km long train. Running 20 000 tonne gross weight trains to provide a 50 million tonne annual payload movement requires 8 to 10 trains daily in each direction. Clearly double track is required as well as extensive facilities at either end for loading, unloading and removing or remarshalling defective rolling stock.

Block trains on mixed traffic railways serve a wider variety of purposes. Transport of bulk coal, fuels or construction materials or container 'liner' services running to fixed schedules make them the freight equivalent of multiple unit passenger sets.

Block train objectives of fast, regular, quick turn-round services between two points have to be met by efficient terminal operations. They also demand strict scheduling to accord with the requirements of other traffic on the same route. On long distance remote lines with more limited traffic, unable to afford modern control facilities, block trains may require special priorities. For instance, in Sudan on a little maintained and under-controlled route, track occupation priority and dedicated locomotives and rolling stock together ensured relatively rapid transport of live cattle from Khartoum to Port Sudan.

Rail wagons for *combined transport* (CT) operations have to be capable of carrying the standard unit loads also transported by road vehicles. Limitations are imposed as weight restrictions on road lorries and as loading gauge and axle load limits on trains.

Inter-modal *multifret* wagons for cross-channel services can carry containers or swap-bodies up to 2.77 m high (Bennett 1991) which apparently embrace all such units as are accommodated by standard road trailers in Britain (Semmens 1992). Most railway flat wagons could carry a 60 ft (18.3 m) length of load, i.e. 3 TEUs, sometimes restricted to 2 TEUs to reduce axle loading. However, conventional railway flat wagons carrying a 2.9 m high ISO (European) container on a platform 1 m above rail level would foul the BR loading gauge.

Lowliner wagons have therefore been designed to avoid this by reducing the deck height to only 0.7 m; this is achieved by providing smaller wheels (RGI March 1990: 171).

Where ample loading gauge is available, as in the open spaces of western USA, and where high axle loads can be carried, double-stacked container trains can be operated.

Piggy-back carriage on railway wagons of complete road vehicles or trailers is similar to roll-on/roll-off operations on ships. It will be a major feature of Eurotunnel's shuttle train services for both freight and passenger vehicles as well as over the longer distances proposed in Europe.

BR clearances do not usually permit 'piggy-back' carriage of complete road vehicles; but the method is proposed for detachable trailers, the most commonly used variation of this type of traffic. In France, *Kangourou* wagons can carry lorries which would otherwise foul even the generous European ('Berne') loading gauge. This is achieved by small wheels and slots in the wagon platform between its axles into which the road vehicles' wheels can fit.

Wagon load and *mixed trains* are still important features of freight operations in some parts of the world, including continental Europe. The location of the terminals, the capability of their cargo handling facilities, the volume of total traffic between key points and the extent to which wagons have to return empty all determine overall delivery times.

If *all freight trains* on any route can be organised to run at a similar length and point-to-point speed, regardless of the type of wagon and merchandise, then the carrying capacity of the track can be increased. Any need for more track route for intensified services might then be avoided and the impact on *land resources* minimised. However, some direct impact of freight trains is felt by *people*.

5.3 ENVIRONMENTAL IMPACT OF FREIGHT TRAINS

A key transport impact of increased rail freight would be reduction of the volume carried by other modes, particularly on congested roads.

Impacts of the trains themselves on the environment through which they run could arise through any pollution or hazard caused by the cargo they carry. However, as with passenger trains, the main perceived impact of freight trains is noise.

5.3.1 Pollution and Operational Hazards

A possible cause of local air pollution could be dust from open wagons carrying coal, iron ore, aggregates or excavated soil. But rarely if ever have specific objections to such traffic been raised. Fine-particled or noxious commodities such as cement or fertiliser are carried in closed wagons.

Derailment or collisions of freight trains carrying hazardous substances pose possible dangers. BR arranged an impressive and satisfactory 160 km/h collision test with a train carrying a nuclear waste flask. Serious accidents with hazardous freight are certainly less common than on the road; obstacles placed on the track are a possible source of derailment but the robustness of containers and isolation of the railway track would normally be expected to afford adequate protection. A more likely danger is fire at sidings where stationary tank wagons are being loaded or unloaded with volatile petroleum products.

From France there have been reports (*New Scientist* 30 January 1992) that the government wants to encourage hazardous goods to be carried by train 'for safety and environmental reasons'; there is even speculation about a new railway line to carry lorries loaded with such goods.

5.3.2 Noise

'BR receive most complaints from residents adjacent to lines where diesel locomotives under heavy load are climbing a gradient or where, for example, they stand for a long time with engines idling at signals' (Institution of Civil Engineers 1990: 37). This observation is more likely to apply to heavy diesel-hauled freight trains, often waiting for a train path, than to faster, more tightly scheduled expresses.

In the same way as for other modes of transport, people expect more stringent noise limitation at night, i.e. either quieter or fewer freight trains. This is a problem where freight traffic is expected to increase, as it may through South London as Channel Tunnel freight builds up.

The noise of freight trains is determined by the type of traction, wagons and track conditions, i.e. the locomotives and the effectiveness of bogie suspension and heavy axle loads on rails and rail joints.

Data for BR coal trains at moderate speed shows noise values substantially *higher* than for much faster passenger trains. The same is apparently true of US freight traffic (Wayson and Bowlby 1989). French reports indicate that freight trains are definitely *quieter* than TGVs but that *some* lineside residents find night freight traffic more disturbing. Evidently there is a great deal of variation in the nature of freight train locomotives and rolling stock and possibly in the circumstances in which noise measurements are made. However, one explanation for the apparent national differences lies in the use of noisier diesel motive power for freight in Britain and the USA. With French electric traction the sound of engines becomes negligible.

The subject of train noise is dealt with in more detail in Chapter 7. Meanwhile it is apparent that electrification, track alignment for fast mixed traffic operations and technical advances in wagon design can be beneficial in redu-

cing the noise of freight trains. Further improvements, for instance in wheel/
suspension systems of wagons, may follow *if* there is sufficient evidence that
the noise of freight trains is a widespread problem.

5.4 FREIGHT RAILWAYS

Freight trains can operate:

- on mixed traffic lines, by sharing tracks with all passenger traffic;
- by using tracks only lightly used by passenger trains;
- by sharing 'slow' tracks with the slower passenger traffic;
- by provision of freight-only lines.

Only if rail freight opportunities or demand require new track capacity
will freight lines create a new threat on land resources.

5.4.1 Mixed Traffic Operations

Passengers require regular, probably 'regular interval', strictly timed services
during the day plus additional peak time services. If freight trains share the
same track they have to fit into available 'train paths' in the non-peak hours;
if the demand is heavy most freight may have to be moved at night.

To achieve the maximum track capacity on any section all trains should
travel at the same speed. Where speed capabilities vary it will be the slowest
trains which dictate the pattern of working.

In spite of modern motive power, braking capacity and permissible speeds,
freight trains could not cope with the *steep gradients* incorporated on some
lengths of TGV railway. The somewhat easier ruling gradient of 1.8 per cent
(1 in 55) of new Italian high speed lines is unsuitable for trains carrying bulk
materials like coal or concrete aggregates. But such gradients are not far
beyond the capability of fast inter-modal freight trains; whilst multiple locomotives already haul 700 m long trains up slopes as steep as 1 in 38 on
trans-Alpine routes.

On existing British routes, with few exceptions, gradients are acceptable
for all types of freight train.

Curvature limits train running speed. Curvature in itself therefore poses no
problem to freight trains where track can carry passengers at greater speeds.
But the appropriate amount of superelevation or 'cant' of the rails differs for
different train speeds. It is possible to design mixed traffic track alignments
by adopting compromise cant angle on horizontal curves. But if heavy
freight and fast passenger trains share the same track, the rate of rail wear
can greatly increase by grinding of wheels on the inner rail by the slower,
heavier trains.

5.4.2 Sharing Routes with Light Passenger Traffic

There are many railway routes where these problems do not arise, for instance on straight track or where fast express passenger services do not run. Typically, the line from Namur in Belgium up the Meuse valley to France is used as a major international freight link. Passenger trains are regular but of branch line standard with services across the central section only in summer. The *environment* is that of a primary freight route, the rumble of long diesel-hauled freight trains reverberating across the valley where the short electric passenger trains are barely noticeable.

5.4.3 Provision of Separate 'Slow' Tracks

There is not much difference in speed between that of modern freight trains and many non-express passenger services—commuter or regional services. Therefore, on main lines, expresses can use one pair of tracks—at 200 km/h —goods and ordinary passengers trains another pair—at up to 120 km/h. Accordingly four-track main lines are common within 50 km or more of major cities like London.

The capacity to carry increased national and European freight throughout the day on certain sections of British two-track mixed traffic lines may be doubted. Goods transport could be limited to non-peak and night traffic but this cannot be achieved throughout the longest distance journeys. Provision or extension of four-track sections may therefore have to be considered. The environmental impacts involved will be those of corridor widening or realignment.

5.4.4 Freight-only Lines

In Europe the majority of freight-only lines are those where passenger services no longer operate but goods traffic justifies retention and maintenance of the track. The vast majority of railway route mileage still in operation in the United States is now freight only.

Other freight railways connect mines or factories to existing networks whilst some have been built over long distances for specific large-scale operations such as moving minerals to ports or industrial plants. The railways already mentioned in Australia or South Africa are outstanding examples.

The construction of railways from mines continues where extraction at new deposits is economic. One South African coal line was recently doubled to increase its capacity to 44 million tonnes per year. This involved duplication of many substantial bridges, tunnels and viaducts. A new alignment was adopted for the new track to reduce a severe gradient for loaded trains.

In eastern Siberia new lines were built for general but primarily freight use and to 'open up' and develop previously remote areas. This development is blamed for exploitation of the 'taiga' forest, claimed by some ecologists to be as serious as cutting down the Amazonian rain forest. Where commercial development damages the environment in remote places, the transport system enabling that development must take a large part of the blame.

In already developed country the implications of land acquisition for completely new freight routes are similar to those of any type of railway. Therefore the maximum use of existing routes and transport corridors may be the best solution for preserving land resources. In Europe this can be achieved by restoration of routes previously abandoned because of falling passenger demand, but only if the land concerned has not already been irrevocably committed to another use.

In densely trafficked regions, like Western Europe, ideas are being floated for provision of completely new freight lines. We have referred earlier in this chapter to French consideration of new freight lines to reduce motorway wear and to avoid motorway duplication. The example examined by SNCF envisages 900 km of new railway built to a large loading gauge for carrying lorries 'piggy-back' between Paris, Dijon and Lyon (RGI July 1991; 437). Whether such a scheme could become a reality will probably depend on:

- convincing evidence not only that the cost of such a railway would be less than that of another motorway, but that its economic performance would be superior;
- confidence that the necessary finance would be forthcoming in the face of other transport priorities;
- evidence that the environmental impact of a new railway would be less severe than that of an alternative motorway (perhaps easy to prove) or of upgrading an existing railway route (more difficult).

Where existing lines are to be adopted for international freight trains, the question of loading gauge compatibility arises. In Britain it has been proposed from time to time that most of the network should be reconstructed to European UIC (or 'Berne') loading gauge for Channel Tunnel freight. BR did not favour such action for any part of the network believing that it would be commercially unviable (Dynes 1990; BR Railfreight 1990). If certain lines, such as from Kent to London or Birmingham, were to be expensively modified to take continental wagons, then the other parts of the island beyond the UIC railheads would be put at a major disadvantage. Preference has therefore been given to provision of wagons that can carry international containers and other unit loads within the narrower British loading gauge. However, only UIC gauge wagons can carry most lorries 'piggy-back'.

A choice between high speed passenger, heavy freight or mixed traffic alignment criteria has to be made at the route selection stage of any new railway. Environmental assessment cannot be completed until that choice has been made and the basic alignment criteria have been determined.

5.5 TERMINALS AND YARDS

Where wagon loads remain an important feature of rail freight and where very wide networks are covered, the need to sort out wagons and marshall trains remains. Western Germany alone operates 22 marshalling yards and 2291 freight terminals (Shannon 1994). There remains a need for marshalling yards in large countries like the USA, Russia, Pakistan or India; but there also remains a need to rationalise the number of such yards to suit their operation to changing freight patterns.

In Britain much less wagon load operation now takes place; many marshalling yards or other extensive siding areas have become obsolete. In Western Europe also the number of points at which long distance freight is sorted has been greatly reduced. The modern need is for a limited number of well-equipped freight terminals which can deal with containers and inter-modal transhipment. The area needed for railway rolling stock, road vehicles, specialist handling equipment and intermediate or regrouping storage facilities is modest compared with that required for the original marshalling yards. The space in obsolete yards can be used for intermediate storage of bulk freight (such as construction materials) or new inter-modal freight depots and the surplus sold off for other types of planned development or adapted for conservation or amenity.

Freight depots must be located where the goods can be received or delivered cost-effectively, i.e. either at rates competitive with road traffic or else as part of an equally competitive combined transport system. Most modern rail freight is in 'block' form, i.e. for delivery of the whole train load from one station to another. Indeed most freight depots are now true 'terminals' replacing the huge number of local goods stations and sidings where individual wagons were shunted, dropped off or collected in pre-Beeching days.

Freight terminals are *not* located in city centres. Much of the traffic is containers or materials for industrial use for which ready road access is essential. Freight depots are sited at suitable transhipment locations or at quarries, coal-mines and factories delivering goods, or power stations and ports receiving them. Nevertheless, many terminals are in suburban areas or surrounded by housing and there can be objections related to frequent deliveries and collection by lorries. Most problems are the same as those which arise in industrial and port development.

Where marshalling yards do still exist, there are certain special noises that arise from shunting operations. These include the piercing squeal of wheel flanges on sharp curves and the use of wagon retarders (externally applied brakes) as wagons descend humps.

5.6 ENVIRONMENTAL ADVANTAGES AND ECONOMIC REALITIES

Most railways have spare capacity whereas some sections of roads are crowded and overloaded and the surrounding air is polluted by vehicle emissions.

The spare capacity is evident on most British railway routes, except in Kent and near certain cities. It is also available on most of the long distance global routes. Nevertheless *new* rolling road freight lines are being considered in Germany and France as well as three or four very long 'base' tunnels under the Alps, mainly for 'piggy-back' traffic and evidently to spare road wear and congestion.

Comparative figures for energy consumption and air pollution are examined in Chapter 8. Different data gives a wide range of answers, from those claiming that the larger articulated lorries can equal rail performance to those giving railways very marked environmental advantage.

Meanwhile only a minor proportion of total freight traffic is carried by rail in Europe because of the disadvantages of railway transport. These are:

- the lack of railway facilities at many points of loading or delivery;
- the higher cost of rail transport for short or medium distances, typically below 300 to 500 km for general merchandise.

The combination of these two factors has different implications for different goods. Agricultural produce, originating at numerous and sometimes remote farms, is economically carried by road for even greater distances; but for coal, metals or construction materials, moved in bulk between railheads, the threshold distance at which rail is economic is considerably less.

The number of railway access points has been greatly reduced because of declining business on branch lines or at intermediate goods stations. The flexibility of road transport cannot be doubted. Its cheap cost arises from medium-term macro-economic and political causes. The cost of fuel is more significant to road transport than to rail. World fuel prices may remain low as long as oil reserves and production capacity can meet demand. Powerful lobbies make most politicians unwilling to transfer the tax burden towards use of fuel, although environmental pressures to reduce global warming may eventually force them to do so; 'carbon taxes' are now on the political agenda.

Some governments are accused of discriminating against rail by constructing new road capacity out of central funds and at no direct cost to the user. The success or failure of privately funded toll roads should show how far this is relevant.

Meanwhile, *some* direct government assistance in improving rail freight infrastructure is available for environmental reasons. In Britain, Freight Facility Grants (1974 Railway Act, Section 8) can pay part of the cost of facilities which will achieve a switch of traffic from road to rail. This incentive to freight handlers is applicable where there are otherwise good reasons for *not* using rail freight and in 'sensitive areas.' It appears that most urban and single carriageway roads and some dual carriageways are 'sensitive' in this respect; it is also government intention to place no limit on financial assistance for such schemes if they are of 'exceptional environmental benefit' (Semmens 1991). Presumably the facilities are new or upgraded depots, sidings and even branch lines; they clearly include the rail element of intermodal schemes.

Government fiscal, economic and environmental policies might therefore influence the balance of road and rail freight traffic. Two other courses of action may also improve this balance.

The first is to encourage industrial development to occupy sites where both road access and rail links can be provided. This can succeed only through a combination of planning controls with a serious intention and capability of railway operators to provide an appropriate freight service.

The second course is already in hand through the development of combined transport (CT). Using road transport of containerised or other easily handled units, CT can deliver to and from road-accessible industrial locations; it can also transfer the loads to railway transport along routes where the speed and capacity of that mode is superior.

Meanwhile, the long-term viability of CT is not yet proven. Its multimodal nature makes it essentially a private sector and multi-enterprise activity. Its success is likely to be dependent upon the formation of a number of competing inter-modal freight companies offering complete services including a praticable framework of prices, cargo documentation and negotiating procedures.

CT also has political problems, particularly across national frontiers, because of legislation and commercial agreements favouring one-operator transport throughout. However, political and environmental determination should be able to overcome these.

Rail freight *can* be commercially viable, particularly over long distances, as is amply demonstrated by practice in western USA. In the former Soviet Union railways carried as much freight (in tonne-km) as the rest of the world combined; these railways are likely to continue to dominate freight traffic in the huge resource-wealthy areas of Siberia or Kazakhstan where alternative transportation is impracticable. But in western Russia, where the cost of providing freight services was hardly relevant under a centralised economy, the imposition of free market forces could lead to a diminution of rail freight and the associated environmental disadvantages which have occurred in the West.

To compete with or recover part of the freight market from trunk road haulage, railways have first to offer widely available frequent services with fast overall transit times and at competitive prices. The extent to which this can be done in free markets is limited—witness the bankruptcies of some US railroad companies and the closure of freight services elsewhere. Successful competition also requires proven reliability of delivery and security of merchandise.

Part II

IMPACTS ON PEOPLE

LIVERPOOL JOHN MOORES UNIVERSITY
Aldham Robarts L.R.C.
TEL. 0151 231 3701/3634

6 Social Impacts and Public Perception

6.1 INTRODUCTION TO IMPACTS ON PEOPLE

New, more frequent, faster or heavier trains can affect living conditions in people's homes or in their working or recreational surroundings; they may even affect the nature of lineside communities. These effects arouse people's interest or concern about the positive and negative features of railways and lead to a need for public consultation when new lines are planned.

This chapter therefore:

- introduces the direct impacts of railways on people's *quality of life* (Section 6.2);
- describes *socio-economic* impacts of railway operation on communities (Section 6.3);
- discusses *public perception* of railways and response to proposals for new rail development (Section 6.4);
- describes how *public consultation* takes place (Section 6.5).

A new railway service is provided to fulfil a social or economic need. It should be planned to achieve this to the best commercial, operational and environmental effect. Environmental planning of new lines is primarily a matter of choice of a practicable route that will best preserve land resources of the sort described in Part III. People's concern lies firstly in the fact that these land resources may include their own property or the facilities they use, and next in disturbance which trains may cause them.

It can be frustrating to a railway planner, endeavouring to fulfil an urgent national transport need, to find that public reaction in the neighbourhood focuses quickly on intangible disturbance factors—which may or may not constitute real problems—rather than on wider benefits and sensitive use of available space. But such public reaction is understandable, particularly where there is uncertainty.

Planners must therefore provide a comprehensive explanation of the levels of disturbance which result from the construction and operation of railways.

6.2 DIRECT IMPACTS ON THE QUALITY OF LIFE

People living near railway lines *may* be sensitive to the passing of trains. Whether they are disturbed, interested or indifferent depends upon individual circumstances.

Common causes of external disturbance by transport systems are noise, structural vibrations and atmospheric pollution whilst in some situations people may even complain of visual intrusion. More temporary but often more disruptive is disturbance caused during construction of the civil engineering works which form new transport infrastructure.

Each of these impacts on people, described in Part II of this book, is related also to some aspect of resources, described in Part III:

- Noise and vibration of trains (explained in Chapter 7) also affect property values, if anticipated disturbance results in 'blight' on sales; see Chapter 12.
- Pollution (Chapter 8) has a similar effect on communities and on heritage and amenity resources (Chapter 14); water pollution can damage wildlife and its habitat (Chapter 13); and air pollution, resulting from the use of fossil fuels, can be reduced by saving energy and by restricting use of non-renewable fuel resources.
- Visual impacts (Chapter 9) are related both to close intrusion and to wider views of scenic resources (see Chapter 15).
- Construction of railways (Chapter 10) makes a temporary demand for extra land space.

Measurements can be made of these causes of human concern. More difficult to quantify is the severity of actual disturbance in each case.

Thus the level of *noise* at various distances from each type of train can be measured and predicted. Methods of assessing people's sensitivity to different types and doses of noise are much more subjective. Research has been undertaken by sample questioning of people as to their reaction and tolerance of certain types and levels of sounds; the ability to sleep at night or to teach in a classroom provides typical standards against which noise levels can be judged. Certainly train noise does not figure highly in any ranking of common sources of noise nuisance, as do road traffic, aircraft, people or dogs. This could be simply because train noise is confined to specific locations which sensitive people can avoid. Research should be directed specifically at lineside residents.

Meanwhile, in situations where train noise is or will be dominant, it is possible to predict what levels will occur, to compare these with any established standards and to prepare data to decide whether and how any mitigation of disturbance might be effected.

Atmospheric pollution cannot be measured as easily as temperature or rainfall. Nor is its impact on people easily assessed, not least because the most dangerous gases may not be the most noticeable. In addition, where air quality can be shown to be poor, it is difficult to allocate the proportion of blame among often numerous and widespread sources of emission. Fortunately, since the end of steam traction, pollution by trains is rarely seen as significant; electric trains cause virtually no direct change in air quality.

Most legislation against air pollution concerns specific emission limits on engines, thermal power generation or chemical processes. Near locations of heavy gaseous (or liquid) emissions—such as diesel locomotive depots or in heavy traffic situations—these emissions should be measured. Then the assessment and evaluation of each railway-derived pollution source should be undertaken in comparison with that caused by other activities at the same locations. Other sources of pollution or nuisance may be more significant; some may have been previously measured and compared with accepted tolerance levels; and measures to limit or alleviate them may already be in force.

Loss or damage to property (people's personal resources) can result from increased disturbance or by physical damage during railway operation or construction. Unlike conservation land, commercially traded property has a market price which reflects its utility to people and their valuation of its living environment.

All these *impacts on people* and their results in social or monetary terms can be assessed by reference to those caused in the past in a wide range of projects involving land development and construction. Adverse impacts of railways can be mitigated:

- by early intimation of probable routes, at least to define those areas where people will *not* be affected;
- by enforcing appropriate standards in manufacture of trains or construction of new lines;
- by specific measures to reduce direct disturbance;
- by compensation for unavoidable disturbance or damage.

People's convenience is a short-term consideration compared with irreversible changes in land use or permanent damage to natural resources. Therefore physical resource issues should be resolved first. Appropriate amelioration of impacts on people and their activities can then be determined.

6.3 SOCIO-ECONOMIC IMPACTS

Railways were conceived as a means of freight transport, initially from mines, later for longer distance distribution, i.e. *economic* activities; but it was people's use of railways which inspired many early schemes. These provided great social as well as economic opportunities. The socio-economic impact of modern railway schemes may be less dramatic but both economic and environmental planning require that it be analysed.

Socio-economic impacts affect the livelihood, earnings, welfare and enjoyment of people. Those related to railway development include:

- *job creation*, either directly in construction or operation of the railway or in commercial activity generated by improved transport;
- changes in the *housing market*, resulting from job creation and from better commuter services;
- *tourism* and *recreation*.

It is important to determine the extent to which positive or negative changes would occur anyway, without any new railway development, due to road transport or other factors in any dynamic or fluctuating regional economy.

Direct *job creation* can be assessed by analysing construction activity and operational requirements. Construction of a 100-km-long new passenger railway, for example, might be expected to generate several thousand jobs at peak, a substantial proportion being filled by people from outside the region. Operation could involve a few hundred permanent jobs initially, increasing to a thousand or more if intermediate stations are involved. New employment can result in changes in local wages, prices and the labour market.

Jobs can also be created by commercial activity based on railway stations and new improved rail services. The mayor of Lille, in France, was so convinced of the economic activity which would be generated by new TGV services between Paris, Brussels and London that he successfully lobbied for a diversion of the Paris–Channel Tunnel route through the centre of Lille; he then became the driving force behind a regional regeneration programme investing £640 million in new commercial development (*The Times* 29 March 1993). On the English side of the Tunnel, the Union Railway route to London, announced in March 1993, led to similar lobbying, related to future economic development, for location of stations in the East Thames Corridor.

It can rightly be argued that such development is fully economic rather than socio-economic in character. However, just as environmental objections to a motorway melt away in the eyes of people whose travel benefits from it, so are prospects of greater economic prosperity a powerful antidote to concern about any inconveniences caused by new railways.

A century ago *housing demand* in Europe was directly related to new commuter railways. In expanding Third World cities this can still be true. Elsewhere the road transport alternatives now available make forecasts more speculative. Judgement, based on careful comparison of previous occurrences in similar socio-economic circumstances, is required to assess such features as:

- new job creation, generated by improved rail services;
- increased commuting to large cities, at the expense of other places which will *not* gain the improved services;
- regional planning policies encouraging or discouraging housing development;
- reduced net populations, in any locations where the perceived adverse impact of the railway might be exceptionally severe.

The last point was raised by a television documentary about the depopulation of a strip of housing in the village of South Darenth in Kent; this occurred when residents accepted BR's offer to buy their houses on the 1989 Channel Tunnel Rail Link route corridor. The houses eventually found occupiers but certainly any new housing schemes in the area were postponed. Whether there was any real long-term effect is still hard to establish. The housing market collapsed nationally and the proposed railway route was altered. Certainly by 1993 house trading appeared to be active in the area.

Recreational or *cultural* pursuits were boosted by Victorian railways, starting with excursions to see the Great Exhibition in London. Daily and weekly holidays to Blackpool and other resorts followed. In the twentieth century, cars severely reduced holiday travel by rail. However, revived country lines promote recreation whilst old railway buildings can be used as new heritage centres (museums), devoted either to railways or to other industrial or historic features of the locality.

6.4 PUBLIC PERCEPTION OF RAILWAYS

6.4.1 The Image of Railways in Britain

Road traffic and the UK economy expanded throughout the late 1980s. Personal wealth and vehicle ownership increased whilst the real price of fuel stagnated or dropped. It is nearly 20 years since there was an energy crisis serious enough to concentrate thought on transport costs and worldwide fuel resources.

In the last 40 years total passenger travel in Britain has burgeoned. Railway travel has remained at about the same level but road use has increased more than three-fold. Much of the increase has been in private car use, multiplying itself about nine times whilst bus traffic has declined. In freight terms there has been a marked decline in rail traffic, at half its post-war levels by 1993 and still falling.

For many years a large proportion of the ordinary motoring public has seen railways either as an uncomfortable weekday necessity or else as an irrelevance to their transport needs.

Freight transportation companies are well aware of the economic advantages of sending all but bulk long distance consignments by road. Fuel is cheap and roads are provided by the government. The powerful road lobby continues to press for the provision of more trunk roads and bypasses around traffic bottlenecks.

Policies and attitudes could change by the end of the century. On the one hand road charges and higher fuel taxes are seriously contemplated; these imposts are seen as environmentally acceptable ways of raising government revenue. At the same time, demand has risen for high speed rail passenger services between cities and rapid transit within them. Increases in rail freight services should follow from the recent opening of the Channel Tunnel, continuing implementation of the Single European Market and the effect of these events on longer distance European and British railway operations.

How does the public view the introduction of new railways or faster or more frequent trains to meet these demands?

Locally, light rail commuter developments are welcomed—because they benefit local people. But major new express lines tend to be treated with grave reservations concerning their impact on the immediate environment.

Nationally, people are quicker to recognise the wide benefits of rail transport. Commonly they envisage freight being removed from roads, leaving more room for cars! But some see a genuine fast and convenient means of business travel.

New main line railways come as something new. In Britain there were none for 72 years until diversion of the East Coast Main Line around the new Selby coalfield in 1982. Few people were even aware of construction of the diversion which passes through sparsely inhabited countryside and was completed before environmental assessment was obligatory. However, since the 1960s many motorways have been built—the impact of which has been similar or more severe than one would expect from railways. Nevertheless, unfamiliarity with high speed trains reinforced by new media attention causes the Channel Tunnel Rail Link (CTRL) to be treated as if it were something unprecedented in contemporary experience.

France followed Japan and led Europe into the modern era of high speed railway building. The first TGV (*Trains à Grand Vitesse*) routes apparently suffered no undue resistance from environmental lobbies, perhaps because of the emptiness of the country traversed. By the 1980s, however, the projected Provençal route was being dubbed a line '*à grand tristesse*' because local antipathy was now stronger.

In Britain, non-railway travellers assumed that similar new routes were not being built in their country because the government or the railways lacked the foresight. In fact, as the increasing army of inter-city travellers began to appreciate, high speed trains could be and are being run without excessive difficulty on existing tracks. The need for any completely new line was therefore deferred until a faster, extra capacity route from London to the Channel Tunnel became a hard necessity. In comparatively thickly populated country at a time when environmental awareness has never been stronger, it is not surprising that such a venture should cause concern as well as enthusiasm.

Sheer necessity for an alternative to road transport may change the image of British railways. Real improvement, such as more frequent higher speed inter-city services and modern rolling stock on secondary and some commuter lines, is going some way towards a more sympathetic attitude. But, whilst subsidies are withdrawn and fares rise, these steps are unlikely to reverse the stereotype of commuter suffering or some vehement criticism of British railway services.

Meanwhile more could be done to demonstrate and publicise the true nature of railway impacts on people and resources.

6.4.2 Local Perception

If a motorway is built through your back garden you will object, probably vociferously. Your neighbours will support you but people living more than a mile away will not; they will look forward to less traffic on the old road and a chance to get away quickly on the new one.

A high speed railway without a local station provides no such obvious benefit. In villages near the CTRL, announcement of tentative route plans in 1988 resulted in the rapid formation of protest groups. Some were new, some existed as earlier anti-motorway groups. Others were serious countryside protection societies with selective and researched concern about particular aspects of possible routes.

Residential groups provide social forums, usually to fight the presence of a new railway and to press for it to be moved elsewhere. In rural surroundings, it is in nobody's immediate interests to support railway development; those who appreciate its national necessity keep a low profile while the storm is raging. The great display of anti-CTRL notices in some villages in Kent in 1989 gave an impression of unanimous opposition to the projected railway. However, this impression was not necessarily correct—count the numbers of residents who did *not* display notices or did *not* write to their MP!

Some local opposition is based on misunderstanding about the impact of a new type of railway and on what one can reasonably get used to. Some is much more reasoned objection and requires logical and scientifically-argued response.

Local reaction may be less negative in urban areas—particularly if there is to be a station providing railway services and jobs. We have seen how commercially-oriented communities are seeking intermediate stations in France and England on the Paris–London route.

Prominent factors in opposition to new railways at local level concern:

- noise and any other direct nuisance to residential life;
- construction activity;

- problems about particular amenities;
- intangible factors such as undefined visual intrusion;
- the effects of all these on the value of, and ability to sell, houses.

However, constructive criticism by informed local people can make a positive contribution to environmentally-sensitive planning. The means for doing this through public consultation are discussed later.

6.4.3 National Perception

National perception is reflected in the views and transport strategies of planning authorities, political parties and major environmental groups—and by the news media who report these views and choose to foster or ignore them.

Government perception of the roles of different transport modes is reflected in their transport and environmental policies. Implementation of these policies depends upon political decisions and, particularly for publicly-owned transport, availability of funds. In democracies it is ultimately the perception of voters which influences decisions.

As road congestion in and around towns increases and if energy price rises alter transport economics, the environmental advantages of trains and buses may be increasingly accepted. By whatever means the role of railways comes to prominence, there remains a need:

- to offer services which travellers and freight forwarders can afford to use and find preferable to competing transport modes;
- to improve rail's image as an efficient operator;
- to publicise the national benefits and, for those seriously affected, to explain the realities of railway disturbance and the mitigation measures which can be introduced.

The British image of railways is a mixed one—enthusiasm for high speed trains and certain other modernised services, despair with several troubled commuter lines, often indifference. In countries like The Netherlands there is a clear awareness of the advantages of railways whilst in continental Europe generally the business advantages of the new high speed lines are clearly recognised. Intense crowding of suburban station platforms and trains reaches a nadir in cities like Bombay; but, judged by stark living standards, the discomfort is accepted as inevitable.

Perhaps no national perception is typical but the situation in Japan may be enlightening. An independent and informal view of trains and national transport policy in that country is included in the appendix to this chapter. Much of the commentary in the appendix could be generally true, if less enthusiastically stated, for some of the better European commuter services. There are obvious differences in culture. Japanese rail services may be particularly reliable and well planned; but they appear to be quite as crowded and less comfortable than their Western equivalents. Yet the official policies and personal appreciation reported from Japan would not be easy to replicate with equally subjective statements from Britain or the Continent.

It may also be noted that, in spite of the pride which the Japanese obviously take in their railways, there was prompt public protest about the noise level of the first *Shinkansen* trains in the 1960s. Action to reduce noise levels on existing and new high speed routes was demanded and taken.

The different attitudes to railways can well be related to responses which are received when new lines or services are proposed.

6.5 PUBLIC CONSULTATION

Public consultation comprises any opportunity to describe proposed projects to the public and for the public to put their views. Consultation with local authorities and affected parties may also be a statutory obligation in railway promotion. It has thus become a feature of the environmental assessment process. Where genuine objections are raised, they must be considered by the planners and answered.

Planning for successful public involvement comprises two elements—contacting and where practicable winning collaboration with local government authorities; and preparation for meeting the wider public.

Local government contacts should be made as early as possible with both elected representatives, who will have to answer to everybody for whatever they achieve or concede, and officials who have to implement planning and environmental regulations and guidelines and are conversant with some of the technical issues.

Before wider public consultation, the promoter may be well advised to decide which individuals or groups he should deal with directly. He should devise a plan and budget for consultation and provision of information extending up to eventual implementation of the project. He should then prepare a preliminary statement of current proposals and the likely environmental impact.

Particularly informative documents are those which contain maps (1/25 000 scale) of the whole route and larger (oblique) aerial photographs of sensitive points, both marked up to show the proposed line of the railway. Maps were produced for each section of the Union Railway route from the Channel Tunnel to London at the time of public consultation from April to July 1993. For the Tunnel itself and its terminals at each end, documents describing the consultation process, the environmental investigations and the action which resulted was published throughout the planning and implementation process and constitute a complete case study (Eurotunnel 1988, 1991b, 1993; Kershaw and McCulloch 1993).

In some cases it may be wise to select some practicable inducement or popular benefit that might be offered to a local community to gain at least its neutrality. The cost of small physical improvements, such as an extension to a village hall, is likely to be modest in comparison with that of attendance at public meetings, undertaking surveys and providing information to what might otherwise be a less receptive community. Compensatory environmental deals are commonplace in many countries for all but the least controversial projects (O'Riordan and O'Riordan 1993: 24).

Objectors usually complain that they were not consulted early enough. Then, afterwards, some will protest that their objections were ignored. There can be considerable skill, both in providing enough information at each stage and in dealing sympathetically but objectively with local opinions. Sympathetic is being seen to have listened to people's concerns; objective is explaining the realities of the sources of concern—such as noise and disturbance— and drawing balanced conclusions as to their significance. There is no way of pleasing all the people all the time; but the experience of people who already live near railways can be included in the evidence that is offered.

Public consultation often takes the form of an exhibition, staffed by the promoter of a scheme, his planners, engineers and environmental advisers. At exhibitions the public are invited to see the alternative schemes which are proposed, to ask questions and to air their objections. The scale of exhibition can vary from a permanent centre connected to a major project such as the Channel Tunnel to temporary displays in public libraries or in trailers at points on a long proposed railway route. Exhibitions should be well publicised, well presented and informative; sometimes questionnaires are sent out to all affected households inviting their views on specific issues. The results of the consultation, much of which is correspondence, are written up in a document to go forward into the next stage of the decision-making process. This document reports on all substantive objections and proposals and comments on why these can or cannot be taken up.

In Britain formal public consultation was introduced for road schemes in 1973 and is now likely to be incorporated in all main line railway developments. The Draft Order/Public Inquiry procedure makes allowance for objectors to air their views. The Inquiry itself involves numbers of expert witnesses. The appointed Inspector, who can regulate the progress of the Inquiry, makes recommendations on the findings but the final decision is a political one, i.e. by the Government's Secretary of State.

This can be compared with procedures for planning and implementing new high speed railways in Europe. The Council for the Protection of Rural England's publication *How Green is your Railway?* (Sullivan 1989) includes a review of the public consultation processes. In Germany and Switzerland

what is notable about the procedures is their thorough nature and the emphasis on the promoting body having to negotiate a solution with local authorities and affected owners.

In France

> procedures are decidedly faster, somewhat superficial, and allow much less scope for opposition; however, in the case of the TGV-Atlantique, the occupants of the two critical areas . . . got what they wanted.

In The Netherlands about one month is allowed for each of the stages of public consultation, which is open to all interested parties. An EIA commission assigns an expert working party to ensure that environmental impact assessment takes all issues and interests into account.

Clearly the circumstances in areas of sparsely inhabited French countryside traversed by some TGV routes are different from those in urban conditions or more densely populated country in parts of Germany or south-east England. However, there is evidently a danger of placing greater emphasis on local agreement than on objective assessment and national environmental interests. This seems less than equitable in promoting the most satisfactory or cost-effective solution to a more than local transportation problem. It can also delay the planning process.

To quote CPRE's publication again

> The recourse to tunnels or cut-and-cover in both Germany and Switzerland as a means of meeting objections is very common. Indeed the problem in Switzerland is knowing where to draw the line—it is sometimes said that if all the requests were met the whole line would end up being in tunnel. There is no technical reason for more than a very few short tunnels on the proposed new Bern–Olten line.

A more recent development for that line arose out of the Swiss parliament's decision (reported in *International Railway Journal* 1994) forcing Swiss Railways (SBB)

> to cut back all of its previous track building proposals, with the exception of the Bern–Olten line, where it is fighting to preserve plans for some sections to be built above ground. Canton governments are attempting to force the railway to construct in tunnel, but the railway says this option is too costly. A decision on the dispute will be made by the transport minister, though the issue is likely to go to the courts if he rules in SBB's favour.

Thus economically unsound and possibly environmentally unsuitable solutions can arise from a relatively straightforward democratic procedure.

Public correspondence about major projects like the Channel Tunnel Rail Link (CTRL) confirmed that the more obdurate objectors will be satisfied with nothing less than a continuous tunnel. The more extreme misunderstandings can only be corrected by clearly presented facts, for instance:

- that the cost of tunnelling is prohibitive and beyond what any investor can afford;

- that the benefits in terms of passenger enjoyment (view from the train) and speed (restricted in tunnels), and hence of rail revenue, will be reduced;
- that the local impact of a surface route is not necessarily serious; evidence must be presented to prove this.

It should not be a fundamental assumption that plans for a new railway will be altered always and only in the direction of concessions to public opinion. For example, the route for CTRL announced in March 1989 indicated a subsurface route throughout south-east London. BR's spokesman said that tremendous advice and criticism had led to sinking one-third of the whole route into tunnel for 'environmental' reasons. Only when the true cost of tunnelling and of the whole CTRL became evident did it become necessary to consider partially above-ground routes through the London suburbs. Whilst the extra cost of constructing tunnels may have always looked excessive to engineers, this might never have been accepted publicly without such a serious initial investigation and cost estimate.

Thus there are severe problems in processing transport development plans through the democratic process. A proper balance must be achieved between genuine but often short-term popular objections and proven long-term goals, like provision of economic infrastructure and preservation of irreplaceable resources.

Very positive achievements can arise out of public involvement, for the environment as a whole not just for individuals. Many of the measures taken to improve the alignment of the M40 motorway extension and to ameliorate its impact on landscape and wildlife were achieved through well-informed public involvement. A major realignment around Otmoor in Oxfordshire was made as a result of the first public inquiry.

The total planning and implementation process for new transport links is a very long one, typically taking 10 to 15 years for a new road in Britain. Of this, less than one-third may be the actual construction period. Industrial and transport interests believe the time lapse is too long. Consultation about routes and the public inquiry process alone can add years. In August 1993 the British government announced that steps would be taken to reduce the planning time for roads and railways. However, it remains questionable as to whether the economic advantages of earlier provision of transport infrastructure investment can outweigh the dangers of hasty environmental planning.

Approved national transport plans related to environmental protection policies might settle general principles in advance. But prior agreement on their detailed application could delay actual projects at least as much as present consultation procedures. Rather, transport plans and environmental policies should be developed in parallel at both national and local levels.

A lot of experience has been gained over the last 20 years by involving the public in road planning. Planners of new railways can learn from this experience and improve their strategy for presentation accordingly. Engineering is the fundamental design process for a new railway. Public consultation is an important element of the environmental assessment which can guide that engineering.

APPENDIX   A View of Mount Fuji: The Image of Railways and National Policy in Japan

Christopher Savage

Eight and a half years of mostly daily commuting by train have provided the basis for a deep admiration for the Japanese railway system. In the eyes of a professional civil engineer and environmentalist the experience has thrown new light on social economics of the environment.

On the 10 minute walk to the station on a crisp winter morning the distant view of the snow-covered summit of Mt Fuji hangs in the sky like a scroll painting between the electric poles that flank the narrow village street. At the station, wives in expensive cars deliver their 'sabiru' ('Saville Row') suited husbands to catch the 0714 to Tokyo. Most of us get seats, if not immediately then three stops later when the school children thin out—and this is why we live as far out as Higashi Zushi—so we can get a seat.

The train has hard seats and little leg room, and soon fills up to overflowing, but not before we have all savoured further glimpses of Mt Fuji, surely the most poetic and inspirational of mountains. Passing through the ancient capital of Kakamura there are other visual delights at all times of year whether it be plum and cherry blossom, or the great variety of beautiful bamboos, or the azaleas in season. Even the suburban and city sky-line appears to be landscaped for the benefit of the traveller.

As the train fills up the air is filled with the aroma of exotic soaps because the Japanese young and old are very clean people who believe in 'asa-sham' or morning shampoo. The train itself is spotless at that time in the morning and station precincts have been washed down earlier. There is no first class but green cars as they are called provide a less crowded environment for the frail or elderly. For most, from janitor to company director, the morning commute is the most pleasant way to travel to the office. It is possible to read a paperback or a shrewdly folded newspaper, but most people snatch a short nap whether strap-hanging or sitting. The evening return journey provides a similar opportunity for a nap.

Only a company president or chairman will arrive at the office in the firm's limousine. There are no company cars as such. In addition to their convenience for the usual variety of travel uses, cars are valued by the young especially for their privacy; Japan is a crowded country where it is common for three generations to share the same accommodation.

So much for the setting, but note the following points:

1. The railways are usually considered to be by far the most pleasant way to travel in Japan. Frequency can be every $2^3/_4$ minutes though 3 minutes is more usual. Variations in passenger volume are taken up by varying the train from 16 carriages at peak to perhaps two at the lowest off peak.
2. Noise mitigating technology such as screens and insulation do not intrude on the views of passengers, whereas passengers in cars can often see nothing of their surroundings on motorways because of the noise barriers.
3. The government has had a consistent transport strategy for many years now to keep cars off the road and to develop the country round the railway system.

Item 3 is the most interesting and important feature because above all Japanese people are proud of their railways and support a government strategy that is really national. Let us therefore see how it works.

Commercial and social planning of cities, towns and villages centre round the railway stations. New and old railway stations are being developed to be the best and smartest shopping centres in town. There are huge underground malls and pedestrian precincts with high quality shops and restaurants at the larger stations, and even the smaller stations have high quality concessionaires as well as the purveyors of delicious low-priced noodles.

Outside the stations are taxi-ranks and bus-stops, but no car parks except deep in the country on a vacant plot of land. Parking is not allowed on the streets; anyone wishing to own a car must prove to the police that they have an off street parking place. Needless to say, new supermarkets located away from the stations provide parking for their clients—but only for their clients.

How then does one do bulky shopping without a car? That too is easy for there are several excellent forwarding agencies that provide a 24-hour delivery service. Even air travellers can confidently consign their checked baggage to the airport by commuter train without fear of arriving late due to traffic congestion.

A secondary effect of the national transport strategy is that property prices reflect nearness to railway stations and distance to central Tokyo or the other major cities.

Another major feature of Japanese railways is the number of private systems even before 'JR' was largely privatised. These private systems have been able to use JR technology under preferential terms and have every fiscal encouragement. They have also been able to vertically integrate transport planning and real estate development more closely than the national railway system could have done.

Three further features make for successful integration of private and public systems:

1. In nearly all cases, interline tickets can be purchased direct from automatic vending machines, and in case of difficulty there are always intelli-

gent booking clerks who can be called to assist.

2. The railway timetables and atlases are excellent; connections really do work because the trains run precisely as advertised even though they sometimes have to wait outside a station in order to arrive on time.

3. There are many imaginative concessionary fares.

Most of the railway systems discussed above are predominantly for commuters. In addition, there are the *Shinkansen* or 'bullet train' lines, and *Narex* (the Narita airport express). These lines are very comfortable and fast without appearing so. Views of mountains and countryside are superb, and it is possible to enjoy all the joys of the changing seasons and the beauties of Japan without getting mud on one's shoes. The main *Shinkansen* lines are scheduled at 3–4 minute intervals; yet whether in the town or countryside the sounds are non-intrusive due to good track engineering and suspension design.

7 Noise and Vibration

7.1 NOISE AS A NUISANCE

Most of the adverse impacts of transport systems on people could be described as nuisance. Only under particularly adverse circumstances can noise or exhaust fumes constitute a direct health hazard or will trains or railway construction pose a threat of injury.

'Nuisance' is defined by the *Oxford Concise Dictionary* as 'anything injurious or obnoxious to the community or member of it for which legal remedy may be had'. The legal reference is interesting in that it effectively excludes trivial annoyance. Legal constraints on nuisance are evidently getting tighter, as are assumptions about people's rights to be undisturbed.

Noise of trains is the most measurable 'nuisance' caused by railways. Noise of French trains is dealt with by a unit in the *Départment Nuisances des Transports*. 'Sound' becomes 'noise' when it is a nuisance, although the terms are used interchangeably; the word 'noise' is preferred technically regardless of its effect. How noise is quantified is explained in the next section. The nature of train noise and its prediction is described in Sections 7.3 and 7.4. Noise impacts, assessed in Section 7.5, comprise general annoyance and specific disturbance. The standards adopted and how noise can best be mitigated are reviewed in Sections 7.6 to 7.8.

Noise has its origins in vibrations. Particles of a vibrating body, such as an engine or rail track, set neighbouring particles in the surrounding medium into motion, transferring physical vibrations to adjoining buildings or sound waves to distant observers.

Most of this chapter is concerned with airborne noise (audible vibrations) whilst Section 7.9 deals more briefly with vibrations through other media (like the ground or structures) and re-radiation of inaudible noise at audible frequencies.

7.2 QUANTIFYING NOISE

Noise can be described firstly by its level or loudness, secondly by its quality or tone and thirdly by its pattern of occurrence.

The *level* of noise is expressed in decibels (dB), a measure of acoustic pressure expressed in a logarithmic scale and proportional to the square of energy. $1\,dB = 2 \times 10^{-5}\,N/m^2$.

A 3 dB increase represents a doubling of energy or twice the sound-generating activity. Two similar trains passing at the same time cause 3 dB more noise than one ($10 \times \log_{10}2 = 3$).

If loudness is 'people's impression of sound intensity', then a 10 dB increase represents a doubling of loudness. A ten-fold increase in traffic ($10 \times \log 10 = 10$) gives the impression that the noise volume has doubled.

A change in sound level of 1 dB from a varying source is generally considered to be just perceptible. Most people can detect a change of 5 dB (ICE 1990: 28). Utter silence is 0 dB but very low background levels below desert conditions (about 10 dB) are comparatively rare. Typical quiet bedroom levels are in the range 20 to 40 dB. Communication in offices or schools becomes difficult above 55 to 60 dB. Health hazards are posed in factories if workers are subjected continuously to 90 dB but many people frequently experience this level for *very short durations* without comment. The threshold of pain is above 120 dB and physical damage to the ears can occur at 150 dB.

The motion of vibrations through air or other media can be measured in terms of amplitude (displacement) and frequency. Pressure level or *loudness* (dB) is related to the amplitude of vibrations; *tone* or musical note and audibility depend on frequency which is measured in hertz (Hz, oscillations per second).

Quality of noise is a complex effect resulting from changes as overtones interfere with and override fundamental tones, for instance in the various noises related to different operating modes of a diesel locomotive. Both the perception of noise by people and the effects of noise spread and attenuation are related to frequency as well as to loudness.

Actual measurements and predictions of noise are usually expressed as dB(A). The 'A' denotes a standard weighted composite of the different frequencies in the noise spectrum (18 to 18 000 Hz) which gives a good correlation between measured noise and human perception of it. Typically, passing train noise falls into the range 100 to 10 000 Hz. The dominant range for wheel/rail contact is 800 to 4000 Hz but diesel locomotives ticking over may emit very low frequency, inaudible noise (6–12 Hz). The frequencies which are responsible for groundborne noise and physical vibration are also low, generally 10 to 200 Hz. Consideration of frequency is important in detailed analysis of particular noise situations. But to avoid unnecessary complexity, comparisons of train noise are expressed primarily in dB(A) data.

The *pattern of occurrence* of a sound varies from a sudden, sharp event, like the blast of a locomotive horn, through the more continuous mildly varying noise of road traffic to the constant monotone inside an aircraft.

Because the noise of many activities, including train operation, is variable and intermittent, a number of ways of expressing noise, instantaneously or over time, are in use.

Maximum noise level (L_{max}) is the highest instantaneous pressure actually recorded, e.g. as the locomotive or power unit of a diesel train passes the observer.

SEL (single event level) is the noise, continuous for one second, giving the same energy as a specific noise event over its whole period; SEL is thus an artificial measure used as a descriptor of a whole train noise event. For a uniform noise lasting more than one second, SEL *exceeds* the actual maximum noise level.

Parameters representing the amount of noise occurring over a longer period include:

- L_{10}, the noise level exceeded for 10 per cent of a period of hours, usually 18 or 24, and expressed as dB(A) $L_{10}(18h)$; the latter is the measure required for UK Department of Transport road environmental assessment frameworks and also for determining eligibility under the 1975 Noise Insulation Regulations.
- L_{90} is the level exceeded for 90 per cent of the time, being the background level between isolated noise events; it is used in rating of industrial noise.
- L_{eq}—equivalent continuous sound level—the notional steady noise level over a stated period giving the same energy as the actual, intermittent noises; now commonly used in assessing aircraft noise in place of the noise and number index (NNI) and in the assessment of industrial noise (BS 4142); for road traffic, $L_{Aeq} = L_{10} - 3\,dB(A)$ approximately.

For assessing the impacts of railway noise, good arguments have been advanced for concentrating analysis on dB(A) L_{eq} (24 h) values (Dept of Transport 1991: 16; Fields and Walker 1982: 177–255). In investigation of particular events, such as the passing of a certain type of train, attention is paid to maximum and SEL values. As heard at a fixed location in constant conditions, a certain type of train at a certain speed will produce a maximum noise level (L_{max}) and a specific amount of sound energy (SEL). The longer term noise level, such as L_{Aeq}, can then be calculated or measured for a number of such trains passing in the time concerned together with any other types of train. L_{Aeq} (24 h) is a logarithmic sum of the single event levels of all the noise events which occur within 24 hours.

$$L_{Aeq} = 10 \log \sum \frac{10^{\frac{SEL}{10}} \times \text{number of trains}}{\text{total period (seconds)}} \, dB(A)$$

Table 7.1 indicates, for comparative purposes, some typical noise levels related to transport. More data for various types of trains is given in Table 7.2 on pages 140 and 141.

LIVERPOOL JOHN MOORES UNIVERSITY
LEARNING SERVICES

Table 7.1. Typical transport noise levels

	Peak noise (dB(A))	Maintained noise level (dB(A))	Position of observer
Passing car	80	–	25 m
Articulated lorry	85–90	–	from
Motorway traffic	–	75–85	vehicles
BR electric 160 km/h	93	–	or trains
TGV SE 270 km/h	99	–	
Aircraft (Boeing 747)	110	–	– 250 m below
Busy road junction	–	60–70	– on sidewalk
	–	50/60	through closed/open windows
Train horn	100–110	–	– at 30 m
Train entering metro underground station	100–105	–	– on platform (not acoustically treated)
Track ballast tamping	–	85	– at 25 m

7.3 NATURE OF TRAIN NOISE

7.3.1 Examples of Train Noise

Figures 7.1 to 7.5 are graphic traces of noise levels (dB(A)) against time for different types of trains. These were recorded on level ground at between 20 and 30 metres from the trains depending on which of four tracks they used.

Figures 7.1(a) and (b) show the noise profile of BR Class 47 2580 hp diesel locomotives hauling seven coaches at different speeds. In (a) the train is accelerating at about 100 km/h causing a maximum noise at the engine of 89.5 dB(A), receding gradually to 84 dB(A) as the carriages follow past. In (b) a similar train is cruising at the maximum speed (150 km/h) permitted for Class 47 locomotives. The engine is less dominant and it is the wheel/rail contact noise which keeps the passing noise of the train within a narrow 90 to 92 dB(A) band. Towards the end of train (b) there is an irregular noise probably associated with some loose attachment to a carriage. The recorded SELs are 95.5 and 97 dB(A) for the slower (a) and faster (b) trains respectively. If 10 of the faster trains were to pass in an hour, then the L_{eq} (1 h) would be 71.4 dB(A), or 150 trains in 24 hours would produce an L_{eq} (24 h) of 69.6 dB(A).

Figures 7.2(a) and (b) show the sound traces of much less powerful three- and six-coach 1960s 'heritage' style diesel multiple units (dmus). A typical dmu emitted L_{max} and SEL values of only 81.5 and 84.5 dB(A) respectively.

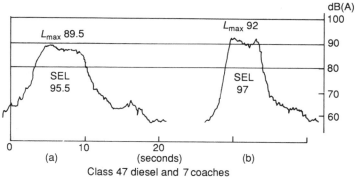

Class 47 diesel and 7 coaches

Figure 7.1. Noise profiles—locomotive-hauled trains: (a) accelerating at 100 km/h; (b) full speed 150 km/h

Ten dmus passing in an hour or 150 in a day would produce L_{eq} for 1 and 24 hours of 58 and 56 dB(A) respectively. Adding their noise to those of the larger, faster 97 dB(A) SEL trains would produce no appreciable difference in the L_{eq} values. These dmus are short and relatively slow (top speed 112 km/h) with diesel engines located under the floor, usually arranged to deliver 200 hp per coach. By 1993 more modern faster dmus—Class 165/166 145 km/h, 350 hp per coach 'turbo' trains—were being introduced, mainly to replace firstly the locomotive-hauled passenger trains of the type shown in Figure 7.1. and then the older dmus. The noise profile of the turbo trains ((c) and (d) on Figure 7.2) is similar to or quieter than that of the older,

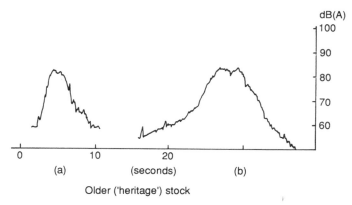

Figure 7.2. Noise profiles—diesel multiple units: (a) (b) older (heritage) stock;

slower dmus. Further comparisons show that the L_{max} and SEL of the turbo trains are at least 10 dB(A) lower than those of the locomotive-hauled trains.

Figure 7.3 shows the different effects of two electric multiple unit trains (emus) passing in opposite directions at another site; the first (farside) four-coach train is running on jointed and the second (nearside) six-coach set on continuously welded track. Allowing for the shorter distance to the latter and the train's greater length, there is no significant difference in the basic noise generation levels; but the four-coach train on jointed track clearly

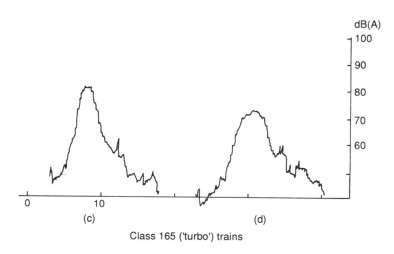

(c) (d) Class 165 ('turbo') trains

(c) (d) Class 165/166 ('turbo') trains

reveals the traditional clickety-clack of the bogies passing over both near and distant joints.

Figure 7.4 shows the distinctive noise profile of an HST 125 high speed train. The 2250 hp diesel (electric generator) engines at both ends are working almost at full power which is normal when these trains are running at near to their top permitted speed of 200 km/h. L_{max} is 96 dB(A) as each engine passes.

The 3300 hp Class 59 diesel locomotive on freight train (a) in Figure 7.5 is also at full power recording some 10 db(A) more than the wagons full of aggregates which it is hauling towards London. The second freight train (b)

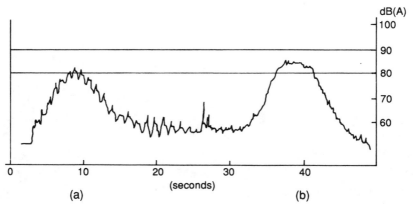

Figure 7.3. Noise profiles—electric multiple units: (a) 4-coach on down (jointed) track; (b) 6-coach on up (continuously-welded) track

is of similar composition; probably because this train is returning westwards empty, the engine is coasting and its noise is barely perceptible above that of the wheel/rail noise of rattling unloaded wagons.

7.3.2 Causes of Train Noise

Train noise can be generated by:

* motive power units; noise from engines and ancillary equipment escaping through exhausts or openings in the casing;
* wheels running on rails;

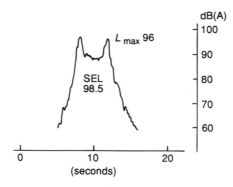

Figure 7.4. Noise profile—InterCity HST 125

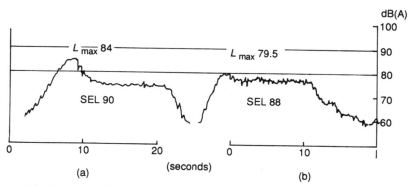

Figure 7.5. Noise profiles—freight trains. Class 59 diesel locomotive hauling 18 ARC aggregate wagons at about 70 km/h: (a) loaded; (b) empty

- aerodynamic effects;
- vibrating structures.

7.3.2.1 Motive power noise

Motive power noise includes that of auxiliary equipment, such as air compressors, radiator cooling fans and heating/air conditioning equipment, which is significant for stationary trains or those running at slow speeds. Diesel engine noise—sounds of reciprocating parts and turbochargers— becomes dominant when full power is being applied, whether for accelerating, maintaining top speed or working up steep gradients. A diesel locomotive causes about 20 dB(A) more noise operating at full throttle than when the engine is idling (Wayson and Bowlby 1989: 21).

Electric engines are much quieter and at speed their sound is not normally discernible from that of rolling stock. Reports of early trials of Japanese *Shinkansen* trains (Wayson and Bowlby 1989: 27) indicated that a substantial element of engine noise was derived from the overhead current collection device; this was discovered in tests of noise barriers which muffled the wheel/ rail contact noise but were not high enough to affect noise emanating from the top of the train. Early design of European high speed electric train pantographs caused noise due to air turbulence. However, this appears to have been successfully eradicated on TGV *Atlantique* trains. In some cases aerodynamic covers have been provided for pantographs. Problems of 'arcing' noise, created when pantographs are momentarily detached from the catenary wire, can be alleviated by electrical connections between different pantographs on the same train (Pyrgidis 1993).

7.3.2.2 *Wheel/rail contact noise*

Rolling noise is roughly related to train velocity—squared for SEL, cubed for L_{max}, i.e. SEL increases by 6 dB(A) (2 log 2) when speed is doubled. Actual noise levels depend upon the condition and type of the rails and wheels as well as on the axle loads and the design of brakes, suspension and bogies. The contact causes characteristic train noise as each pair of wheels crosses open rail joints, but is quieter and more uniform on the continuous welded rail which is now common on fast lines. In either case wear on a rail increases its roughness and hence the noise level. Rail corrugations are mainly caused by the action of iron tread-braked wheels. Variation in the standard of the support system and of rail wear and maintenance can account for differences of 10 dB(A) between running on track that has severely deteriorated and track in good condition.

Typically on smooth British track, the older Mark I/II coaches travelling at 160 km/h cause a peak noise of 94 dB(A) at 25 m. Disc-braked Mark III stock causes only 84 dB(A) in similar conditions; nor does it cause so much rail wear. But BR data shows that if the track *is* worn—say with 0.1 mm deep corrugations—then both types of stock will emit 105 dB(A) (Stanworth 1987: 14/8).

Track conditions are at least as relevant to the noise of heavy freight trains as to that of faster but lighter passenger traffic. Heavy wear also results on curved track if freight trains traverse it at slower speeds than those for which the cant is provided. Thus the cause and degree of track wear are critical elements; in forecasting noise levels for future traffic, perfect track conditions should not be assumed.

Improvements in rolling stock wheels and suspension mean that higher speeds do not necessarily mean louder noise compared with older stock. The latest BR electric IC 225s and diesel HST 125s are reported to run at 200 km/h at L_{max} noise levels at 92/93 dB(A), similar to those of WCML electric trains at 160 km/h. Note that the intermediate coaches of any HST 125 (see Figure 7.4) emit less than 90 dB(A) at close to 200 km/h, but the sub-150 km/h Mark II coaches in Figure 7.1 cause between 90 and 92 dB(A). Whilst the 1980s TGV produced noise in the high 90s dB(A) at 270 km/h, design criteria for the 300 km/h TGV *Atlantique* and Eurostar stock stipulate that the noise shall not exceed 93 dB(A). These trains utilise articulated bogies following the precedent set by Spanish *Talgo* stock whose smaller and fewer wheels account for a noise level apparently less than 80 dB(A) at 160 km/h. Still using conventional bogies, the 160 km/h Wessex Electrics are claimed to be about 10 dB(A) quieter than the older 144 km/h Southern emus.

Design of passenger coaches is aimed at a smooth, comfortable ride. Carriage suspension is designed to limit the vibrations at audible frequencies which pass up into the body of the coach. Achievement of these objectives has resulted in additional benefits of noise reduction outside. No such incentive exists in the design of goods wagons. Nor is it economically attractive to provide for smoother suspension than safe operation requires.

However, solutions are being investigated to provide smoother, and hence quieter running, suspension for goods wagons. For instance, cross-bracing in three-piece wagon bogies has been developed by BR to improve stability and prevent 'hunting' on curves. But so far this is considered suitable for application mainly to tighter curves than are common in Britain. Improvements will be needed if the noise of freight trains becomes a recognised environmental hazard.

Wheel design for quieter running has to take into account the resonant response of the wheel rings, the size of the wheels and the type of brakes. Wheel ring resonance, transmitted torque or the occurrence of 'wheel flats' are complex subjects but ones where 100 years of operating experience must be of value. Research has been undertaken and models developed to relate noise and vibration to track conditions (Cox 1994). It has already been seen that small wheels have been introduced on some wagons in order to provide more superficial cargo space; reduction in the size of wheels can also be effective in reducing noise of rail/wheel contact.

As we have seen, greatly reduced wheel and rail wear, and hence noise, results from use of disc brakes. But the use of disc brakes on freight trains is almost unknown outside Britain and there will be brake compatibility problems on international trains until economic and operational considerations make agreement on a uniform approach possible.

7.3.2.3 Aerodynamic noise

Aerodynamic noise is insignificant in open situations except at exceptionally high speed. Such noise is believed to be proportional to the sixth power of the velocity. Tests with magnetic levitation trains, operating without engine, wheel rail contact or catenary, indicate that aerodynamic noise is not a significant cause of disturbance below 300 or 400 km/h.

7.3.2.4 Vibrating structures

The noisiest vibrating structures are steel bridges. In general, augmentation of noise during the passage of trains over structures is in the range 1 to 9 dB(A). However, where the track is fixed directly to the bridge girders, increases of up to 20 dB can be generated. Such large increases can be avoided by suitable design of track bed and fastenings.

7.3.3 Other Railway Noise

In addition to the basic noise as the train passes, there are various special effects, usually at particular locations:

Table 7.2. Railway noise levels (dB(A) 25 m from track centre line)

Speed (km/h)	Type of train	L_{max} (free field)	SEL	Typical L_{Aeq} 24 h (façade)		Source
70	**BR** aggregates	83	89	–	250 m long	A–E
70	**BR** aggregates	83	94	–	500 m long	A–E
80	**BR** Intermodal freight	84	93	–	600 m long	DNV
80	French Freight	86*		64*	450 m long	CPRE
97	US—Three Metros	78–81*	–	–	average 3 cities	W&B
100	**BR** dmu	80	83	–		A–E
100	French parcels train	89*	–	64*		CPRE
120	**BR** Intermodal freight	87	95	–		DNV
144	**BR** Southern 8-coach electric 4-coach	82–93	92	72	170 trains/day	BR
144	**BR** Loco + 5–8 coaches	82/85	88/90	–		BR/A–E
		89	95	–	Class 47 diesel	A–E
160	Eurostar	87/88	93/97	71	2013 CTRL Traffic	BR
160	West Coast Main Line	93	–	76	180 trains/day	BR
200	High Speed Train (HST 125)	94	97	–	Diesel	A–E
200	TGV SE	92				IRJ
200	ICE	86/82				W&B/IRJ
200	Talgo Pendular	82				IRJ

225	Eurostar	93	97	74	2013 CTRL traffic	BR
250	ICE V	85				IRJ
250	TGV SE	95				IRJ
250	TGV	96*		68*		CPRE
270	TGV SE	99/97				BR/W&B
300	Eurostar/TGV-A	96			Specification	BR
300	TGV SE	97				W&B
300	Maglev	84				W&B
300	ICE	93				W&B
300	ICE V	89				W&B
400	ICE V	102				IRJ
400	TGV-A	100				W&B
400	Transrapid Maglev	100				IRJ

Sources
BR   = British Rail Channel Tunnel Link—Noise Brochures, 1989. Range of values in Explanatory Note and Papers A1, A3.
DNV = based on DNV Technica railway noise (façade) calculation data.
CPRE = Sullivan (CPRE 1989).
IRJ = *International Railway Journal* (June 1989)
W&B = Wayson and Bowlby (1989).
A–E = measurements by the author and S. Ellis.
* Data adjusted as equivalent at 25 m distance.

- accelerating diesel engines as trains leave stations or after signal stops or speed restrictions, or high power output climbing gradients;
- at points and crossings;
- 'swoosh' of air pressure as trains leave tunnels or pass under bridges—this effect can be minimised by careful design of structure openings;
- noise emerging from tunnel vent shafts;
- hooters/horns.

Other noise incidents are associated with track maintenance trains. These include ballast cleaning and tamping, occurring once or twice a year on well-used routes and occupying an hour or so at night at any one location. Rail grinding operations also take place occasionally.

Under certain circumstances detonators may still be used in foggy conditions. Off-track noises related to railways may also be heard, such as public announcement systems, cargo movements and road traffic at stations.

Electric railways are generally quieter and cleaner than those using diesel traction because power is generated at a remote source. But substations are located at intervals at the lineside to regulate the voltage supplied. The noise of these substations, primarily due to large cooling fans, gives average continuous levels of 56 dB(A) at property lines (Wayson and Bowlby 1989: 30)

Table 7.2 illustrates some published, measured and deduced data for the noise of a variety of modern trains at different speeds. There are some differences among the values quoted but these are probably less than the variations which actually occur due to track wear and other circumstances. Therefore all values should be taken as indicative rather than specific. Generally, however, variation in the louder noises—such as from HSTs—is slight, whilst there may be differences of 5 to 10 dB(A) in the recorded noise of apparently similar quieter trains, e.g. lower powered dmus or emus in the 70 to 80 dB(A) range.

7.4 NOISE MEASUREMENT AND PREDICTION

7.4.1 Measurements

The purposes of noise measurement in investigating railway impacts are as follows:

1. To record the actual noise generated by trains of different sorts under various operating conditions; this information is essential for predicting train noise.
2. To record the ambient (existing) noise at locations, such as outside houses, in gardens or streets, where new railway development threatens to introduce new and perhaps dominant noise. L_{Aeq} or noise incident data can then be compared with the predicted new noise levels; this informa-

tion may be relevant in assessing the impact of introduced train noise into particular situations.

Sound pressures can be measured using portable equipment. Instruments can be set to record particular frequency ranges, e.g. in dB(A) for those frequencies most sensitive to the human ear. They can record instantaneous and continuous noise levels and provide data for calculation of SELs, L_{eq}s or other parameters.

There are a number of important qualifications about the circumstances of noise measurements which must be recorded and considered when making subsequent predictions. The position of the observer and the distance from the track is clearly critical. Meteorological conditions (wind, snow or temperature inversions) can be important if noise propagation is thereby affected. Ideally, measurements should be made in still weather at a distance from the track approximating to the standard (25 m), on level ground and where there is only low background noise.

Basic data about the permanent way should be recorded, if only to confirm the type of rail jointing, sleepers and ballast. Information as to the state of wear of the track, if obtainable, can be very relevant.

Information about trains that must be recorded can vary from a simple note of its type, e.g. an HST 125 of standard composition or details of the type of locomotive and (full or empty) rolling stock on a goods train. In all cases it is important to ascertain the length and hence the speed of the train; for some trains at close quarters the time taken for the train to pass can be deduced from a trace of the sound measurement. However, it may be difficult to identify on a chart the actual time of passing of relatively quiet trains in which wheel/rail noise is dominant. Calibration of speed calculations is therefore desirable.

British Rail and some other railway companies have collected comprehensive information about train noise on their own systems. This data is generally sufficient for predictions of the noise impacts of future traffic on the same systems. However, measurements are still likely to be needed in many countries where noise issues have not previously been scientifically investigated.

7.4.2 Prediction

Forecasting of future noise levels will be required at all places (sites) where people (receptors) could be disturbed or affected by train noise. The necessary steps in prediction are:

- assembly or measurement of standard reference data about the noise generated by each type of train;
- calculation as to how the noise is propagated towards the various receptors/sites;

- adjustment for receptor/site conditions;
- calculation of L_{Aeq} or other total data.

7.4.3 Reference Noise Data for Trains

Noise levels generated by each type of train are recorded at, or adjusted to, a standard distance from the track. The values used in predictions (L_{max}, SEL, etc.) should be for the same track conditions and train speed and length. Where these differ, corrections to the data should be applied—for instance for speed, for the number of coaches or wagons and to allow for timber sleepers or steel bridges. Corrections may have to be made according to which track the trains will use, especially where noise is absorbed by ballast on other tracks.

7.4.4 Propagation

A simple model has been described (Hemsworth 1986) for predicting railway noise levels, i.e.

$$L_d = L_o - L_s - L_a - L_g - L_e$$

where

L_d is the unknown level at distance d from the source,
L_o is the known noise level at a fixed (close) distance from the source,
L_s is the attenuation due to geometric spreading,
L_a is the attenuation due to air absorption,
L_g is the attenuation due to ground absorption,
L_e is the increased or reduced attenuation due to geographical features such as cuttings and embankments.

L_o is the reference noise level already discussed which will be reduced over distance by the various forms of attenuation L_s, L_a, L_g and L_e.

Geometric spreading (L_s) accounts for the diminution of sound at a distance from the train. It is generally assumed that noise decreases proportionally to the logarithm of the distance. However, the actual constants in the equation depend on the frequency of sounds generated as well as the length of the train. It can be estimated that if the noise received from a *point* source (like a locomotive) is reduced by 3 dB(A) each time the distance from that source is doubled, then the equivalent reduction for an infinitely long *line* source, like a very long freight train, is 6 dB(A). In fact Walker (1989) found that noise level *typically* falls by about 20 dB(A) as the distance from the track increases from 25 to 200 m. These values probably include elements of air (L_a) and ground (L_g) absorption as well as geometric spread (L_s).

Data relating noise attenuation to distance has to be adjusted where the angle of view from the receptor site to the section of track concerned is less than 180 degrees or where it is at a skew.

Air absorption (L_a) depends on the frequency of the sound and the ambient temperature and humidity. The attenuation is slight; at 100 m the air might absorb 0.5 dB(A) from a diesel locomotive, slightly more from rolling stock.

Ground attenuation (L_g) results from ground impedence and from absorption by ground cover. It depends not only upon the nature and topography of land crossed by sound waves but also on the frequency of that sound. For instance data quoted by Hemsworth (1987: 15/8) indicates that vegetation may be 10 times as effective in reducing high frequency (8000 Hz) noise as it is for that below 100 Hz. Dense shrubs are three and a half times more effective in absorbing sound than smooth surfaces such as grass.

Measurements of noise at a distance across level ground record $(L_s + L_a + L_g)$ and each element $(L_s$ or L_a or $L_g)$ can only be differentiated in theoretical calculations. L_g is the most variable and its condition the most difficult to specify. If measurements and prediction formulae reflect low absorbent, grassland conditions, then assumed values of L_g will be low. In typical conditions, therefore, L_g may be conservative in calculating total noise reduction. If dense vegetation exists between the source and receptor site then the effect of additional ground attenuation might be estimated; but if the intervening surface is reflecting and acoustically hard, such as concrete or water, then no ground attenuation L_g should be allowed.

Excess attenuation (L_e) accounts for other means whereby noise propagation may be affected, i.e. by barriers, such as rows of houses and fences, or where the track is elevated on embankment or hidden in cutting. Corrections may be calculated as a separate value in each case and adjusted for the standard (close to the track) level L_o. Alternatively, ground and track contours can be incorporated in a mathematical model for calculating L_s, L_a, L_g or total attenuation.

Based on data quoted by Hemsworth (1987: 15/10), L_e could represent an increase of 10 dB(A) for track on a 250-m-long embankment over 6 m high, a decrease of 10 dB(A) in a 7 m cutting, and a similar decrease if a row of double storey houses obstructs the view.

7.4.5 Receptor Site Conditions

For comparison of effects outside different buildings, noise data is usually stated as at 'façade' level. This may require an adjustment (addition) to 'free field' levels to allow for reflection from walls or nearby ground surfaces.

7.4.6 Calculation of Total Train Noise Levels

L_{max} or SELs can be calculated for each receptor site and each train or for each element of the train. Data can then be digitised so that calculations can be undertaken rapidly for different train circumstances and different receptor sites.

SELs can be added logarithmically for all trains of all types passing in a period to give 24 hour, day or night L_{Aeq} values.

7.4.7 Prediction Methodology

A range of formulae is available for calculation of the various elements of train noise—either in detail for each element (e.g. Hemsworth 1987) or in simpler versions produced for public use and related to particular trains (e.g. BR CTRL 1989). Such formulae are generally semi-empirical, i.e. based on observation as well as theory.

Investigators or planners seriously interested in achieving objective data about particular situations should review all the available scientifically-established relationships and should qualify their results by stating the precise circumstances and assumptions. It should be noted that many formulae relate to noise pressures measured on the A-scale of frequencies without differentiating between actual frequency bands or between structural or ground-borne vibration and airborne sound propagation.

Figure 7.6 is a typical methodology for predicting noise from a railway. Various approaches are possible for the detailed calculations at each stage according to which formulae are used for each element of noise generation and projection, what corrections are to be applied and in what sequence the calculations are to be made.

In Britain comprehensive prediction methodology has been drafted by the Department of Transport (1993b) which may be applied in Britain to support legislation concerning noise insulation near new railway lines (mentioned in Section 7.6 below).

7.5 NOISE IMPACTS

7.5.1 Disturbance or Toleration

A considerable amount of research has been undertaken since 1983 on people's response to train noise. This section reviews the issues involved and Section 7.6 describes standards which are being set to determine when action might be taken.

The impacts of noise on people can be examined as follows:

1. By subjective survey—how much will people tolerate? This approach

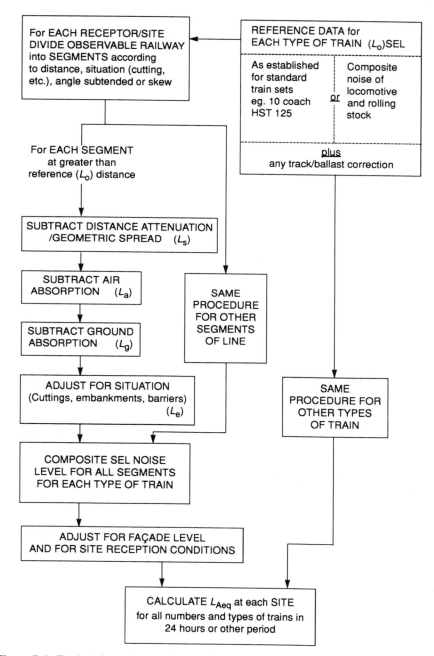

Figure 7.6. Typical flow diagram for predicting noise from railways

shows great scatter in results when 24 hour L_{Aeq} data is plotted against annoyance by railway noise (Walker 1986); it may exaggerate long-term disturbance if the people questioned are unfamiliar with train noise and apprehensive about its effect.

2. By comparison with existing situations; this may underestimate initial discomfort because people in the compared (railway) situation are already used to train noise.

In fact there is some initial evidence (Dept of Transport 1991: 25) that the anticipated 'habituation' (getting used) to noise does not occur in the first 18 months of new transport operations. Light should be thrown on the subject as more data about relationships between perception and familiarity is made available for the TGV or other routes constructed through *suburban* areas.

There is a wide range of views on what levels of noise are acceptable. On the cautionary side local authorities have established noise limits, sometimes as low as 55 dB(A) L_{Aeq}, above which they refuse permission for residential development. At the other extreme, French lineside residents told the BBC in July 1993 that they were unaware of the passing of TGVs. (This was evidently not deliberate understatement because the same people said they *did* notice the passing of night freight trains.) Visitors from Rainham in Essex, taken to France to assuage their fears about Eurostar trains on the Union Railway, remained unconvinced.

7.5.2 Thresholds of Tolerable Noise

Tolerable noise levels must be examined with respect to:

- maximum noise (L_{max}), for instance levels above which speech or sleep are seriously interrupted;
- equivalent continuous sound levels, such as L_{Aeq}, over longer periods.

7.5.2.1 Dangerous Noise

Physical damage to ears can occur at sound pressures around 150 dB(A). Such noise levels are not associated with railway operations. There is a finite risk of disability associated with noise levels greater than 85 dB(A) lasting over all or most of a working day for 10 years (Dept of Transport 1991: 9).

No sudden railway noise is likely to be loud enough to damage anybody's hearing. Standing about or working on railway tracks can of course be dangerous but it is the *quietness* of an approaching train which causes the need for a look-out. It was possibly because his hearing had deteriorated with age that the Great Western locomotive engineer G.J. Churchward was killed by one of his own locomotives at Swindon. On the other hand, George Huskisson MP must have been totally confused by the whole unfamiliar situation

100 years earlier in 1829 when he was knocked down by an engine at the Rainhill Trials.

7.5.2.2 Disturbing and Annoying Noise

Disturbance affects specific activities. Annoyance is more general aggravation.

There is controversy as to what constitutes serious disturbance, particularly in defining the level at which speech—indoors or out—is interrupted. Most people can communicate against a background of 35 to 40 dB(A) but may find speech difficult in the range 50 (indoors) to 65 dB(A) (outside). Teachers report difficulties in classroom communication against background levels greater than 60 dB(A), a level commonly associated with crowed offices or with a busy street heard through open windows. There is also argument as to what duration or numbers of interruptions are tolerable. Shutting the window in order to speak while Concorde passes over is different from keeping it shut due to continuous traffic. Informed objectors may demand evidence about the extent to which conversation or audible messages are affected. Comprehensive explanations are likely to involve analysis of the nature and sound frequency of the interfering noise as well as its volume.

TEST (1991: 143) quote data showing that 50 per cent of the population of the UK is exposed to road transport noise exceeding 55 dB(A) and 25 per cent to that exceeding 60 dB(A). L_{10} values quoted by the Department of Transport (1983: B.1.9.3) are close equivalents whilst TEST's data for Japan, the noisiest country listed, shows that 80 per cent of people there are exposed to 55 dB(A), 58 per cent to 60 dB(A) and 31 per cent to 65 dB(A). It is suggested that if a quarter of the British population is used to 60 dB(A), then it is not unreasonable to adopt that level as an acceptable measure of a moderately noisy environment. Certainly there are some people normally exposed to particularly quiet outside levels who might object to 60 dB(A) as a normal level and would express annoyance when first exposed to it. But, in view of the researched proportions, 55 to 60 dB(A) represent reasonable levels for minimum noise exposure below which consideration need only be given to effects on exceptionally sensitive environments. The Department of the Environment (1993) indicates that, below an outside noise level of 55 dB L_{Aeq} in the day and 42 dB L_{Aeq} at night, noise should not normally be a material consideration.

Opinions differ as to whether an overall 24-hour L_{Aeq} can adequately cover disturbance at night or whether special limits should be applied to sleeping conditions. Limits quoted vary but 20 repetitions in a night of 55 dB(A) indoors or 85 dB(A) outside is typical. These could be peak rather than SEL levels and it is difficult to estimate equivalent L_{Aeq} values. However, the L_{Aeq} equivalent to the indoor value would be too low to be meaningful whilst the outdoor equivalent would be nearer to the day than to the night values envisaged above by the Department of the Environment. Recent studies by the Civil Aviation Administration appear to indicate that sleep disturbance occurs at a higher level than originally thought.

There is considerable evidence but continuing argument about the effects on people of different types of transport noise. DNV Technica (1990:8) presented data showing percentage response/L_{Aeq} curves for railways, roads and aircraft in which railway noise above 60 dB(A) L_{Aeq} is consistently considered less annoying than the same noise level from roads or aircraft.

There are well-researched claims that tolerance of all-day train noise levels is higher than for roads. Conclusions of such research indicate that the tolerance level is commonly between 4 and 9 dB(A) L_{Aeq} greater to trains than to road traffic in the 60 to 70 dB range. Very little difference in perception is reported at lower levels (Dept of Transport 1991:21). This seems very credible in Britain where for many years homes have been built close to railways. Social attitudes to conventional (pre-*Shinkansen* 'bullet') trains were different in Japan and may be changing in Europe.

What sort of criteria can be used for deciding whether a noise impact is significant and, if so, what can be done about it? Three types of threshold can be used:

1. a noise level over a period, for instance an L_{Aeq} (24 h) value, above which some form of mitigation should be considered;
2. increases of a certain increment (dB) above previous ambient or railway noise levels;
3. maximum tolerable peak level, e.g. causing sudden shock in daytime or waking from sleep at night.

Judgement on what criteria to apply, and at what levels, must be made in the light of how much weight should be given to various receptors' reaction and to the practicability of achieving targets.

7.6 PLANNING STANDARDS

Standards can be set for either noise generation or noise reception levels. *Maximum train noise* (L_{max}) *generated at source*, measured at 25 or 30 m distance, has been subject to such upper limits as:

- 90 dB(A) on US diesel locomotives built after 1979;
- 85 dB(A) for rolling stock wheel noise on Danish railways;
- 96 dB(A) in the specification for Eurostar 300 km/h trains.

No EC directives have yet been issued relating to the noise emitted by locomotives or trains.

7.6.1 Housing

Noise *reception standards* can be set to define levels of acceptability, to limit housing development or to determine whether insulation should be provided. Typical standards for 24 hour L_{Aeq} noise at house façades, are as follows:

- 70 dB(A) for new railways in Japan;
- 69 dB(A) for SNCF TGVs, reduced to 64 dB(A) in the latest guidelines for application in residential areas;
- 65 dB(A) incorporated in the parliamentary bill for the London Docklands Light Railway's Beckton extension or 60 dB(A) for Tyneside Metro (noise insulation criteria);
- 55 to 65 dB(A), being limits of local noise level above which some county councils refuse planning permission for housing development.

Such planning standards are set on a district by district basis. They might take into account the inclinations of those potential house purchasers who, through experience or otherwise, will choose to live near railways. If standards imposed were unnecessarily high, they might prevent housing development where it might otherwise be very suitable. Enforcement of severe

Figure 7.7. Diesel HST 125 passing housing built about 1990, 15 years *after* these trains were introduced. At 25 m from the train maximum 'free field' noise level at 200 km/h is about 95 dB(A); Single Event Level (SEL) about 97 dB(A); slightly higher 'façade' noise levels would be recorded against walls. The 18-hour L_{Aeq} noise level at 25 m from this line is estimated at about 74 dB(A)

limitations could increase demand for housing space elsewhere, threatening more valuable green field land resources. However, the particular needs and characteristics of different types of housing will no doubt continue to play a part in land use planning. Blocks of flats without individual garden space and with multiple window glazing have recently been built very close to HST routes where 70 dB L_{Aeq} façade noise levels can be expected (Figure 7.7).

In British Department of Transport practice, threshold indicators are to be used as guidelines rather than rigid standards. To quote SACTRA (1986), 'We do not recommend the use of single values as "standards" which, if exceeded, would require action.' Nevertheless, noise insulation regulations impose a duty to insulate dwellings against traffic noise from new roads (Dept of Transport 1991: 13). The standard applicable to new roads in Britain entitles noise insulation treatment where increased façade noise levels exceeds 68 dB(A) L_{10} (18 h).

In comparing all-day noise standards for road and rail traffic, British Rail (BR CTRL 1989) argued that 68 dB(A) L_{10} (18 h) for road traffic is equivalent in annoyance effect to 70 dB(A) L_{Aeq} (24 h) for railways. Meanwhile Kent County Council (DNV Technica 1990: 13) proposed different rail noise standards at different times of day. Kent's standards are mathematically equivalent to a L_{Aeq} (24 h) of 63 dB(A) (ICE 1990: 29).

To address the issue specifically, the British Government set up a committee in March 1990 to recommend national noise insulation standards for the operation of new railway lines which would equitably relate to the standard set by regulation for new highways.

The Department of Transport Committee, chaired by Dr C.G.B. Mitchell, reported early in 1991. The report confirmed earlier conclusions that L_{Aeq} (24 h) was the most suitable standard measure for (intermittent) train noise. For road noise, L_{Aeq} (24 h) was equivalent to 3 to 4 dB(A) less than L_{A10} (18 h), i.e. the standard for roads is equivalent to 64 to 65 dB(A) L_{Aeq} (24 h).

The Committee examined a wide range of evidence about disturbance and annoyance caused by railway noise and recommended a standard façade level of 66 dB(A) L_{Aeq} (24 h) for situations equivalent to those justifying insulation in the roads standard. In addition a separate, lower standard of 61 dB(A) L_{Aeq} applicable over eight night hours was recommended. Note that the standards do not apply to existing railways and only to those new or altered railways which create more noise than exists from railways already in the area. The Committee's recommended standard was subsequently altered by the Secretary of State to 68 dB L_{Aeq} (0600–2400) and 63 dB L_{Aeq} (0000–0600). This appears to make a clear departure from the commonly considered view that the 24 hour L_{Aeq} adequately represents annoyance due to railway noise.

Data on the differences between *day* and *night* railway operation noise (L_{Aeq}) levels varies widely. On one route investigated by BR (Dept of Transport 1991:6) actual night-time L_{Aeq} was only 1 dB(A) less than during the day. On another route the reduction at night over the quietest six hour period (midnight to 0600) was between 7 and 16 dB L_{Aeq} according to the site. These latter values are comparable with those recorded on roads. The overall desirable difference of 5 dB(A) between 24 hour and night values proposed by the Mitchell Committee would seem to pose a problem on the first route but not on the second. The number and type of trains running at night is of course a matter of operational planning.

Calculations of the 18 hour L_{Aeq} façade level were made by the author in 1991 based on a limited number of measurements beside the (Western) Paddington to Reading line. It was estimated to be 74 dB(A) at 25 m for a mixture of traffic including 200 km/h HST 125s, locomotive-hauled passenger trains, freight trains of various types and 1960s vintage dmus. The Department of Transport's proposed standard of 68 dB(A) L_{Aeq} (18 h) was probably experienced under these conditions at about 75 m from the track.

In 1993 the locomotive-hauled passenger trains were largely replaced by much quieter 'turbo' trains. HSTs remain the dominant traffic but new measurements indicate that the SEL of turbo trains is 10 dB(A) less than the trains which they have replaced. If all the locomotive-hauled trains were replaced, the 18 hour L_{Aeq} for the traffic is reduced by about 1 dB(A). Only if the HSTs were replaced by a completely different type of train, probably after electrification, could any more substantial reduction be expected.

The noise and occurrence of freight trains vary too widely to make a firm estimate on how conditions on this line would relate to the proposed night-time limit of 63 dB(A) L_{Aeq} (0000–0600). However, if freight traffic were to build up so that substantial trains (emitting 95 dB(A) SEL) were to pass at night every 30 minutes in each direction (24 total in six hours) then the L_{Aeq} (0000–0600) would be 65 dB(A) free field, say 68 dB(A) façade level equivalent at 25 m from the track or 63 dB(A) at about 60 m distance.

No wide acceptance is evident of *change in noise level* as a principal threshold for action. However, minimum increases in either total or railway-related noise levels have been proposed as *qualifying* criteria, i.e. as additional conditions to be satisfied before applying absolute noise level standards.

7.6.2 Non-Residential Property

The imposition of standards or design of noise mitigation measures is concerned primarily with *residential* property. Public concern is much more related to what happens at home than to where people work, shop or play. So different attitudes and noise standards may apply to non-residential receptors.

Figures have been proposed for 'acceptable background noises' inside buildings. Typical are 35–40 dB(A) for 'private' offices, slightly more for open-plan offices or classrooms and 40–50 dB(A) in restaurants. Corresponding outside levels could be 10 to 20 dB(A) higher. *Some* people with experience of noisier backgrounds in offices and restaurants would exhibit higher noise tolerance; whilst questionnaires to hospital patients have indicated little preference for lower exterior noise levels than those existing outside other public buildings. Offices and schools are noisy places because of the behaviour of their inhabitants; hospitals are quiet for the same reason. Exterior noise is usually secondary.

It is concluded that non-residential buildings require no special treatment in environmental assessment. Any special requirements for low noise levels for particular activities should be catered for by design or insulation where they occur. On the other hand, for certain activities higher noise levels may be tolerated at work than would be acceptable at home.

7.6.3 Fauna

Most research indicates that domestic or wild animals or birds do *not* suffer any ill effect from aircraft or train noise and that they adapt quickly when new noises are introduced. General proof of this is evident by looking out of the window of a TGV. However, the following qualifications are worth making:

1. Horses exhibit 'startle response' for longer than other animals when in confined quarters (where they may need similar insulation to humans).
2. In quiet countryside, construction work should be planned to minimise *sudden* (start-up) noise in the early morning or evening in the wildlife breeding season.

7.6.4 Noise inside Trains

The subject of noise inside trains has been investigated thoroughly. Such noise is generated primarily by wheel/rail interaction vibrated through the suspension and the body of the carriage. Where traction is provided by underfloor diesel engines in diesel motor units (dmus) or when break pumps operate in electric emus, these also contribute to inside noise levels.

In older, conventional main line carriages the noise inside, quoted at 75–95 dB(A), is rather higher than in motor cars (65–85 dB(A)) and similar to that in buses or passenger aircraft.

However, in modern air-conditioned high speed stock the level is lower than all these, commonly in the range 60 to 70 dB(A)—sometimes slightly higher at the carriage ends over the bogies. Noise of only 59 to 65 dB(A) was predicted, achieved and apparently appreciated by passengers in new 160 km/h stock in Denmark (Pederson 1986). On faster German ICE trains internal noise is reported as 63 to 72 dB(A) depending on the speed and location within the carriage (Wayson and Bowlby 1989: 21). The noise inside a TGV has been described as a quiet 'purr'. Whilst luxury is often equated with quiet (Hardy and Jones 1986), some people claim that passengers need an optimum background noise—sufficient to relate to the sense of movement and to mask the noise of fellow passengers. The ideal noise level may be that sufficient to prevent private conversation being easily picked up by strangers without actually making talking difficult. The most annoying noise commonly complained of in trains is that of personal radios (Bovey 1986) rather than of the train itself.

It is concluded that noise inside trains is not commonly a serious cause of disturbance. This aspect of the welfare of passengers can be left in the hands of the railway companies who wish to attract customers.

7.7 NOISE MITIGATION

Adverse effects of noise can be countered by measures to reduce it at source, by creating barriers to attenuate its propagation over distance or by insulating affected buildings from exterior sound.

7.7.1 Reducing Noise at Source

Noise *generation* on the *rails* can be reduced by:

- design of wheels, suspension and brakes to reduce wear;
- regular grinding of rails to remove corrugations;
- avoidance of rail discontinuities in sensitive areas;
- resilient track mountings or fastenings, especially on elevated structures or in tunnels.

In addition to rail grinding measures, wear—and hence noise—between wheel and rail can be alleviated by regular reprofiling of wheel flanges (Pyrgidis 1993) or by optimal design of wheel flanges (Leary 1990). Some types of slab track produce up to 10 dB more noise than ballasted track. Removal of rail corrugations can reduce wheel noise by 10 dB(A) or more.

Wheel damping is not considered cost effective although wheel number and size reduction and suspension improvements are. Disc-braked rolling stock is undoubtedly quieter and less likely to cause rail corrugations than that fitted with iron tread brakes. Lighter construction, such as the light alloys used in the latest *Shinkansen* 300 sets in Japan, is also understood to help noise reduction.

Considerable progress has been made in construction of quieter smoother wheels and suspension for passenger trains to the extent that there is some doubt whether there is any scope for further improvement. The same may be true of track maintenance. However, these improvements have by no means been universally introduced on passenger trains and great scope remains for designing quieter freight wagons. Stricter limits on *engine noise* could be imposed but this is likely to prove impracticable without a complete change to electric traction.

7.7.2 Attenuation of Noise projected from the Railway

Barriers can be erected either at the lineside or near the affected property. Noise barriers are made of concrete, stone, metal, plastic, wood or composite materials or can be earth bunds. They can be fences or part of structural roofs, walls or floors (see Figures 7.8 and 7.9). Their effectiveness depends on the height, mass and absorbent quality of the barrier. Typically it is claimed that a structure 2 m high, just low enough for nearside train passengers to see over, will reduce noise levels by up to 10 dB(A) (ICE 1990: 37). But noise at *low frequencies*, below 500 Hz, is not as effectively blocked by barriers as is higher frequency noise (Wayson and Bowlby 1989: 27).

Elevated sections of railway are particularly likely to cause high noise impact. The sound of Japanese *Shinkansen* trains on bridges is reduced by 7 dB(A) by installing shielding *beneath* the track. If additional barriers are erected *between* the tracks this apparently improves the reduction to as much as 12 dB(A) (Remington et al. 1987: 17/6).

Thus variations can be made in barrier shape and configuration. Actual design of noise barriers should take the following issues into account:

- that adequate mass is provided;
- capacity to absorb noise or to reflect it; generally absorptive barriers give between 3 and 4 dB(A) additional attenuation over that from reflective noise barriers;

Figure 7.8. Noise barrier on French-*Atlantique* TGV route (Photograph by courtesy of Graham Parry of DNV Technical Ltd)

Figure 7.9. Noise barrier between tracks, Tyne and Wear metro (Photograph by Courtesy of Graham Parry of DNV Technical Ltd)

- operational factors, such as whether track workers can hide inside or outside the barriers when trains pass;
- the visual appearance, from the train and within the landscape, of potentially ugly walls and fences.

The negative visual impact of high noise barriers can be seen in reports by passengers on the German *Neubaustrecke* lines that 'the view from the windows is severely curtailed by noise barriers' (RGI 1990: 581). Bray (1993) states

> installing noise barriers involves a trade-off between visual intrusion and noise reduction. Noise barriers have traditionally been totally ugly affairs. Many people would object if London's railways began to be obscured by featureless concrete slabs—particularly in South London where sections of elevated railway are part of the area's character.

Where wheel/rail interaction is the main source of train noise, a relatively low 'see-over' barrier may be practicable, particularly if a narrow inward cantilever roof is provided to reduce sound 'spills' over the top. However, this will not attenuate any higher source noise such as might be caused by overhead electric current collection equipment.

Cost and performance are key issues in making decisions about the need for and composition of noise barriers. Visual appearance must then be taken into account in determining what is an appropriate type and height of barrier.

7.7.3 Noise Insulation of Buildings

Soundproof buildings can be designed specially to avoid noise and vibration from various sources including railways. Birmingham's Symphony Hall, opened in 1991, was designed to an ideal shape and size and incorporated a wealth of facilities—movable ceiling canopy, doors and sound-absorbing panels and reverberation chambers—all adjustable to achieve acoustic perfection. Air conditioning was particularly quiet and the whole building was mounted on rubber blocks to isolate it from noise and vibration emanating from New Street railway station (Hawkes 1991). The Hall was sited near the station as a convenience to its patrons and was designed to ensure that train noise was no problem.

The effectiveness as noise insulators of double glazed windows varies according to the type applied. Noise levels indoors, even near an open window, are perceptibly lower—by 5 to 15 dB(A)—than those outside where the noise is usually measured and predicted. Closed windows may increase this difference to 20–25 dB(A) whilst suitably spaced and sealed double glazing may raise the total insulation to 35 dB(A) (DNV Technica 1990).

In general, the wider the spacing between panes in double glazing the more effective is the sound insulation. However, there are certain ranges of frequency—usually 50 to 300 Hz—in which resonance in the cavity may reduce or even neutralize the insulating effect (Welsh Office 1992). This implies that window insulation is more effective for high frequency noise generated by high speed, particularly electric, trains than for low frequency road traffic or possibly diesel-hauled freight trains.

7.8 EVALUATION OF NOISE IMPACT AND OF MITIGATION MEASURES

7.8.1 Why should we be Concerned about Railway Noise?

Railway noise levels have not changed much in the second half of the twentieth century. Whereas some faster, more frequent high-powered diesel expresses have increased the local lineside disturbance, this has been balanced by substantial transfer of freight services to the roads.

Nevertheless it is feared by some parties that noise problems may be caused by higher speed expresses or a resurgence of goods traffic when the Channel Tunnel opens.

The effects of higher speeds are often balanced by quieter rolling stock hauled by electric engines, whilst rapid transit systems are usually less noisy than the road traffic they are designed to replace. But increased noise of freight traffic could be significant in urban areas like South London as rail freight to and from Europe increases.

Current environmental attitudes and legislation require that problems be examined, even if the actual circumstances are less severe than they used to be. People feel more comfortable if they know that the means of reducing noise are available and can be implemented. Just as important to the investor in railway development is that the costs of reducing noise should be justified.

7.8.2 Economic Evaluation

Any proposal to mitigate noise impact must be justified in terms of its main and secondary benefits and costs or disbenefits.

Various attempts have been made at putting monetary value on noise pollution damage, e.g. at national level, US $200 million (£130 million) in The Netherlands in 1986 (about 0.03 per cent of GNP), or $10 billion (£6.5 billion) in Germany in 1985 (2 per cent of GNP). It is not surprising that these figures (from Pearce et al. 1989: 57, 58) do not match some other estimates. Reference to national calculations has been made here only to indicate their imprecise nature. Certainly there is a great variation in the criteria used. It is concluded that there is no accepted way of pricing disturbance or annoyance directly in terms of any increased dose of noise.

Means of costing noise pollution from particular sources which can be considered are:

- what people are actually willing to pay to avoid the nuisance, i.e. the difference between house prices, where they are situated by new roads or under aircraft flight paths, compared with those elsewhere;
- 'defensive' expenditure compensating for the annoyance, such as the cost of double glazing for partial (indoor) noise reduction.

House price or similar valuation methods require large amounts of data and there have not been enough new railways to provide this. House prices near and far from *existing* railways could be examined but it is doubtful whether there would be any significant difference, i.e. whether there was a measurable disbenefit of railway noise.

Defensive expenditure is a more practicable means of assessing the cost of noise disturbance or at least the *benefit* of avoiding it. Parity is introduced into the assessment if a standard of disturbance is set which must not be exceeded.

One might, for example, allot a value of £1000 per person or per household where a 65 to 70 dB(A) L_{Aeq} is predicted or £10 000 if the level exceeds 70 dB(A). These values might be a measure of insulation costs or of compensation or of house prices but would be applicable wherever these noise levels were reached. If, along a stretch of line 100 m long, there were five houses which would experience over 70 dB(A) at their track-side façade plus a further ten behind receiving over 65 dB(A), then the allotted cost of noise would be £60 000. The sum for all the houses or residents represents a price on noise 'disbenefits' which can be added into any estimate of total project economic costs. A 2-m-high noise barrier along the 100 m stretch might eliminate all noise disbenefits at a cost of £50 000 (£250 per m^2). Whether the cost of insulation or compensation could be less than this amount will be relevant in the actual choice of mitigation measure. If there were more houses a lineside barrier becomes more economic; if there were fewer then its cost may not be justified.

Secondary costs or benefits may also arise from noise mitigation measures. For instance, a high opaque barrier or deliberate dropping of the track level into cut-and-cover has a detrimental effect on the view from the train and, therefore, in some cases, on passenger enjoyment and perhaps revenue.

7.8.3 Railway Noise in the Total Environment

A valuable resource which is at risk is the quiet or peace in any area. In Chapter 11 there is discussion as to how quiet or 'unspoilt' countryside can be defined and on how to quantify the effect on this resource of any new transport developments, at least in comparative or order-of-magnitude terms.

There is a clear need to define at least comparative effects of transport systems on levels of peacefulness in urban, suburban and rural land areas.

7.9 VIBRATION

7.9.1 Nature of Vibration

Defined as 'rapid motion to and fro', vibration is more physical than noise. It causes perceptible shaking, usually described in terms of 'peak particle movement'. It is transmitted and dampened through the ground but can be magnified where it resonates with the natural frequency of a structure or rises through a flexibly-framed building.

Vibration is closely related to noise:

- in that noise has its origin in vibration—both are wave movements in surrounding media, noise mainly through the air, vibration through structures or the ground;
- through re-radiated noise ('rumble') caused by low frequency ground-borne vibration, converted at a change of medium into the audible range; vibration in buildings can be generated by low frequency airborne sound as well as by ground pulses.

Note that low frequency sound/vibration waves may be inaudible (below 18 Hz) or barely perceptible sounds and therefore are often described as vibration. In general, detectable vibrations and structure-borne sound occur at frequencies below 45 Hz.

Particle movement of vibration is commonly quoted, as here, in terms of peak particle velocity (mm/s). Other related parameters of vibration are frequency, amplitude (or 'displacement'), and acceleration.

Vibration dose values (VDVs), defined in BS 6472 and based mainly on acceleration measurements, provide a vibration equivalent of noise $L_{Aeq}s$.

Re-radiated noise, measured in dB(A), is likely to be masked by directly radiated airborne noise from trains on the surface. Problems with re-radiated noise arise generally in tunnels or deep retained cuttings where sensitivity to such noise often exceeds that to pure vibration.

7.9.2 Vibration caused by Trains

Diesel engines generate mainly low frequency sound and this could be re-radiated where it reaches buildings. However, the main variations transmitted to the ground arise from the forces between the wheels and the permanent way.

Design of the suspension and bogies of passenger trains, particularly electric multiple-units, has made them smoother as well as quieter. Heavy freight wagons are more likely causes of vibration.

Train-related vibration is propagated through unresilient structures, such as steel bridges, or through the ground, especially by underground railways. New vibration problems can arise in existing situations from:

- heavier axle loads on longer and, on new connections to Europe, more frequent freight trains;
- construction and operation of new lines tunnelled through urban areas.

7.9.3 Impacts of Vibration

The impacts of vibration are potentially serious in the following circumstances:

- for structures subjected to violent shaking, above, say, peak particle velocity of 3 mm/s for historic houses, 10 mm/s for residential and 20 mm/s for commercial buildings (DIN 4150);
- for human comfort and sleeping, if movements are plainly perceptible (above 3 mm/s);
- at lesser particle velocities for musical recording and surgical or other equipment particularly sensitive to vibration.

Typical acceptability criteria are:

- 0.3–2.0 mm/s for *vibration*, less for sensitive equipment; for buildings there should be measures to limit vibrations at the structure's natural resonance frequencies;

- 35–40 dB (A) L_{Aeq} as the maximum home indoor level of *re-radiated noise* during passage of any train; lower levels may be required for concert halls; rather more can be tolerated in commercial buildings.

Re-radiated noise from tunnels is not likely to be significant at radial distances greater than 20 m (in clay) to 40 m (in chalk).

New techniques are now being developed for assessment of human sensitivity to variable, intermittent vibration events. Standards (BSI 6472, ISO 2631) prescribe an equal perception value (EPV), related to frequency and particle velocity, for assessing human reactions to vibration. It may also be possible to correlate these reactions to other parameters such as VDVs.

Note that human concern is often for the safety or integrity of the buildings which people own or in which they detect the vibrations. This concern is sometimes unfounded. Structural damage to buildings is assumed *not* to occur below about 200 times any humanly-detectable magnitude (measured as amplitudes at the same frequency). Some architectural damage (superficial cracking) may occur at 5 mm/s peak particle velocity.

7.9.4 Calculation and Prediction

The complex nature of railway-induced vibrations and of wave propagation in the ground has been investigated in detail by others.

Vibration is propagated by compression waves, shear waves and surface ('Rayleigh') energy waves. Two typical situations arise on railways:

1. On main lines, carrying fast or heavy trains, in the open air: vibration, typically around 10 Hz frequency, is radiated via Rayleigh waves of 5 to 50 m wavelength (compression and shear waves decaying to nothing within a short distance).
2. In tunnels: compression and shear waves (commonly 30–150 Hz, typically with 50 Hz peak 'rumble') travel through the intervening material at a velocity proportional to the square root of elasticity divided by density. Attenuation is greater in most soft materials than in rock but vibration may also be muffled at rock interfaces.

Analysis of the propagation of vibration from the track to potential areas of disturbance can be undertaken through:

- establishing the 'site laws' of the ground conditions (geology, density, ground water) to determine how vibration is attenuated with distance in different soils and at strata interfaces;
- calculation of vibration transmission through structures supporting the track or through affected buildings, for which there is extensive theory and experimental data.

Rough estimates or comparisons can be sufficient to estimate firstly whether a vibration problem exists. If these give a positive indication, then more precise calculations are needed.

Until recently there was a great shortage of reliable data about the generation and propagation of vibrations and re-radiated noise through the various soil types and strata interfaces commonly encountered. Investigations for the Channel Tunnel Rail Link and other new railways may have rectified this. Based on this experience, empirical relationships will have been further developed. The related formulae and graphs will no doubt be valuable in making preliminary investigations of potential problems. But, in undertaking detailed calculations, close attention should be given to actual ground materials, wavelength of vibrations and the resonance characteristics of affected structures. Any new *assessment* techniques should be developed in parallel with appropriate measuring equipment.

7.9.5 Vibration Reduction

Vibration can be reduced at source:

- by improved design of suspension and bogies of rolling stock;
- by vibration-absorbing resilient track.

Major improvements to high speed rolling stock have already been achieved, mostly to alleviate internal and external noise, by reducing suspension stiffness and unsprung weight. On lighter urban rapid transit stock, exceptionally smooth running has resulted from use of pneumatic rubber tyres or air-cushion suspension; (unsprung bogie weights could even be too light for such trains). The main scope for improvement now lies in the suspension of heavy freight wagons, if the cost of such improvements is exceeded by that of any disturbance which their vibrations cause.

Vibration absorption at the track can be achieved by supporting its mass on a resilient spring. Where the cost can be justified, 'suspended slab' track can provide this absorption, at least for frequencies above 30 Hz. Less costly partial solutions lie in thicker ballast, sleeper soffit pads or insertion of a resilient mat beneath the ballast. Where maintenance cost or space constraints obviate the use of ballast, as in some tunnels, the high vibration transmission capacity of rigid track support slabs can be mitigated by inclusion of elastomeric layers. In all cases the value of the solution depends upon the frequency range at which it is effective. For some very low frequency situations there may be no practicable means of alleviating vibration.

An equivalent to sound insulation for the receptor of vibration and ground-borne noise is flexible suspension for sensitive equipment. The natural resonance of whole buildings cannot easily be adjusted. But, in rare cases where ground vibration is potentially serious to structures, passive protection could be provided by excavation of trenches of adequate depth (at least one-third of the vibration wavelength). These trenches could be open or filled with slurry in the way widely used for temporary excavation support or for permanent containment of hazardous wastes.

Where the available measures for vibration reduction are ineffective (for re-radiated noise) or prohibitively expensive (for isolating structures), then there remains a choice of buying the properties for demolition or other use or paying compensation for inevitable disturbance.

New buildings can be designed specifically to resist known vibration problems by attention to:

- natural resonant frequencies, e.g. in the spacing of floor joists;
- the mass of the structure;
- the means by which the structure is coupled to its foundations.

8 Pollution

8.1 TYPES OF POLLUTION

Pollution is making the environment foul or impure; the term is commonly applied to contamination of air, water or land. The following are recognised physical forms of pollution:

- *Emission of gases* or small particles into the atmosphere creating dirt, impairing the quality of air breathed, causing regional pollution in such forms as acid rain and ultimately influencing long-term climatic conditions.
- *Liquid effluents* discharged as sewage or industrial or agricultural runoff, adding toxic metals, chemicals or bacteria to surface or ground water and thus contaminating supplies downstream.
- *Solid wastes* as poisonous chemicals or radioactive material or, more often, as inert waste in such quantities as to cover large land areas, rendering them unusable for other purposes.

Air pollution was a widespread consequence of, the industrial revolution. It could be readily recognised by the sulphurous smell and visible smoke and grime common in many industrial cities until the mid-twentieth century. In Eastern and Central Europe it has continued to the present day where high sulphur coal or lignite still provides the basic fuel of power generation, industry and domestic heating. In many Western countries, coal still dominates power production but has been succeeded as an industrial, heating and transport fuel by other fossil fuels—petroleum products or natural gas. These cause air pollution of a type which is often less visible and probably more regional or continental than local in impact. The changeover to these fuels has been accompanied by rapid increases in usage whilst they remain cheap and abundant.

Water pollution is an immediate danger to human health if a source of drinking water is affected. This danger, rather than damage to aquatic life forms, has been the spur to developments in public health engineering—treatment of water and waste water—as well as in legislation prohibiting the discharge of untreated pollutants.

Land pollution can be caused by dumping of toxic or dangerous materials rendering the land hazardous or impossible to use for other purposes. Treatment may be expensive and may not be practicable if the location of the dangerous material is unknown or widely distributed. More often land is devastated by much greater quantities of generally inert material such as arises in mining operations. China clay working in Cornwall has left a 'moon-scape' of pits and waste heaps from which a clay suspension enters river estuaries, producing sterile sea bed conditions close to the shore. Dereliction of the land is thus often associated with water pollution.

Ill-planned coal and lignite mining in Eastern Europe has concentrated on high production directed by central planners rather than on sustainable use of land resources which were considered a local matter of no immediate consequence. Not only are land areas put out of use by open-cast pits and spoil heaps but subsurface mining may cause subsidence, impeding surface drainage, breaking pipelines and undermining the foundations of railway structures. Perhaps most seriously of all, pumping of some deep mines produces saline water which renders useless the river water into which it is discharged. Expensive desalination plants have been installed in Poland to start tackling this problem but the economics of such operations are unlikely to justify such mining in a free market economy.

Pollution caused by transport is mainly related to the exhausts of engines burning fossil fuels such as in diesel locomotives or lorries. But railways are increasingly reliant on electricity which can be environmentally cleaner than direct fuel combustion. Where railway traction depends on electrical energy, then the problems are those of the power stations. Wherever it is abundant, coal—with its associated problems of air, water and land pollution—is likely to remain an important fuel for electricity generation.

8.2 ATMOSPHERIC POLLUTION CAUSED BY TRANSPORT

This section describes the effects on air quality of the main pollutants and presents comparative data for rail and road transport.

In dry conditions, dust may arise from the movement of road vehicles but rarely from trains. Most emissions into the air originate as waste products of combustion, for instance from train, vehicle or aircraft engines or from associated electricity generating stations. The emitted gases and particles can cause health problems or nuisance, especially near the source; they can attack buildings, crops, natural vegetation and aquatic life through the formation and deposition of acid, sometimes at a considerable distance; and build-up of some gases is believed to contribute to the 'greenhouse effect' which may alter the climate of the whole earth.

8.2.1 Direct Effects of Air Pollution

Local air pollution can be caused by dust or by smoke or gas from a combustion or chemical process. Emissions from the engines of diverse, vehicular sources are detectable in high concentrations; these occur at particular locations such as the entrance to a tunnel or where traffic builds up and waits at a crossing or station.

Pollutants commonly involved are the following:

- Suspended particulate matter—a broad term covering any finely divided solid or liquid that is dispersed into the air; the most familiar example in urban environments is 'smoke' which refers to visible emissions of incompletely combusted particulate matter in suspension with soot or unburnt fuel and lubricants; particulates are very noticeable by sight and smell and are irritant to eyes, nose and throat; they are often emitted together with other pollutants.

- Carbon monoxide (CO) results from incomplete combustion; it is odourless, invisible and toxic, affecting blood and absorption of oxygen; 50 per cent of urban people may be exposed to concentrations in excess of World Health Organisation short-term guideline levels (TEST 1991: 101).

- Sulphur dioxide (SO_2) results from combustion of fossil fuels, particularly oil and coal which contain sulphur compounds; it can be a strong irritant to eyes and mucous membranes; with particulates it can form sulphuric acid (H_2SO_4) in lungs or as the main constituent in acid rain.

- Nitrogen oxides (mainly NO_2 and NO, the group being referred to as NO_x); like SO_2 these are formed by the combustion of fossil fuels, particularly diesel; NO_x is a source of longer distance pollution through acid rain; associated with photo-chemical smog, nitrogen dioxide is highly toxic, affecting the respiratory tract, and increases sensitivity to dust and pollen.

- Hydrocarbons (HC) are found in exhaust emissions or arise from spills and leaks of liquid or gaseous fuels; various health hazards, mainly synergistic, are related to the wide range of compounds associated with petroleum products; hydrocarbons contribute to pollutants known as volatile organic compounds (VOCs).

- Other hazardous pollutants such as lead (in petrol) and other heavy metals, asbestos (from vehicle brake linings) or benzene (in hydrocarbon fuels).

The above pollutants are listed in a manner which is inevitably arbitrary and over-simplified. Fuller treatises are available (e.g. TEST 1991: 98–105). All these emissions are arguably injurious to health. The danger of some of them is not fully proven nor are all the thresholds determined at which their effects become significant. A few are known to be carcinogens (inducing cancer) or mutagens (producing genetic alterations), whilst some are only suspect; many uncertainties exist regarding the causes and significance of hazardous levels. Not only is the data about health and mortality mostly related to the effects of a multiplicity of causes but pollutants like CO, SO_2 or NO_x make their most adverse impacts in combination with other gasses whilst some carcinogens adhere to otherwise relatively harmless particulates.

Secondary pollutants are mixtures or compounds formed subsequently from primary pollutants. For instance, O_3 as atmospheric ozone, which is not detectable by human senses, is formed by the reaction in sunlight of NO_x with hydrocarbons and oxygen (O_2). The irritation and susceptibility of people to high O_3 levels justifies publication of measurements of ambient atmospheric ozone in air quality reports or predictions.

It is possible to estimate typical or average levels of polluting gases at various distances from certain types of transport sources as is done for noise levels. However, air pollution, both as a local phenomenon and as dispersed elsewhere, is largely related to incidents or 'episodes'. Dust, smog or gaseous effects arise mainly as a result of particular weather conditions but may also be due to occasional or unplanned industrial operations or to traffic congestion. 'Smog', as a combination of natural fog and coal-derived smoke, was responsible for some very serious health hazard episodes in London and elsewhere in the early 1950s. Los Angeles smogs are a somewhat different variety resulting from sunlight and ozone production in photosynthesis with the exhaust gases of road vehicles.

8.2.2 Regional or Long Distance Air Pollution

For 200 years emissions from static coal burning and other industrial plants have been fed through tall ('high stack') chimneys to dissipate the smoke from the immediate area. In recent years there has been widespread recognition of the damage which can be done at long distance by acid-carrying water vapour and similar wind-driven phenomena. Even though the output of coal-burning emissions may have fallen in Western countries, a considerable increase in long distance pollution from fossil fuel combustion has taken place—now arising from numerous small mobile transport units as well as more identifiable high stack point sources.

The United Nations Economic Commission for Europe has sponsored a programme in which a great deal of information about the source and deposition of pollutants throughout Europe has been gathered and, with meteorological information, fed into mathematical models of long distance movements of various gases including SO_2 and NO_x. This programme has been entitled 'trans-boundary' in that it can readily reveal the amounts of pollution being carried by the prevailing winds across international frontiers.

Acid rain is a phenomenon which results in the deposition—directly (dry) in fog or in precipitation (wet)—of airborne acids (HNO_3, H_2SO_4) to the detriment of people, water, land, vegetation and structures. These toxic and corrosive acids are formed by chemical and photosynthetic combination with oxygen and water vapour of sulphur dioxide (SO_2) and nitrogen oxides (NO_x). These oxides result from burnt fuel emissions as well as from natural activity like volcanic eruptions. The 'rain' falls up to 1000 km from the source.

Global impacts of air pollution are the 'greenhouse effect', caused by certain gases and possibly affecting the world's climate, and damage to the stratospheric 'ozone layer', potentially a health hazard.

The ozone layer acts as a shield against excessive ultraviolet (UV) radiation. Its thickness varies seasonally and over different parts of the globe. In some areas the ozone has evidently become much thinner and this may increase UV radiation risks. A major cause of ozone depletion has been release of chlorofluorocarbons (CFCs) and the supposed cure is to eliminate their use. The matter is marginally relevant to railway operation in so far as CFCs may still be used in air-conditioned trains, refrigerated wagons or warehouses. Other beneficial measures quoted (Pyrgidis 1993) include reduction of materials (such as pyrenium) which emit polluting gas from electric trains and substation transformers.

The greenhouse effect is an *increase* in the atmospheric gases which trap heat from the sun to the extent that more heat is trapped and the atmosphere increases in temperature, i.e. global warming.

The gas primarily responsible for the greenhouse effect is believed to be carbon dioxide (CO_2). Except as an asphyxiant, carbon dioxide is not otherwise a harmful gas, indeed it is a very common one. Global CO_2 levels are known to have increased since the industrial revolution. One reason for the increase is believed to be deforestation; there has been a decrease in the amount of vegetation which can absorb naturally or artificially generated CO_2. Another reason is the increased burning of fossil fuels.

Directly polluting particulates and gases like SO_2 or NO_x can be reduced at source. Processes vary from relatively cheap but only partially effective screening or pre-treatment of coal, through improved combustion processes to more expensive and environmentally contentious removal from flue gases. No such 'technical fix' is available for eliminating CO_2. The only way to reduce man-made carbon dioxide is to burn less carbon. Whether or not this can make any significant improvement to global climate is debatable. However, such a reduction is desirable:

- because it would automatically ensure a reduction in emissions of the directly harmful gases;
- to conserve global fuel resources.

Therefore it is environmentally desirable to reduce consumption of fossil fuels in transport. In planning railways this is more important than reduction of pollution. Nevertheless, no environmental assessment for any form of transport development would be complete without identification of such pollution as will occur. Emissions to the air in railway operation emanate almost entirely from the sources of motive power.

Little is new about local, regional and global pollution. Smoke abatement legislation was enacted in Britain in 1875 but attempts to reduce coal burning in London go back 700 years, R.A. Smith commented on the level of sulphuric acid in Manchester rain in 1852. John Tyndall gave warning of excess CO_2 and a 'greenhouse' effect in 1863 (Allaby 1986: 10).

8.2.3 Comparative Data and Significant Pollution

Tables 8.1 and 8.2 present comparative data, taken from several sources, about energy use and emissions from railway and road transport, and indicate the wide variations that exist.

The data for trains in kilojoules of energy or grams of emission per kilometre are of the same order of magnitude whether measured in numbers of

Table 8.1. Transport energy consumption

Type of vehicle train	Energy use	Source of data
Passenger traffic	(kJ/passenger-km)	
Cars	2580	Whitelegg (1993: 57)
Cars, 1.4–2.0 1, 1.5 person	1990	Farrington (1992: 64)
Buses	680	Whitelegg (1993: 57)
Buses, single decker	870	Farrington (1992: 64)
Buses, double decker	520	(33% full)
All railways	1270	Whitelegg (1993: 57)
Light rail/trams	1020	Whitelegg (1993: 57)
Diesel trains (sub-200 km/h)	4131 ⎫	Wayson and Bowlby (1989)
Electric trains (sub-200 km/h)	3541 ⎪	(in range 135–5760)
Shinkansen	567 ⎬	(50% load)
TGV	440 ⎭	(50% load)
Inter-city (160 km/h electric)	480 ⎫	Farrington
Inter-city (200 km/h electric)	650 ⎪	(1984: 64)
Inter-city (200 km/h diesel)	590 ⎬	(60% full)
Dmu super-sprinter diesel	550 ⎭	
Freight traffic	(kJ/ton-km)	
All road freight	2890	Whitelegg (1993: 58)
All road freight	730–1850	Bevilacqua (1978)
Rigid lorry, 4 axle 20 t		EEC (1992)
70% load	2220 ⎫	*The Impact of*
100% load	1550 ⎪	*Transport on the*
Articulated Lorry, 5 axle 38 t		*Environment,*
70% load	990 ⎬	Brussels, COM 95,
100% load	690 ⎪	20 Feb 1992
Rail, bulk	600 ⎪	
Rail, wagon load	1000 ⎭	
All rail freight	677	Whitelegg (1993: 58)
All rail freight	220–440	Bevilacqua (1978)

Table 8.2. Typical emissions from transport

Type of vehicle or train	CO$_2$	VOC	NO$_x$	SO$_x$/SO$_2$	CO	HC	Source
Passenger traffic	(g/passenger-km)						
Cars	180	2.2	2.1		11		Whitelegg 1993 (Germany)
Cars and taxis	127		1.2		5.6	0.6	TEST 1991, Table 4–18 (UK)
Cars							
1 pers, average			1.9	0.07	24	3.15	TEST 1991, Table 4–29 (Canada)
2.4 p., average			0.8	0.03	10	1.31	
3.6 p., rush hour			0.5	0.03	6.6	0.88	
1 pers, commute			1.3		9.4	1.30	TEST 1991, Table 4–28 (USA)
Buses	35		0.4		0.3	0.06	TEST 1991, Table 4–18 (UK)
Buses	48	0.3	0.8		0.3		Whitelegg 1993:57 (Ger)
Diesel bus							
20 p., average			0.7	0.09	1.5	0.11	TEST 1991, Table 4–29 (Canada)
37 p., rush hour			0.4	0.05	0.8	0.06	
Transit bus			1.0		1.9	0.12	TEST 1991, Table 4–28 (USA)
All road diesel (passenger)	42	0.8	1.6	0.08	7		TEST 1991, Table 4–17 (UK)
All road passengers			1.7		9.3	1.10	TEST 1991, Table 4–21 (Germany)
All rail passengers	79	0.3	0.5		0.1		Whitelegg 1993:57 (Germany)
BR (NSE)	84		0.4	1.0	0.02	0.01	TEST 1991, Table 4–20 (UK)
Diesel rail	98		1.0	0.16	0.12	0.07	
Electric rail	82		0.4	1.1	0.01	0.002	
Tram/metro	61	0.1	0.2		0.01		Whitelegg 1993:57 (Germany)

Continued

Table 8.2. *Continued*

Type of vehicle or Train	CO_2	VOC	NO_x	SO_x /SO_2	CO	HC	Source
Freight traffic	(g/tonne-km)						
All road freight	207	1.1	3.6		2.4		Whitelegg 1993:58 (Germany)
All road freight	275		4.7	0.23	2.6	0.4	TEST 1991, Table 4–18 (UK)
All road freight	220		3.6	0.32	1.6	0.81	TEST 1991,
Lorries, <50 t	255		4.1	0.18	1.9	1.25	Table 4–22
Lorries, >50 t	140		3.0	0.07	0.25	0.32	(Germany)
Diesel rd freight	36	0.7	1.4		6.0		TEST, Tab. 4–17 (UK)
All rail freight	41	0.1	0.2		0.05		Whitelegg 1993:58 (Germany)
All rail freight	40		0.7	0.13	0.12	0.07	TEST, 1991,
Diesel rail freight	38		0.8	0.06	0.13	0.08	Table 4–20
Elec. rail freight	50		0.2	0.66	0.01	0.001	(UK)

passengers or tonnes of freight. This is a convenient coincidence. In any serious analysis it is necessary to consider what parameters apply to each type, capacity and method of operation of passenger and freight trains.

The energy consumption figures depend entirely on what basic assumptions about operating conditions are used for the calculations. For instance, although high speed trains are very highly powered in comparison with more conventional trains, their long non-stop journeys and high occupancy make them surprisingly efficient.

The data for emissions of particular pollutants are even more widespread because these are usually calculated using:

- data for energy consumption which, as Table 8.1 shows, varies widely;
- conversion factors for then calculating proportional amounts; these factors vary according to circumstances and are often controversial since energy producers and industrial enterprises are keen that the level of their emissions is not exaggerated.

Table 8.3 is an attempt to present *comparative* data about energy use and atmospheric emissions from road and rail transport. It should be emphasised that these are order-of-magnitude figures intended only to show where railway contributions are significant.

Table 8.3. Summary—typical transport energy use and emissions. Order-of-magnitude comparative data based on data in Tables 8.1, 8.2 and 8.4

Transport Mode	Energy	CO_2	NO_x	SO_2	CO	HC	VOC
	(KJ/ passenger- km)	(g/passenger-km)					
Road passenger							
Cars	2000	150	2	0.05	10	1.5	2
Buses	800	40	1.0	0.1	0.5	0.1	0.5
Rail passenger							
All trains	800	80	0.6	0.3	0.2	0.2	0.3
Diesel trains	800	80	1.5	0.2	0.2	0.1	0.5
Electric trains	800	80	0.5	1.0	0.02	0.001	0.001
Road freight	(kJ/t-km)			(g/tonne-km)			
All road freight	2000	250	4	0.3	2	0.5	1.0
Large lorries	1000	100	3	0.2	0.2	0.3	–
Rail freight							
All rail freight	700	40	0.3	0.3	0.2	0.05	0.1
Diesel	–	40	0.7	0.1	0.15	0.1	0.1
Electric	–	40	0.2	1.0	0.01	0	0.01

Of emissions relevant to local air pollution and health, CO and hydrocarbons/VOCs are caused predominantly by petrol-engined road vehicles (cars). Diesel trains and lorries emit much less per passenger-km or tonne-km although they contribute particulates. Electric trains do not cause local pollution except near any fossil-fuel burning power stations without emission controls which may supply them.

Electric trains contribute a small proportion of NO_x at generation; diesel trains and buses possibly emit rather more; but the main source of NO_x is evidently cars and lorries.

Besides NO_x, the other long-distance pollutant is SO_2. The only significant transport source is electric trains, again if high sulphur fuels are burnt uncontrolled in power stations.

If carbon dioxide causes climatic change, then most motorised transport contributes. If CO_2 impacts are adverse then, since CO_2 emission is roughly proportional to fossil fuel burnt, the most fuel efficient engines (in trains, buses and large lorries) and non-fossil power sources should be encouraged.

8.3 GASEOUS EMISSIONS FROM TRAINS

Coal-generated smoke in steam days was a significant part of industrial atmospheric pollution. Up to the middle of the twentieth century a majority of the British population lived in a smoke-laden atmosphere which would have been intolerable by modern standards.

The amount of coal consumption has been substantially reduced, totally on Western European and North American railways. In Britain, total coal consumption in 1957 was 213 million tonnes of which direct railway use accounted for 11 million tonnes. Current figures are about one-third and one-hundredth respectively. Concern has switched from local industrial air pollution to more widespread acid rain and the global effects of burning all types of fossil fuel. But the contribution of railways to either form of air pollution is now undoubtedly less significant.

8.3.1 Coal and Steam

Although steam traction is now only of historic interest in Western Europe, it still has a part to play in China and has not been entirely phased out in Eastern Europe; it is certainly necessary to consider briefly coal smoke which was for a century the *main* environmental impact of railways.

Steam locomotives emit smoke particulates in the vicinity and contribute sulphur and nitrogen oxides and CO_2 to the wider atmosphere in essentially the same form as do coal-burning power stations. However, major differences in quantity arise in that locomotives are small units, emitting heavy bursts of smoke when they are consuming coal most rapidly to build up or maintain a head of steam. Even when working at full pressure steam engines are only about 10 per cent thermally efficient, compared with 20 per cent for petrol engines and 30 to 40 per cent for diesel or electric traction. Nor can sophisticated flue gas scrubbing devices be applied to clean up locomotive exhaust.

As a result the level of pollutants emerging from a steam locomotive is much higher, per unit of energy produced, than from a 'clean' power station. In the absence of direct data for steam locomotives it is uncertain as to what are the constituent proportions of their smoke and therefore how significant is any threat to the environment compared, for instance, with diesel emissions. All that can be said with reasonable certainty is that these proportions would exceed any modern limits applied to industrial processes or power stations.

In Britain steam locomotives are exempt from prosecution under nuisance legislation because of the pleasure they give. The generation which grew up with steam trains is well aware of the smuts and soot which they produce and of the smell of smoke. It may be that this very familiarity as well as the absence of advice to the contrary, bred confidence that *occasional* doses of such smoke have no adverse affects on health. The same might be said today of hydrocarbons pervading the atmosphere at motor racing events.

Where steam locomotives are still in widespread use it is suggested that the quantity of air pollutants which they emit should be considered in the light of:

• the relative quantity (i.e. volume concentration) of emissions in the vicinity in which they work;
• the likelihood that, in the long run, they will be superseded by other types of locomotive or electric traction.

8.3.2 Diesel

Diesel locomotives and rail motor units produce the same sort of emissions as do road lorries, including carbon monoxide, nitrogen oxides, hydrocarbons and carbon-based particulates. However, the railway contribution is less than 1 per cent in each case in total national terms. Not only do railways carry only a small proportion of total transport movements but, broadly, they move twice as many passengers and freight *per fuel unit* as road transport. Exceptions to the last are buses and the largest lorries, but the proportion of rail diesel emissions is reduced as railways are electrified.

Diesel engine particulates have been classified as a probable carcinogen and are blamed as a cause of respiratory illness among city children. These emissions are associated with the bursts of black smoke seen when a diesel engine starts up from cold.

Among gases associated with diesel fumes, hydrocarbons and CO are emitted in relatively high concentrations when cold or when idling or running at low output. Therefore diesel pollution is greater in stop/go operations, such as shunting, than in long fast trips such as diesel HST 125s undertake. Emissions of all polluting gases, except particulates, SO_2 and NO_x, are generally lower from diesel than from petrol engines. In addition, diesel fuel has never required the addition of any lead additives.

Particulates, sulphur and nitrogen oxides can be reduced by water scrubbers even at small static generators. Catalytic processes can be applied to convert CO to less harmful CO_2 and to eliminate odours and aldehydes derived from hydrocarbons. Some measures of this type should be feasible for application to the mobile generators in diesel-electric locomotives, should such steps be necessary. Many types of emission treatment involve increasing certain potential pollutants in order to decrease others which are judged more serious. Trade-offs, for instance between NO_x and particulates, have to be planned to achieve the optimum balance.

A local source of air pollution arises when the engine of a diesel locomotive is kept idling for long periods rather than shutting them down. In the USA 'layover protection systems' are advertised which permit the safe shutting-down of engines, resulting in savings in fuel and reductions in emissions of HC, CO, NO_x and particulates. These reductions are claimed as considerable at the throttle settings required for idling in very cold overnight conditions.

8.3.3 Electric Traction

The only real contribution of electric transport to atmospheric pollution occurs where the electricity is generated at fossil fuel-burning power stations. Reduction in pollutant emissions from these power stations is the responsibility of the electricity generating planning authorities. Reductions can be achieved firstly by energy conservation, then by pre-treatment/cleaning of fuels (e.g. washing, physical separation or chemical techniques) and improved combustion (e.g. atmospheric or pressurised fluidised bed boilers, or integrated coal gasification combined cycle processes); if these measures are not sufficient, there are 'end-of-pipe' treatments of exhaust gases such as flue gas desulphurisation (FGD) for SO_2 and fabric filters or electrostatic precipitators for particulates.

In Britain power is increasingly being generated by burning natural gas— for 'environmental' (less SO_2 and NO_x) as well as economic reasons. However, this 'secondary' use of gas is regarded by energy planners as a wasteful use of a relatively scarce natural resource.

In France or Switzerland most rail traffic is electrically propelled and most electricity comes from nuclear or hydro-electric power stations. Air pollution at such stations is essentially zero.

Other atmospheric effects of electric railways which are occasionally quoted relate to emissions arising from the high speed contact of pantographs on wires or, for any railway, wear between wheels and rails. These effects are in fact negligible.

Electromagnetic radiation and catenary arcing problems have occasionally been cited in relation to high tension power lines, but rarely for railways. Effects of motors or line equipment on communications systems or on other electrical equipment are occasionally mentioned.

8.3.4 Comparative Pollution from Diesel and Electric Traction

There are so many forms of air pollution, ways of measuring them and variations in train operating conditions that it is misleading to quote typical figures without qualifications. The comparative figures quoted in Table 8.4 are taken from a wide range of primary and derived data. They are representative only but do provide evidence as to the order of magnitude involved.

There are very wide variations in the data quoted in Table 8.4 because of the different assumptions made about the composition of the fuels used, the emissions caused and operating pattern for each type of engine, and the train loading factors. But the clear inference is that electric train transport is cleaner in terms of all emissions to the air except particulates, SO_2 and CO_2. Other factors in pollution derived from electricity generation should also be considered:

- Particulates can be removed at the power station and environmental legislation could ensure that this is done in future.
- Sulphur emissions can also be removed at power stations, albeit by relatively expensive processes which *may* be implemented if plans for national SO_2 reductions require it.
- It is by no means certain that global warming, which is the *only* known physical impact of CO_2, is related directly to particular CO_2 emissions rather than to more complex mixed effects.

Figure 8.1 summarises diagrammatically the possible connections between railways and pollution.

8.4 ASSESSMENT OF AIR POLLUTION

Adverse impacts of air pollutants occur:

- as nuisance—smell of fumes, layers of dirt and reduced visibility because of smoke;
- as health hazards;
- through damage to materials, buildings, agricultural production and land and water resources.

Table 8.4. Atmospheric emissions from trains

	Particulate matter	CO	HC	NO2	SOx	CO2
(a)*						
Passengers	(g per passenger-km)					
Conventional diesel	0.33	1.70	1.24c	4.87		
Conventional electric using coal generation	4.76	0.13	0.067c	1.00		
TGV-type electric using coal generation with 'controls'	1.08	0.030	0.015c	0.23		
	0.011			0.19		
using fuel oil with 'controls'	0.025	0.013	0.003c	0.26		
using natural gas with 'controls'	0.004	0.006	0.000c	0.25		
Shinkansen (coal gen.)	1.40	0.039	0.019c	0.29		
(b)†						
(i) Passengers				NOx		
Diesel—[TEST Table 4–17 UK, average]		0.49	0.34v	1.33	0.15x	38
—[TEST Table 4–18 UK, average]		0.62	0.26	2.07		83
—[TEST Table 4–20 Inter-city]		0.13	0.07t	0.88	0.15	92
—[TEST Table 4–20 Provincial]		0.12	0.06t	1.56	0.18	111
—[TEST Table 4–20 Network SE]		0.12	0.07t	0.99	0.16	98
Electric—[TEST Table 4–18, UK average]		0.030	0.001	0.49		117
—[TEST Table 4–20, Inter-city]		0.010	0.001t	0.32	0.90	68
—[TEST Table 4–20, Provincial]		0.017	0.002t	0.56	1.56	118
—[TEST Table 4–20, Network SE]		0.012	0.002t	0.39	1.09	82
—[TEST Table 4–20, RT, Vancouver]		0.001		0.01		

Tram and metro [Whitelegg, Germany]		0.01	0.15v	0.15	—	61
All Trains [4–21, Germany]	0.08	0.06	0.43	0.43		
All Trains [Whitelegg, Ger.]		0.13	0.31v	0.46	0.2	79
All Inter-city [4–20, UK]		0.08	0.04t	0.63	0.49	81

(ii) Freight

(g per tonne-km)

Diesel—[TEST Table 4–17]		0.17	0.12v	0.46	0.052x	13
—[TEST Table 4–18]		0.02	0.09	0.72		29
—[TEST Table 4–20]		0.13	0.077t	0.76	0.06v	38
Electric—[TEST Table 4–18]		0.010	0.003	0.17		41
—[TEST Table 4–20]		0.008	0.001t	0.24	0.66v	50
All freight [TEST Table 4–21/2 Germany]	0.05	0.08	0.03	0.27	0.25	49
All freight [Whitelegg, Ger.]		0.05	0.08v	0.2		41

Notes

Hydrocarbons (HC): c denotes measured as CH_4, v as VOC, t as all volatile organics including methane and solid hydrocarbons.

Sulphur oxides: x denotes measured as SO_x; all others measured as SO_2.

Sources:

* Data from Wayson and Bowlby (1989).

This 'conventional' (less than 200 km/h) train data is calculated using mean values from a very wide range of *energy use* data, e.g. from 88 to 3777 kilojoules (kJ) per passenger-km supplied to electric trains. The energy data quoted allows for power station energy conversion efficiency and is then multiplied by a US Environmental Protection Agency conversion factor for each type of traction. It is not certain how the basic data for TGVs and *Shinkansen* trains were calculated.

The power station factors assume no environmental controls. The 'with controls' generation emissions calculated for TGVs assumes removal of 99% particulates and 15% NO_x. Note that for actual TGVs (in France) a large proportion of the electricity may be derived as nuclear or hydro-electric power in which case there is *no significant air pollution and no CO_2 emission*.

† Data from TEST (1991) and Whitelegg (1993).

This data has been abstracted from tables which were themselves abstracted from official publications or the results of independent investigations in the UK and Germany. The TEST report gives a comprehensive explanation of the difficulties in assembling comparative data for transport related air pollution.

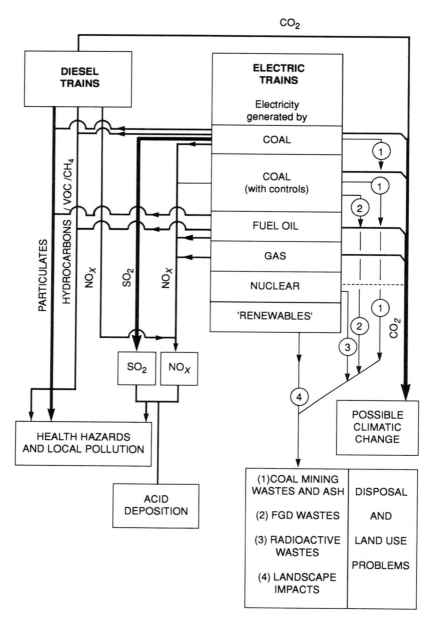

Figure 8.1. Rail transport—possible causes of pollution

Relationships between particular air pollutants and consequent damage to health or resources are difficult to establish with accuracy or without contention. The best that can be done is:

- to compare causes of pollution and ambient conditions with those which have occurred in similar situations elsewhere;
- to predict the ambient pollution levels from the quantity of emissions so that specialists can determine their significance and make comparisons with baseline and air quality standards.

Ambient conditions at locations beside a railway or near a diesel locomotive depot can be predicted using mathematical models for particular situations and under various wind and atmospheric conditions. This has been undertaken at road tunnel entrances but is likely to be effective only if such models can be calibrated by authenticated measured data. To quote Wayson and Bowlby (1989), dealing with high speed railway systems:

> Health effects should be calculated based on concentration, which is a function of release rates and dispersion. Dispersion depends on the geometrics of location, weather parameters, reactivity of release and background concentrations. Accordingly, the analyst must use an appropriate dispersion model if more than a comparison of total pollutants is required.

There is an analogy between air quality modelling and noise prediction discussed in Chapter 7. But the factors in empirical relationships derived for sound propagation are much less complex than the cocktail of visible and invisible gases emitted by motors or generators; and atmospheric conditions are less crucial in noise transmission than in the formation and movement of airborne chemicals.

Prevention of high levels of *local* pollution has often been identified as the most urgent need in controlling industrial air emissions. This is because:

- the direct impacts on health and environmental conditions are most apparent locally;
- main sources of local pollution can be readily identified.

Predictions of air quantity near new roads in Britain are based on relationships, established by the Department of Transport, between types, numbers and speed of vehicles and resulting concentrations of CO, HC and NO_x.

It is unlikely that diesel locomotives could produce what the Department of Transport would consider an air pollution problem because of the much smaller number likely to be concentrated at one point, the relatively low output of CO and HC by diesel engines compared with petrol vehicles, and the greater separation of most railway stations or yards from houses or communal areas. Nevertheless, environmental concern requires that a technical explanation of the *impacts* should be given and in specific instances this should be supported by estimates of the *quantity* of polluting emissions.

Statutory emission standards can be applied in specifications for the manufacture and operation of diesel locomotives of electricity generating stations. But, beyond banning *visible* exhaust emissions, quality regulations are difficult to enforce on mobile units.

Predictions of emissions should be based on the best primary data available for particular types of locomotive or power station. Estimates should be provided for emissions which are in general use for transport planning and comparisons (such as CO and HC) and for pollutants for which there are statutory limits or international guidelines for local or national application (e.g. CO, NO_2, SO_2, particulates).

If the railway planner makes clear the order of magnitude of the emissions he anticipates it is up to electricity power planners, transport priority setters and long-term fuel price fixers to consider the wider environmental implications.

8.5 LIQUID EFFLUENT AND SOLID WASTE

There is no need to examine in detail here the disposal of liquid wastes from moving trains. But questions are sure to be asked, so the matter justifies serious if brief attention in the environmental appraisal of any particular train operation. Basic points are as follows:

1. Train lavatories discharging directly on to the track have raised no apparent objections from railway workmen or lineside residents in the past.
2. On many modern high speed trains (certainly TGV, Eurostar and BR Mark IV coaches) washroom effluents are sealed-in and disposed of at terminals in the same way as from aircraft.

Apart from trains, railways can be associated with water or land pollution through other aspects of operations, such as activities at depots or in electricity generation, or through railway construction. Spillage of fuels and lubricants may occur at stations, locomotive depots and rolling stock servicing sidings. Fuel storage tanks can be protected by bunds against accidents or spillage and drainage to sumps provided at fuelling, maintenance or standing points; precautions can also be applied to mobile and stationary equipment used in railway construction. Whenever materials are stored in the open, measures must be taken to protect the external drainage system and ground water aquifers from rainwater runoff polluted by these materials.

The effects of gaseous exhausts from electricity generating stations have already been described. Steam power stations fired by fossil fuels produce ash which has to be disposed of, preferably as fill or lightweight aggregate rather than by wasting land space. Nuclear powered steam power stations produce various grades of radioactive waste for which disposal is more problematical and costly. Land occupied by nuclear processes is also unlikely to be available for any other use for 100 years or more. All steam power stations require cooling water and precautions must be taken to avoid contaminants in return flow. Hydro-electric stations also make use of water resources; the reservoirs with which they are usually associated provide opportunity for fuller optimal use of water but involve the inundation of significant land areas.

9 Visual Impacts

9.1 INTRODUCTION

There are four ways in which railways can give visual pleasure or displeasure:

1. Reactions which arise from the sight of trains: stimulation or distraction of lineside residents and visitors—relaxing, working, playing and watching games, or visiting sites of cultural interest;
2. Obstruction of views by railway structures.
3. The clash or blend of railway infrastructure and trains in the wider landscape.
4. The attractiveness or interest of the view from the train.

The appearance of trains (1) and railways (2) is described in Section 9.2 and the way in which their visual impact can be assessed in Section 9.3. Seen close at hand, high embankments or bridges carrying railway lines may be visual obstacles, dealt with in Section 9.4. In the wider landscape (3) lines and trains may be less prominent but may nevertheless intrude on the scenery (see Section 9.5).

The view from the train (4) is considered in Section 9.6. It has been observed that good landscapes are experienced by moving through them (Preece 1991: 72). If there has to be a railway, it should lose no opportunity to provide its passengers with the visual delight that is human appreciation of scenery. Engineering features require that the view be occasionally obscured, for example in tunnels; an inventory of views which are available to passengers should be part of any visual assessment.

The significance of railways in scenic landscape has to be examined with the character of that landscape. The visual impact of a new railway is first on *scenic land resources* and thence on *people's reaction*. Therefore a whole chapter (15) in Part III is devoted to the place of railways in various landscapes.

Scenic landscape is also an *amenity*. The way in which amenity values are affected by industrial or transport infrastructure is considered in Chapter 14. Massive heaps of mining waste or extensive railway marshalling yards may be an affront to the lover of hillsides or country fields. But canals and railway structures like Ribblehead Viaduct or St Pancras station have become admired features. Whether new railways become eyesores or part of heritage landscape depends upon their design now and the perception of observers in the future.

9.2 ASPECTS OF RAILWAYS

Features of railway infrastructure and operations which could affect existing scenery are trains, earthworks, structures, track and lineside accessories.

General description of railway structures and of particular types of trains has been given in Chapters 3 to 5. Here it is necessary to examine their *appearance*.

9.2.1 Trains

The passage of a train is an event each time it occurs. Unlike the continuous passing of main road traffic, trains run only intermittently; but anyone working for a long period near the line will become nearly as unaware of the passage of trains, as of cars or aeroplanes. If his concentration is disturbed at all it will be by noise before sight. The occasional or casual onlooker, on the other hand, is likely to take more definite interest, to see what it is that he can also hear. For many, the sight of a train passing is a pleasurable sensation. The fleeting glance of passengers can also be stimulating, occasionally even resulting in waved greetings.

The uniform characteristics of modern trains are such as to render their appearance generally unexceptional. Even to the technically-minded, electric or diesel multiple units are common and therefore unremarkable once they become familiar. A range of different types of locomotive-hauled freight trains can attract more interest. However, there is one sharp and intentional visual impact of modern British trains—their bright yellow front end which gives warning of their approach. People who like trains (and yellow as a colour) find this attractive; others justifiably claim that it is a sharp intrusion into green landscape. The matter is one of no great concern to sustainable environment.

In more distant views, trains are merely one of many widespread mobile or static man-made features. People seek good viewpoints to obtain sight of everything within a wide landscape. Items of interest include not only broad semi-natural features such as woodland or river meadows but also the variety of man-made structures, some of which the observer recognises from his knowledge at closer quarters. These structures may be large bright buildings or wide paved car parks, liked or loathed according to taste, or less conspicuous features such as railway lines, difficult to detect except when a train is passing.

As to the place of trains in closer views, they were already noticed with interest by artists in the period before photography. Trains feature in wild weather landscapes (Turner, Cox), on bridges (van Gogh, Doré, Monet) or at stations (Monet, Pisarro). Most of these pictures were painted by impressionists, not to record mechanical details but to present trains as adjuncts of scenery.

Such is human reaction to seeing trains. There is even less evidence that animals are disturbed by the sight of trains than they are by the sound. In the early days, screens were sometimes erected between railways and highways to prevent horses being frightened by the sight of steam locomotives. But these screens proved unnecessary.

9.2.2 Earthworks and Structures

Earthworks are built to carry railways across undulating country where trains could not negotiate the sharp curves or steep gradients of a route at natural ground level. Hilly country demands more earthwork for any railway which crosses it but is at the same time usually more scenic than flat land. Therefore railways intrude more into the better scenery! Cuttings or embankments necessarily cut into or supplant the natural shape of the landform. A railway may run with the lie of the land, for example along the side of a broad, long hillside, or against it, across the spurs at the foot of an incised escarpment.

Cuttings are not easily visible except from above or where they pierce a high horizon. The latter occurs more often on motorway routes, which can reach such heights, than on railways which are more restricted by gradients and usually have to tunnel beneath any steep escarpment.

High earth *embankments* may both obstruct the view up a valley and intrude on its features. A *viaduct* or high bridge in the same situation may be less obstructive in that you can see under it; but a viaduct will be at least as visible and potentially as intrusive. An embankment may be disguised by trees whereas a high viaduct or railway building is necessarily prominent; *architecture* determines their visual acceptability. Morgan (1971), discussing early railway viaducts, calls them 'a prototype for a type of structure which was to be repeated a thousand times by the end of the [nineteenth] century . . . and which remains a characteristic of the English landscape as do her church spires and last hedgerows'. Commenting further on early railways, Turnock (1990: 69) observes that 'although the railway was considered by some people to be an unwarranted intrusion in the landscape . . . it was merely one more change after the building of canals and enclosing of the fields'.

9.2.3 Ballasted Track and Line Equipment

Permanent way is of a width and texture generally less striking in its contrast with the countryside than is a wide smooth motorway pavement. Major and less compatible accessories of electric railways are *gantries* or masts from which overhead power line catenaries are suspended. However, advances in design have resulted in lighter, less obtrusive overhead equipment. Comparison of the massive 1960s West Coast Main Line suspension frameworks with much lighter later versions shows this clearly. On some high speed railways recently completed in Europe it has been a particular requirement that power line suspension equipment should be unobtrusive on viaducts.

To put overhead electrification into perspective, the structures are different but surely no more obtrusive than the hundreds of wires which were suspended between telegraph poles that formerly lined up beside all railway lines. In another context, the towers of high voltage power transmission lines are a higher, more intrusive and common form of linear intrusion into scenery.

A recent and contentious fixed feature of high speed roads or railways is the *noise barrier*. Many fences provided recently on roads as noise barriers are visually unattractive (from both sides). Use of more sympathetic materials or colours is called for. Where noise barrier heights can be restricted, say to about 2 m to absorb much of the wheel noise, this is much less intrusive—from inside and outside the train. If higher barriers are necessary, transparent ones may be appropriate.

Where houses are *not* separated from railways by barriers, there is a different visual 'impact' between trains and people. This is *intrusion on privacy* by train passengers looking *into* houses. Objections to this intrusion may be raised by parties seeking to oppose new trains. Its validity as a serious long-term objection can be discounted by the precedent of existing urban railway lines or double decker bus routes.

9.3 APPRAISAL OF VISUAL IMPACT

9.3.1 Elements of Visual Appraisal

Figure 9.1 shows the steps in identifying visual impacts and taking them into account in planning railways. Starting from consideration of a preliminary route line, two approaches are proposed. The first is to define the views which *people* commonly observe and to investigate how new developments will obscure or alter those views. This type of appraisal is reviewed here.

The second approach is relevant where the route has to cross land of acknowledged *scenic resource* value. It involves consideration of the nature of various types of scenery and of the actual visual impact of existing railways in each, locally and regionally. This approach is developed in Chapter 15.

Will a railway detract from a view? In a rural scene, will it create a visual barrier, a conspicuous discontinuity or an unwelcome contrast? In an urban

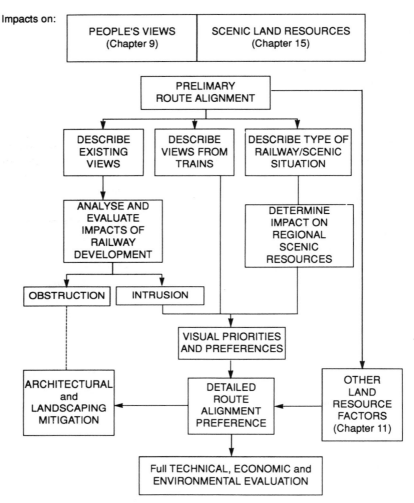

Figure 9.1. Assessment of visual impact

setting, how will it blend with industrial or building development that has taken place in the past or is likely to do so in the future?

Who will be aware of visual changes? What are the available options and which will be seen as the best solution by future generations?

A suitable *methodology* for appraisal of visual impacts of railways can answer these questions. It can comprise the following steps for any proposed route or route section:

1. Description of existing views and scenery, emphasising features that are attractive and potentially long-lasting.

2. Identification of the impacts—ways in which railway development will change these views; preparation of information illustrating the changes.
3. Analysis and evaluation of the visual impacts; determination of their significance.
4. Definition of visual priorities and preferences; criteria for visually acceptable route alignment and proposals for mitigation of adverse effects.
5. Preparation of a visual/scenic impact statement.

9.3.2 Description of Existing Views

The nature of views can be recorded:

- objectively by defining specific 'areas of visual influence' contained within existing visual boundaries, by listing the various types of visible features and by making some measure of their abundance;
- more subjectively, in terms of scenic characteristics;
- visually, using photographs of particular views; and—for general scenic assessment—by indications on available published or specially prepared maps showing contours and land use.

9.3.3 Identification and Illustration of Visual Impacts

Visual impact very close to a new railway bridge or embankment can be described as *obstruction* if a substantial part of the previous view is excluded.

At a greater distance, trains and railway features may affect the character of the view without materially obstructing it. The effect is then defined as *intrusion*. In Figure 9.2 the viaduct *intrudes* on valley scenery whilst the embankment in Figure 9.3 *obstructs* what could once be seen behind it.

The line and levels of a proposed railway can be marked on a large scale contoured map together with the resulting crossings, cuttings and embankments, all as part of the initial planning and outline design process. Observations and calculations can then be made to determine which areas of land can be seen from track or train level and vice versa. These areas can then be mapped as 'visual envelopes'. (Similar measurements, calculations and perhaps mapping may be needed in order to predict train noise levels in the vicinity.)

Visual envelopes may be used to locate viewpoints where visual obstruction or intrusion might be significant. They may also be contoured in gradations to indicate the magnitude of visual effects, such as by the angle subtended by a visible railway line or its distance away. The data can be studied to define 'zones of visual intrusion' (Preece 1991:93) related to specific railway structures. These zones can then be examined to see how intrusion affects the areas of visual influence that already exist.

Figure 9.2. Railway viaduct at Eynsford, Kent. An architectural triumph, like most tall viaducts, this structure nevertheless intrudes on the view of the river Darent

Illustration of the quality of future views can be presented by:

- fixed pictorial representations such as *photo-montage* in which a representation of the new project is superimposed on photographs of the current scene; this is an ideal form of presentation for inclusion in a document or for obtaining sample judgements;
- more sophisticated and expensive ground-level movie techniques known as '*travelling matte*' such as were developed by the Transport and Road Research Laboratory (Watkins 1981: 136) to include simulated moving road traffic in the views; the cost of showing trains at speed, rather than in still views, is only likely to be justified for particularly comprehensive demonstrations, probably with simulated sound added;
- movie or conventional *aerial photography* in which the railway track is superimposed on the aerial plan view by photo-montage techniques; since this shows views not normally seen, it is probably better suited to assessment of impacts of rail construction on physical land resources rather than on scenery;
- construction of *models*, only likely to be justified in solving or demonstrating planning problems in situations where visual impacts are only one of the environmental factors.

Figure 9.3. This Great Central Railway bridge was built in 1899 across the then quiet valley of the River Misbourne in Buckinghamshire. More than 80 years later the M25 motorway was ingeniously threaded between the bridge's piers. Before the motorway, the railway caused *intrusion* in the valley view at the bridge section and *obstruction* where this view was fully blocked by embankments. On the latter, trees have now grown, forming a new and generally attractive landscape element through which trains can only be seen in winter

9.3.4 Analysis and Evaluation of Impacts

Analysis starts with the systematic description of existing views and continues with appropriate presentation of data about visual impacts. Presentation of 'before' and 'after' views provides visible evidence of what will occur. Judgement on the severity of impacts can be made in terms of popular reaction to the change and of the numbers of people who will notice it.

People's reactions to the introduction of new structures into their familiar views vary from the plainly reasonable—such as insistence on a pleasing appearance for a bridge—to implacable opposition to any sight of the proposed development and a demand that it be hidden from view.

Opinions as to whether the change is welcome or constitutes 'a blot on the landscape' have to be based largely on subjective judgement but comparison with existing situations elsewhere is a good guide. Evaluation of the nature and gravity of visual impact requires:

- analysis or classification of the types of scenic situation which the railway will affect; comparison with views of well-established railways in similar scenery;

- expert, objective or subjective comment on anticipated, or of similar existing visual situations and on their perceived significance;
- estimation of the regional implications of intrusion into scenery, for example in terms of the proportion of similar scenery which is left undisturbed.

The steps in assessing visual intrusion are examined further in Section 9.5.

How much visual intrusion will be noticed depends on such criteria as

- the number of dwellings from which a view can be enjoyed;
- the nature of viewpoints, e.g. private gardens, village streets, hilltops, recreation grounds or other accessible amenity land;
- the extent to which each view is already noticed by residents, workers or ramblers.

9.3.5 Visual Priorities and Mitigation

Action must be taken to preserve or enhance those views which give most pleasure. Possible approaches to minimising visual obstruction and intrusion are indicated on Figure 9.4. These comprise:

- route alignment, which is seldom likely to be changed for purely visual reasons but may be modified to avoid related damage to scenic land resources;

VISUAL PROXIMITY	Close and Prominent	Middle Distance	Far Distance
	Direct impacts on people (Chapter 9)	Impacts on scenery (Chapter 15)	
TYPE OF VISUAL IMPACT	OBSTRUCTION	INTRUSION	
		Private or local	Wide landscape
		Change in character	Loss of remoteness
POSSIBLE MITIGATION MEASURES	Route alignment and adjustment		
	Partially clear the view with open viaduct or transparent barriers	Appropriate design of structures	
		Soft or hard } Landscaping { Tree or shrub planting: Earthwork	

Figure 9.4. Visual impact and its mitigation

- provision of slender or open structures as far as this can be related to functional design;
- architecturally acceptable appearance and soft or hard 'landscaping', taking into account also economic, construction and ecological factors.

These approaches are discussed under route selection in Chapter 11, railway architecture in Chapter 14 and scenic mitigation in Chapter 15.

The built environment on the earth's surface, including transport infrastructure, is continually being expanded at the expense of ecological resources. Conservation of the latter is related to scenery, both directly and in people's perception. It is therefore important that railways should be aligned, their structures designed and their surroundings affected in a manner which is *ecocentric* rather than purely *technocentric*. The former implies the greatest sympathy with the more natural forms of landscape; the latter emphasises functional performance. The two are quite different but not necessarily incompatible.

9.3.6 Visual/Scenic Impact Statement

A section of any environmental statement should describe the scenery and the visual obstruction and intrusion which will take place. In expressing preferences for particular route options, conclusions will have to be related to other environmental factors, particularly land-take. In proposing measures to mitigate the visual impacts or to enhance the ultimate scene, practical factors in engineering design and construction may be significant.

9.4 VISUAL OBSTRUCTION

Visual obstruction is very real. The *features* of landscape which can be seen from certain viewpoints before and after erection of a structure can be specifically defined (see Figure 9.5). The *magnitude* of obstruction can be measured in terms of the solid angle subtended. For instance, the UK Department of Transport's *Manual of Environmental Appraisal* (1983: B2) defined degrees of obstructive severity based either on the distance to and height of an embankment, or on more precise angular measurements for a less uniform structure.

The *nature* or *significance* of the obstruction depends upon:

- how the appearance of the new structure compares with the previous view in the same direction;
- whether sunlight is actually denied by the shadow of the new structure;
- what can be seen *through* it;
- whether vegetation can soften the obstruction.

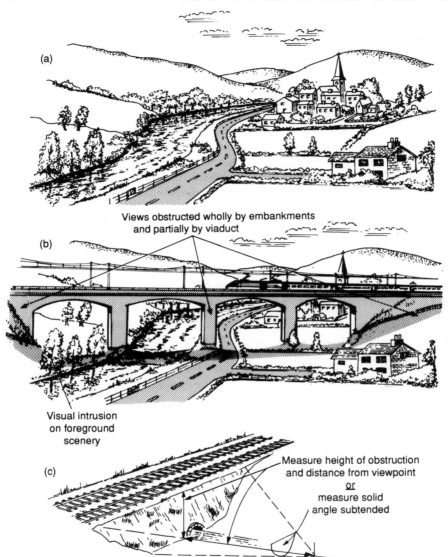

Figure 9.5. Visual obstruction: (a) view that will be obstructed; (b) effect of proposed construction; (c) measurement of visual obstruction

9.5 VISUAL INTRUSION

Assessment of visual intrusion involves analysis of the merits of the scenery affected and determination of the significance of railway intrusion into that scenery.

9.5.1 Analysis of Scenerey

Elements of scenery are *natural* land-form, geology and vegetation; *man-managed* woods, fields and streams; wildlife—flora and fauna—and their *habitat*; man-made *structures*—roads, bridges, buildings and quarries.

Description of scenic features along a route can be, first, in terms of land-form, including slopes, rock outcrops and water features; next, of vegetative ground cover and habitat, and then of human habitation and all related man-made infrastructure. Seasonal variations in the appearance of certain features must be identified.

Classification or ranking of scenery has been attempted by many parties. Official classifications, such as Areas of Outstanding Natural Beauty (AONBs), reflect official but popular recognition and imply a degree of statutory protection. Other approaches are necessary to classify the much more common areas of land which are considered visually attractive but have no official protection. Some, such as the US Bureau of Land Management's *Visual Resource Management Program*, define classes of land in categories from those so sensitive that no change can be allowed to those in which no large-scale development is likely to degrade the current situation. In between lie categories in which changes may be permissible subject to management (or planning) measures in effecting them.

Some detailed analysis systems award points for particular features or characteristics; these may incorporate individual or group preferences for different types of scenery. Some writers have referred to the inadequacy for particular applications of any single visual assessment method. 'A technique for assessing scenic quality should be selected with criteria relevant to the *problem at hand* as no single methodology can fulfil all relevant criteria' (Watkins 1981: 129).

9.5.2 Significance of Railway Intrusion on Scenery

Typical impacts of railways on the quality of scenery are:

- severance of natural features like rivers, cliffs or open hillsides;
- loss of any such features lying directly on the line;
- creation of discontinuities in semi-natural or ancient man-made aspects of the landscape like woods and hedgerows;
- erection of modern structures whose appearance might be judged discordant.

Practice in assessment of the visual impact of roads has some relevance; roads have been extensively studied in this respect and are similar in linear character to railways. Watkins (1981: 129) draws attention to three approaches to visual impact assessment, i.e. assessment by experts, objective assessment and subjective assessment. Each has a possible part to play in a methodology devised for application to railways.

Expert assessment is undertaken by specialists such as landscape architects or scenic arbiters such as the Royal Fine Arts Commission. These have built up experience in the treatment they see proper for development projects. Their judgements are based on well-structured arguments but may still be subjective in character. However, their assessments may be respected by the public *because* they are acknowledged experts.

Objective assessment attempts to establish factual information. Numerical data is valuable, particularly for comparison of alternatives. But any value judgement by outside experts is liable to meet fierce cross-examination from amenity groups.

Probably the most effective objective approach is analysis of proposed visible features and then comment on the acceptability of similar features in existing railway structures. Typical characteristics analysed may be ground steepness, vegetative cover, 'greenness', 'wetness' and indications of contrast. Objective data about natural and man-made features can include presence, absence, relative scarcity, number and magnitude.

Subjective assessment, whether by experts or members of the interested public, necessarily involves personal preference. Travel and guide books define classic examples of esteemed scenery. Sampled preferences can be based on actual sight or photographs taken from viewpoints, homes, cars or trains. The assessment should be by people who are aware of scenery and by visitors as well as nearby residents. On their home ground people's preference may be weighted in favour of what they are used to and initially against anything new, which is seen as intrusive.

A *rational approach*, making use of all three types of scenic assessment, would be for experts to identify the scenic resources, to use objective analysis in recording (photographically), quantifying or comparing features of the scenery, and for subjective popular preference to be sampled to validate the expert and objective conclusions.

The wealth of *scenic situations* in which railways are already present should first be examined to identify parallel cases. Description and assessment *techniques* then used can be those preferred by the experts who will undertake the assessment and those best suited to linear developments like a railway and to the type of scenery traversed.

Chapter 15 describes and analyses examples of existing railway intrusion in various types of scenic land resources. It also discusses measures which might be adopted to relieve these intrusions.

9.6 THE VIEW FROM THE TRAIN

Scenic views have long been recognised as an attraction of railway journeys. As long ago as 1910 the Midland Railway advertised its route to Scotland as that offering the finest scenery.

Sir Bob Reid, BR Chairman, referring to the low average speed which Eurostar trains can achieve on existing lines in south-east England, wrote to *The Times* in 1992 that 'there are advantages of going through Kent slowly. It is one of the most beautiful parts of the country. Passengers will enjoy going through it at a leisurely pace.' If this remark were to be taken seriously, there would be a corresponding *disadvantage* in travelling through Kent at speed when the CTRL is eventually built! It is more likely that passengers will be disappointed by the sections of the new route which are in tunnel.

Finer scenery than that in Kent can be seen from some trains—mountains, lakes and forests in Switzerland and the Rocky Mountains, or views of the sea-shore in South Devon, western Scotland and along the Mediterranean.

More commonplace but still visually interesting urban sights and activity used to be seen from any remaining and some new overhead city railways.

Environmental and cost constraints may no longer permit railways to be built along scenic coasts, up gorges or even above city streets. But plenty of fine scenery can still be seen from track which is at ground level or on embankment. Early TGV routes proved this. Meanwhile, however, new high speed lines in Germany are being planned with great emphasis on reducing external awareness even to the extent of sinking the tracks below ground level. This is certainly to the detriment of any view for passengers.

Views from the train can be defined by a reverse process of the visual envelopes described in Section 9.3. In steep topography, these views can be materially different in character from those seen outside the railway boundary, depending on how they are enhanced or obscured by the height of trains above or below natural ground level. Pictorial synthesis of views from embankments will be difficult because photography may not be practical from the height of an as-yet-unbuilt structure. A description of views available to passengers may have to suffice.

In flatter country photographic evidence of what can be seen from trains may be more feasible. Attractive features are often rural scenery and sometimes even industrial activity. Some people find residential areas less interesting – except where church spires or the gables and chimneys of older buildings protrude above more mundane features.

An inventory should be prepared of views available on each side of the the line. This should determine whether the views are of general or particular interest, spectacular or unexceptional, clear or obstructed. Measures should be considered which could improve the quantity and quality of these views, particularly where there are other operational, economic or environmental issues which make a choice of options necessary. Such measures include:

- slightly raising the vertical alignment if this can improve visibility in shallow cuttings;
- permitting at least glimpses through trees planted for landscaping;

- improving the appearance of any lineside earthworks where these are the main feature which can be seen;
- limiting the height of noise barriers or, if necessary, making them transparent; in mountain country, providing side openings in snow (avalanche protection) sheds.

Numerous volumes have been published in the past describing views from main line routes for railway passengers. It would be a significant omission if the major visual features of the passengers' view were not included in environmental appraisal of a new route.

10 Construction

10.1 BUILDING A RAILWAY

10.1.1 Characteristics of Railway Construction

Apart from its sight and sound, railway operation between stations is comparatively isolated from local surroundings. Like many civil engineering works, the impact of a railway on its neighbourhood can be more directly noticeable during its construction.

Because railways and roads are both much longer than wide, there are many similarities in their construction. But there are also clear differences:

1. The width of two-track railway formation is only one-third of that of a six-lane motorway; however, because the alignment is less flexible, the height of cutting and embankment needed in undulating country may be more for a railway; even so, the amount of construction activity per kilometre is usually less than half that for a motorway.
2. Because of the narrow width of the level formation, there is less working space; therefore the sites of construction depots, storage yards and sometimes haul roads have to be located outside railway rights of way—as does working space around structures or access to tunnels.
3. Speed of railway construction along the alignment can be faster at each stage because:
   - the volume of earthwork and the width of bridges are less;
   - track-laying and ballasting can be completed more quickly, in kilometres per day, than road pavement construction;
   - electrification and ancillary trackside equipment can also be erected quickly, using rail-mounted equipment.

First we shall examine how environmental factors can be included in the construction planning (Section 10.2). Potential impacts on workers on the site are primarily construction method and safety matters. This chapter is concerned mainly with effects of the construction process on the well-being of people outside. Most of the recognised impacts of railway construction activities, equipment and traffic fall on people. However, the external effects on flora and fauna and on land and material resources must also be considered.

10.1.2 Human and Social Impacts

When early railways were constructed by gangs of itinerant manual workers ('navigators'), the influx of the latter had considerable social impact on rural people and activities. This impact comprised:

- a large temporary population of single workers with relatively high wages and requiring drink, food and lodging;
- entrepreneurs arriving to provide these commodities profitably;
- local inhabitants' concern and need for protection against the occasional recreational excesses of the workers.

In urban surroundings the environment created by railway construction was certainly noticed and recorded in the literature of the day. Charles Dickens (1848) mentions buildings being demolished, undermined or shaken and access cut off or rendered hazardous by fire, ashes and water among the earthquake-like chaos of excavated pits and ponds.

Today the scale of these impacts is much reduced—by the smaller numbers of, now mainly local, workers involved; by the need for better organisation of operations using large machines within tight space constraints and quality control requirements; and not least by legislation and statutory procedures protecting people and property.

Townspeople's first concern is still protection of their property from damage by construction equipment; a second concern may well be traffic and a third, noise. Experience from road construction (Dept of Transport 1983: B.8.3.2) shows that at least half the people living within 50 m of the site boundary are seriously bothered by construction nuisance. This suggests that objective appraisal should consider impacts on all properties within 100 m of construction sites or haulage routes.

Matters of potential concern to people living near railway construction can be related to the techniques, equipment and space needed for the construction processes, viz.:

- excavation and fill: dust, mud, material stockpiles, borrow pits and waste disposal (see Section 10.3);
- underground works: vibration and settlement over tunnels; interference with ground water (Section 10.4);
- structures: methods of construction and their needs for space (Section 10.5);
- water and drainage: effects on water supplies and habitat (Section 10.6);
- hazardous ground and materials: protection from release of contaminants during construction or thereafter (Section 10.7);
- plant and machinery: noise and emissions (Section 10.8);
- work sites: extra space required temporarily (Section 10.9);
- construction traffic: noise, dust, mud, congestion and accidents on public roads or temporary haul roads (Section 10.10).

In all major construction, the severity of public impacts has to be assessed in the light of their duration.

10.1.3 Impacts of Construction on Flora and Fauna

Large animals are controlled by humans and can be protected against construction activities. *Small wild animals and birds* are relatively indifferent to noise. They are more sensitive to dust and water pollution and may abandon their young if their nests are physically touched.

Small plants and shrubs can be replanted if they are sufficiently valuable and if soil conditions are right. Even *substantial trees* have been successfully moved. Partial reconstruction of entire habitats, if it is practicable and affordable, calls for detailed specification, careful timing and subsequent management, and expert supervision. In spacious development, such as around stations in suburban situations, a number of trees can be preserved *in situ* by careful planning of buildings, parking areas and boundaries at temporary work sites.

10.1.4 Impacts on Land and Material Resources

Temporary land-take is often needed for construction depots, haul roads and around the sites of structures. This is because of the narrow width of the railway wayleave which allows much less working space than, for instance, motorway construction.

However, land taken for temporary operations need *not* be directly on the alignment; its location can be chosen:

- where best suited for access from the public road system whilst adjacent to the railway structure concerned or with ready access to the line itself;
- where least damage will be caused and where the site can be readily restored to use, e.g. on arable ground or spare paved land.

Needs for construction materials are similar to those for roads in respect of use for structural concrete aggregates. The railway equivalent of bitumen or concrete road pavements is mined stone ballast, concrete or wooden sleepers and steel rails.

10.2 CONSTRUCTION PLANNING

A construction plan must be based on the particular design features of the railway and on practicable means of construction. The design should already incorporate long-term environmental safeguards. So the construction plan will give first priority to the most cost-effective solution.

A good second priority will nevertheless be given to environmental protection—in methods of working, type and operation of equipment; in dealing with drainage, effluents and emissions; and in subsequent site restoration.

The works will usually be built by a construction contractor, at a price tendered to meet a final design. Therefore it is a responsibility of the designer of the railway to incorporate environmental safeguards into the construction contract conditions and specification. He should define essential requirements in the construction plan and the contractor should complete the detail. Thus the designer's specification contains clauses dealing with construction methods, protection of property and land or water conservation features. His drawings should define site boundaries and existing features or utilities which have to be incorporated; although the acquisition of land and the precise location and avoidance of obstacles may be responsibilities devolving on the contractor.

Contract conditions can impose limitations on noise, working hours, access to sensitive areas and use of roads and haul routes. However, restrictions should not be so inflexible as to impose working methods that are either unnecessarily costly or are not compatible with *all* environmental factors.

In Britain, Parliament or the government authorises railway development itself; but responsibility for approval of the construction processes, day and night noise levels, temporary works and traffic regulation is often delegated to local planning authorities. Some environmental constraints on construction already exist in national legislation and local planning orders; but some measures still need to be negotiated between the promoter or his designer and the statutory planning authorities or local government Environmental Health Officers. Other details will have to be negotiated by the contractor himself.

Local authorities have powers to deal with pollution and disturbance from construction, for instance to make contractors provide noise insulation or temporary access. The cost of temporary disturbance to road traffic can also be brought into the contractor's construction plan and work price; indeed the time for road occupation can be minimised if public road space is rented to contractors while they occupy it.

Figure 10.1 shows schematically how environmental assessment can be brought into the planning process. Figure 10.2 shows how specific activities and their impacts can be dealt with in the framework of a construction plan.

A construction plan should include:

- a network programme for all construction activities and events in building the railway;
- plans and quality assurance procedures for particular operations (such as earthwork, dealing with water, making concrete or transporting materials);
- schedules of manpower, materials and equipment needed for each operation;
- a plan for utilisation of space including location of work sites and haul roads; occupancy requirements for existing rail track;
- a plan for safety and prevention of pollution, on and off the site.

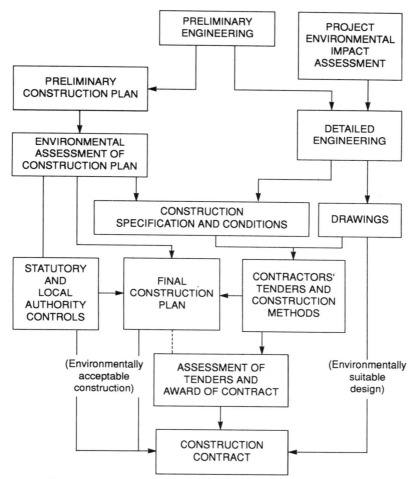

Figure 10.1. How environmental assessment can affect construction

10.3 EXCAVATION, FILL AND FORMATION

Earth and rock moving processes take a large proportion of the time and working effort required in construction of roads and railways. They also use up land, mineral and fuel resources. The main component of earthwork is excavation of cuttings and tunnels and provision and placing of fill in embankments and track formation. Processes include open excavation in various materials, tunnelling, compacted fill, stockpiling between operations, disposal of surplus material and specialised excavation of foundations for structures such as bridges. Placement of track materials completes formation of the railway itself.

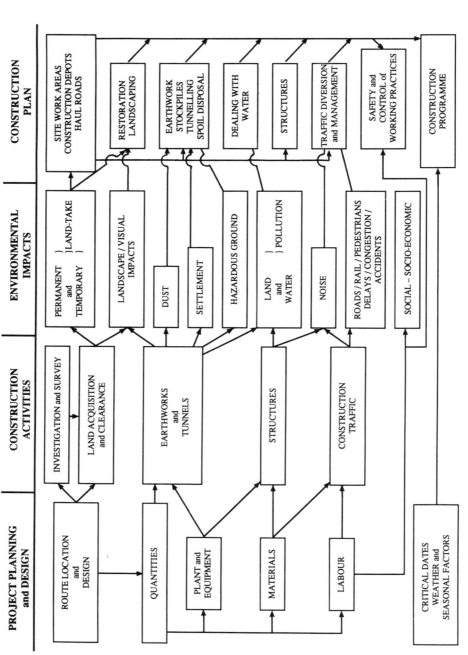

Figure 10.2. Environmental impacts in construction planning

10.3.1 Open Excavation in Soft Material

Clays, silts and sands can be removed by scrapers or excavators whilst soft rocks, gravels or chalk may need preliminary ripping. Environmental hazards arise from dust in dry weather and mud in wet; mitigation can be attempted by watering haul roads and cleaning vehicles respectively. Note that dust can cause disturbance as serious to ecological habitat or to sensitive crops as it is to people. EC directives and derived national legislation place limits on dust as well as noise generation.

Topsoil stripping and preliminary excavation can cause unnecessary damage to near-surface archaeological features or to valuable trees or vegetation on the fringes of working areas. Inexpensive precautions can usually be taken, for instance to protect trees during construction (Preece 1991: 257–262).

10.3.2 Fill

Fill is compacted in embankments by rollers of types suited to the materials being placed. In dry weather, dust is allayed because water has to be added anyway to provide the moisture necessary for compaction. In very wet weather, fill and compaction may have to be stopped because of the excess water. Sand fill is compacted by vibration, finer material by weight at an optimum moisture content; part of the vibration or weight can be provided by the large earth-moving plant depositing fill material before rolling. On the other hand, compaction should be avoided on topsoil replacement or on other non-structural fill placed for landscaping and planting.

Where old embankments are being widened, special measures may be necessary to prevent differential settlement in the existing earthwork. Introduction of a light type of fill may be practicable. Fill material may be derived from cutting excavation but in some circumstances, as across extensive flat country, it may have to be dug from special borrow pits. In the distant past and even today in relatively unpopulated country these borrow pits can be located adjacent to the route; often they are part of a modified lineside drainage system; but structural, environmental or land resource requirements may dictate that material must be brought from more distant, approved quarries at extra cost.

Environmental factors in planning the placement of fill concern:

- use of surplus material in 'hard landscaping' (earth-shaping);
- use of materials suited to any vegetation which is to be created, e.g. *not* topsoil for wild flowers;
- the suitability of compacted or loosely placed soil for subsequent planting; and the time of year at which planting should take place;
- the possible effects of wildlife on the stability of the earthworks, e.g. tree roots or burrowing animals.

10.3.3 Open Excavation in Rock

Excavation of hard material requires blasting. Drilling prior to blasting causes dust which can be reduced by spraying water. The actual explosion of a rock bench can cause a sudden, considerable and dense cloud of dust as well as harder missiles. In sensitive situations, suitably designed reusable protective blankets can prevent such clouds reaching far outside the blasting area. Timely warning of the loud and potentially alarming noise of explosions should be given to people in the neighbourhood. If frequent, blasting should take place at the same times daily. Rock excavation for new or widened railway lines normally takes place in cuttings, i.e. below normal ground level. Vibration may be a hazard to sensitive equipment in the vicinity and special warning precautions may have to be taken.

10.3.4 Stockpiling

Stockpiling of earth may be necessary between its excavation and its placement in fill. As far as possible excavated material is moved directly—by scrapers, dump trucks or conveyor belts—from the source to the final destination. However, sometimes the distance involved, variations in the quality of the material or other practical considerations will not permit this; for instance, it may be necessary to allow wet material to dry out before it can be placed as fill. Stockpiling requires space and can cause the same dust or mud problems as excavation; wind may blow dry material off the top of stockpiles. Different stockpiles may be required for different materials and end uses, e.g. for incorporation in load-bearing fill, for landscaping or for complete removal.

10.3.5 Disposal of Excess Material

Disposal is required of excess excavated material which is not suitable for fill or where cut exceeds fill. In classic design, earth taken from cuttings is used up as fill in embankments. In practice, there may be engineering, environmental or local planning requirements which result in an excess of cut. The cheapest way of dealing with the surplus material may be to use it as additional landscaped fill beside the railway formation itself, as far as land ownership rights and environmental factors allow. Planning its disposal elsewhere must take account of the location of disposal sites and the means of transporting it there.

Sites and methods for disposal of surplus material will require their own environmental assessments. The sites may be old quarries or derelict areas where landfill is needed for reclamation. The method of disposal must be safe; for example landfill slopes must be stable and build-up of potentially explosive methane gases must be prevented. Occasionally the excess material

may be useful for other civil engineering works or even industrial use—
gravel for concrete, clay for bricks and pipes, chalk for cement.

10.3.6 Permanent Way

Placement of track is akin to road pavement construction, particularly if
concrete 'slab' track is installed. An expensive alternative to ballast and sleepers, shallow slab track is used on some lengths of Japanese *Shinkansen*
extensions and, occasionally, in Britain to gain valuable headroom for overhead electrification through existing tunnels or under low bridges.

More commonly, where ballast is used, a source of uniformly graded hard
rock is required. Over fine clay a filter of transitional, potentially dusty
material, may have to be provided to prevent the clay clogging and impairing the drainage quality of the hard ballast.

If a railway line becomes obsolete, ballast is less of an obstacle than slab
track to restoration or removal. Some secondary lines were built on only a
shallow layer of rock ballast or on one of less robust material. This has
meant that, when no longer required, those lines more easily provide the
firm base but discreet appearance for a rural byway.

10.4 UNDERGROUND WORKS

Hazards of tunnelling are very real. But it is the safety of the workers themselves which is usually at risk, rarely the general public.

Tunnelling is performed by boring machines or by drilling and blasting
according to the material and the length of tunnels. At or near tunnel portals the noise and dust of these activities may be evident outside, although
effects may be reduced in deep cuttings. Sometimes the exhaust fans of ventilation systems add to the noise background.

Noise disturbance is less likely to arise from excavation inside the tunnel.
However, particular and potentially damaging effects can arise from

- *vibration*: where impacts are potentially serious, avoid the most sensitive
  times of day but otherwise advance the face as quickly as possible;
- *settlement*: risks can be reduced by continuous excavation and prompt
  placing of tunnel lining.

Both vibration and settlement are related to the nature of the ground surrounding and overlying the tunnel. Their effects should be calculated during
design and construction planning by geotechnical engineers familiar with the
particular type of ground.

Settlement above tunnels is most serious for buildings, particularly older
ones, located immediately above or very close to the line. The foundations of
the Grade I listed Mansion House in the City of London had to be under-
pinned in 1870, about 100 years after its construction; subsequently the

building suffered some cracking during underground railway construction early in the twentieth century. When the Docklands Light Railway was extended under the Mansion House in 1990 settlement had to be countered by horizontal tie bars within the building itself. A number of independent detection devices were then used to monitor any movement during railway construction.

Cut-and-cover tunnels are a common feature of approaches to city terminals and of metro railways. They are traditionally built as surface operations, being enclosed by constructing the roof last. But in locations where the surface space is particularly sensitive to construction activity or urgently needed for transport or commercial use, then the techniques can be reversed, building the walls and roof slab of the tunnel first and then completing the main excavation and installation within the tunnel itself.

Disturbance of ground water by underground excavation can affect its level and availability for water supply, either at wells or after flow emerges on the surface. Where an aquiclude—intervening impervious material between separate aquifers—is pierced, flow between the aquifers can take place, leading to leakage from the higher level and possibly to changes in water quality. However, the risks of underground works and workers being inundated by sudden flows of water are sufficiently serious that any tunnel excavation precautions take ground water characteristics fully into account.

Apart from occasional settlement or vibration risks, construction of bored tunnels has no impact on the land resources above them. They do however require extra working space at the portals and usually suitable sites elsewhere for disposal of unwanted excavated spoil.

Conventional bored tunnelling beneath wide waterways can be difficult or impossible in geological conditions less homogeneous or impermeable than, for instance, the chalk between Folkestone and Calais. Where shallow alluvial material underlies the channel to be crossed, construction of a submerged tube tunnel may be more feasible. A trench is dredged in the channel bed and precast box units placed on a carefully prepared foundation and later connected and dewatered. Such tunnels were successfully constructed for the San Francisco (BART) and Rotterdam metro systems and in Britain for the Conway estuary road crossing. The main environmental requirement in such construction is location of an adequate land and shore area for construction of the tunnel units.

10.5 STRUCTURES

Noise and vibration from processes like pile driving or dynamic compaction are features of structural foundation construction; so are precautions to prevent differential settlement in adjacent buildings. The main opportunity to limit such effects is in the choice of the construction method and equipment.

We have seen that the amount of railway land-take is determined by the width of earthworks as well as that of the tracks. Where there are buildings or other obstacles occupying space needed for cuttings or embankments, the latter can be obviated by incorporating suitable structures. Demolition and rebuilding to release additional areas for cut or fill are unnecessary if earth or rock slopes can be replaced by near-vertical retaining walls or steep stabilised banks. A variety of modern techniques is available, including different types of diaphragm walls and soil reinforcement. Low technology forms like gabions (wire cages filled with local boulders) may also be eminently suited to construction of walls in a visibly acceptable form.

Techniques for rapid construction or replacement of bridge superstructure have been pioneered largely in railway work so as to minimise the time during which rail occupation is required. These techniques include rolling in precast structures above or beneath the tracks; and setting by crane of precast 'model railway tunnel' sections through which traffic can run whilst embankment construction continues overhead.

The main environmental aspect in planning structural works is location of suitable work site areas. These require ample space for construction and assembly and for access but their siting must avoid sensitive land areas such as riverine strips or wetland commonly found under bridges.

10.6 WATER AND DRAINAGE

Crossing of rivers and natural drainage routes has a major influence on the initial route planning of any railway. Restrictions on gradients and curves require that an alignment shall run over, beside and even under watercourses.

Construction activity involves potentially serious difficulties if natural drainage has to be temporarily interrupted. Therefore an important part of any construction plan is that providing for 'dealing with water'. This has to cover:

- measures to accommodate watercourses diverted around bridges or culverts during construction; the risk and consequences of floods which may occur whilst temporary arrangements are in place must be determined;
- measures to drain storm runoff from earthworks or foundations under construction; temporary drains or pumping may be involved;
- provision of water needed for construction purposes such as mixing concrete, compaction or dust alleviation;
- identification of any pollution which may occur to any water users and of any necessary measures to isolate and treat water which might become contaminated;
- measures to prevent damage to aquatic habitat.

Surface or ground water can be contaminated by runoff through chalk or clay leachate in stockpiles as well as by spillage of fuels and lubricants. Areas of temporary hard standing can be laid on a plastic membrane surrounded by drains and connected to a sump whence water is removed or diluted in lagoons. Local government, conservation groups and the rivers authorities have to be consulted about the disposal of runoff and the quality of effluents.

As well as measurements of the pH level and the concentration of chemicals or suspended solids, there are biological ways of monitoring effluent quality, for instance by studying how invertebrates react to changes in water quality. Observations can trigger alarms at which effluents have to be disposed of by special means.

10.7 HAZARDOUS GROUND AND MATERIALS

Construction through hazardous ground is no recent phenomenon. The extension of the Caledonian Railway from St Rollox to a new terminal at Buchanan Street in Glasgow had to be burrowed through chemical waste in 1849 (Turnock 1990: 71).

National legislation has to be considered in planning how to construct railways through land or ground water that is already polluted and may be dangerous to disturb. It is important to examine all visible and historic evidence to decide whether there is any real potential hazard so that further investigations can be arranged and expert advice taken. Most of these hazards will be encountered in land which has been previously used for industrial activities or for disposal of wastes. Apart from the release of noxious gas or radioactive or contaminating substances, the main concern is fire or explosion.

Poisonous chemicals are likely to be encountered where railway construction takes place through disused industrial land or indeed on the site of old railway sidings or engine sheds. Study of the local industrial archaeology and close inspection of the ground and its flora may reveal indications of hazardous conditions.

In country areas, waste disposal tips encountered on the route can range from unauthorised dumps to large organised landfill sites. At the former any possibility of dangerous chemicals or asbestos must be investigated whilst at the latter, the chances of disturbance releasing methane must be considered. Where contaminated soil is removed for railway construction it has to be disposed of at a suitably prepared site—probably in an excavated hole, impermeably lined, drained safely to a sump and with facilities for dealing with gaseous emissions.

10.8 PLANT AND MACHINERY

Construction plant can damage the environment through atmospheric or aquatic pollution and noise.

Air pollution is caused by diesel fumes from excavating machinery and haulage vehicles. Of the emissions hazardous to health in the vicinity, the proportions of carbon monoxide (CO), nitrogen oxides (NO_x) and lead from diesel exhausts are substantially less than from petrol engines. However, emissions of particulate matter, much of it visible as smoke, are much greater from diesel-driven earthmoving plant and dump trucks.

Other visible emissions such as oils can be thrown into the air in certain processes. For instance, diesel pile drivers spew oil liberally, particularly in a high wind, causing damage to clothing of passers-by or on washing lines.

Fuel should be stored within bunds. Machinery and vehicles should be refuelled upstream of drains where effluent is collected for discharge to safe sumps. Storm water and other runoff, which might be contaminated by running through waste cement, asphalt and toxic grouts or slurries, must be tested as to the quality and alkali content of the effluent. Treatment of contaminated water may even be necessary if its discharge to ground water aquifers or to surface streams cannot be avoided. Construction run off is thus a potential source of *aquatic pollution*. Others are temporary blocking of watercourses or excessive extraction of water for construction purposes.

Dust resulting from earthworks has already been mentioned. Other sources of particulates may be cement or concrete aggregate at concrete mixing plants or in stockpiles—also dust from rock crushing equipment if crushing is done on site. Mishandled or leaking compressed air lines can also cause dust storms. Conveyors can be covered to prevent fine material being blown off.

Noise levels are subject to legislation or negotiation with local authority Environmental Health Officers. British Standard (BS) 5228 (1984) on noise control on construction and demolition sites is also relevant. However, construction is a relatively short-term event. It should only be necessary to take special steps to reduce noise where it has definite adverse effects on outside activity or rest, not where it causes a low level of only temporary annoyance. Restrictions on night working or muffling requirements can be applied to particular machinery in sensitive situations. Unavoidable high noise levels may justify insulation measures in houses. In cases of extreme disturbance temporary rehousing may be appropriate.

Explosions are the loudest but not most frequent sound. Common sources of mechanical noise are:

- pile driving (also causing ground vibration as do vibrating rollers);
- aggregate plants, concrete mixers;
- jack-hammers and rock drilling (95 dB(A) unsilenced at 20 m);
- concrete vibrators;
- excavators, scrapers and bulldozers;
- conveyor belts (90–95 dB(A) at 10 m);
- vehicles of all types; alarms such as dump truck reverse gear warning horns.

It should be noted that noise, particularly of vehicles and moving machinery, serves as a warning and that no muffling of traffic noise should be allowed to impair this safety function.

Much of the most serious impact of noise and atmospheric pollution falls directly on the machine operators or other site workers. These are occupational hazards, dealt with as such by recommendations (CIRIA 1984) and by noise at work regulations. Recent industrial legislation concerns three noise 'Action Levels'. The first two are daily personal noise exposure levels (85 and 90 dB(A)) which trigger various requirements for information and for personal or structural noise protection measures. The third Action Level is a peak sound pressure (200 Pa) and related especially to sudden very loud noises.

Some concern has been aired in the construction industry that responsibility for appropriate action is difficult to allocate where several contractors are working on a confined site in what may already be a noisy urban environment. However, such problems are not typical of new railway construction or well-organised single activities like bridge replacement. Railway constructors should have no great difficulty in full compliance. Like accidents, noise on site can be treated as a safety matter. It can be planned at the same time as construction environmental impact control. Previous writers (Watkins 1981) have dealt in detail with calculations for estimating the noise impacts of road construction and many of these are applicable also to railway building.

The polluting emissions and noise we have described are similar to those which occur in any construction project. The net pollutive impact may be equally unlikely to cause more than temporary concern.

10.9 WORK SITES

The corridors of land required permanently for railways are comparatively narrow. Therefore additional land may be required temporarily, both for haul and access roads and for depots for storage, processing of materials and parking and maintenance of vehicles.

On any exploited land which is to be returned to other use, restoration is required after construction is complete. This should reinstate the original situation unless there is opportunity for its positive enhancement.

Restoration has to be planned from the outset in terms of separate and suitable stockpiling of different types of soil, particularly topsoil. Plans must allow for each material to be placed in the location and in the sequence which true restoration demands. Restrictions on height and side slopes of stockpiles are necessary to prevent the break-up of soil units, as are some limitations on placement during wet conditions.

In urban and underground railway construction there may be insufficient space for surface work sites. Where no undeveloped land sites are available, sections of public amenity land may have to be sought temporarily or, in exceptional cases, buildings demolished. The latter was evidently a potential necessity at one site (Warwick Gardens) where an underground route was being planned by British Rail for the 1989 Channel Tunnel Rail Link route under South London. The 1993 route passes under or near the former Great Eastern Railway complex at Stratford and should provide better opportunities for locating surface construction sites.

Construction activity is interesting and many people enjoy watching it. Therefore there is no point in hiding it behind opaque fencing, except for security reasons or to contain continuous loud noise. Rather, viewpoints can be provided at points convenient to public access and to operations.

Explanation of what is going on is an important factor in presentation of the railway developer's image and in mitigation of temporary hostility which may have arisen locally. If high hoardings are essential for practical reasons, they can be designed to look attractive and to suit the local environment. Acceptable examples of colourful hoardings have been erected around building sites in London.

New construction by a railway is also of particular interest to its potential customers. Clear notices seen by passengers approaching London Paddington told them that they were seeing the completion of the North Pole maintenance depot for Channel Tunnel trains—information of definite topical interest.

Night working requires lights on all mobile machinery and on fixed standards, similar to those in football grounds, for extensive work sites; the light diffused around the vicinity may be similar to that around stadiums.

10.10 CONSTRUCTION TRAFFIC

Railways already play a role in the long distance carriage of construction materials. BR's Rail Freight Construction division carries aggregates and cement on a regular basis as well as steel and precast concrete. During construction of the Channel Tunnel, 1.8 million tonnes of precast concrete sections and other materials were carried annually by rail.

In actual railway construction, once the earthwork has been completed, ballast and track material can be brought up by rail, advancing as the track moves forward; electrification equipment and other lineside accessories can follow by the same means.

Occasionally, excavated earth or embankment fill materials may also be brought in or removed by rail, either on the new line itself or on an existing adjacent track. But the economics of short distance haulage usually dictate that the majority of earth and rock fill or spoil and some other construction materials is carried by road vehicles or trailers, using both temporary haul roads and public highways.

Contractor's haul roads usually run as near alongside to the railway as is practicable. Except where the track area is itself wide enough to use, therefore, temporary land-take and measures to evade, protect or provide access to buildings adjacent to the line are necessary. Where the temporary nature of haul roads precludes the expense of hard surfacing, construction traffic can cause dust problems in dry weather. These problems should be allayed by spraying water from bowsers.

Public highways have to be used for deliveries of materials and for other off-side traffic. In some locations controlled on-site operations may have to share short stretches of road with ordinary road vehicles. The inconvenience of extra traffic has to be offset against the cost and extra impact on local property and activities of still more temporary haul roads. Where earth-moving operations make use of public roads in wet weather, 'wheel spinners' can be used to prevent mud being carried on to roads.

On narrow country lanes the main impact of traffic is in removal of hedgerows to widen the road. The side of the lane to be widened and the location of passing places should be chosen so as to prevent loss of the most valuable hedge lengths.

Environmental impacts of construction road traffic affecting people are:

- accidents—collisions between vehicles or with pedestrians;
- traffic congestion and delays to road users caused by the large extra vehicles and by necessary traffic control measures;
- dislocation of access along certain roads;
- annoyance to householders from increased noise of traffic, especially on roads to which normal traffic is temporarily diverted;
- splashing, mud or dust from site vehicles on public roads, especially annoying to pedestrians.

Accidents are the most serious of these impacts. Indeed the most common types of fatal accident in civil engineering are those involving motor vehicles. Incidents should be prevented by precautions at least as tight as those taken for normal road traffic. Potential temporary black spots should be clearly identified, based on experience at other sites and after analysis of particular circumstances.

In Britain, in accordance with the 'polluter pays' principle, public utility corporations and contractors can be charged for disruptions caused by highway construction ('streetworks'). Railway constructors can be expected to pay a rental for use of any access where their traffic or operations causes public inconvenience.

Part III

IMPACTS ON RESOURCES

11 Resource Use and Route Selection

RESOURCE USE

11.1 TRANSPORT AND RESOURCES

Transport operations require *energy* to propel vehicles. Transport systems use *materials* in their equipment and occupy *land*.

Energy is commonly derived from *fossil fuels*, convenient for carriage in liquid form on the vehicles they propel but a finite resource more plentiful in some localities than others. Sustainable planning requires that using up a non-renewable fuel for transporting people or goods must be justified in the light of the long-term prospects for availability to future generations of that fuel or a substitute.

Meanwhile a railway operator has to buy his energy at the market price, whether it is diesel oil or electricity. His choice of traction depends on the anticipated future cost of these sources of energy rather than on their sustainability. For electric traction the choice of fuel, and hence the cost of generation, lies with the producer of energy. For diesel the price is determined by the international oil market. Both can be affected by environmental and other taxes or subsidies which governments choose to impose.

Indications from Chapter 8 are that trains use roughly the same amount of energy as buses to carry passengers. Both are claimed to be about three times as efficient in energy use as cars. Railways carry freight more efficiently than all road transport except perhaps the largest lorries. Appropriate use of resources to provide power is the subject of current work on the economics and environmental impacts of fossil fuel production and electricity generation. This book's main concern about resources is therefore focused on the rail transport system itself.

Physically, railway systems comprise:

- trains, power supply, signals and control apparatus;
- buildings and facilities used for transfer of passengers and freight to and from trains;
- the permanent way on which trains run;

- the land space which is occupied by the buildings, track and its earth-work.

These features have been described in Part I.

The manufacture of trains and equipment requires energy and mined raw materials. The price of these resources is dictated by international demand; their long-term sustainability depends on the size of remaining exploitable reserves. Again the consequences of over-exploitation must be left ultimately to the planners and politicians who can influence the supply. Conserving the energy resources used in obtaining raw materials and manufacturing trains, rails, sleepers, ballast and structures can only be influenced by fiscal or legislative controls such as taxing resources or restricting their use for any purposes. Similar measures can influence disposal or recycling of obsolete rolling stock and equipment.

Land resources for new railways are more directly within the control of railway planners. Use of these resources must compete with commercial demand including industry, other transport infrastructure, mining, agriculture and forestry as well as with conservation of habitat, water resources and open space.

The next three sections of this chapter therefore introduce land and water resources, man's activities which compete for them, and the scarcer tracts which are so far relatively unspoilt. Section 11.5 then describes the historic pattern of railway route development and Sections 11.6 and 11.7 deal with route selection for new and realigned railways in ways suited to the preservation of the most valuable land resources.

11.2 LAND RESOURCES

The width of land occupied by most two-track railways on level ground is about 15 metres; that for a canal or single carriageway main road is similar whilst motorways are much wider. But cuttings and embankments may make the total formation width of a railway considerably more than that of a contour-hugging canal or minor road. The land space occupied by a two-track railway is typically 2 ha/km—rather less across flat country, more in steep terrain. In addition, intermediate stations, with goods yards converted into parking space, may each occupy another hectare. Several London terminus stations cover 5 ha whilst marshalling yards occupy 20 ha or more. The latter are less essential to modern freight operations but very substantial space is needed for road/rail transfer operations. At the English end of the Channel Tunnel the Dollands Moor freight train sidings occupy 9 ha but the adjacent Cheriton road/rail shuttle terminal requires no less than 140 ha (see also Figure 15.10). Even that space is tightly planned because of the high landscape or commercial value of the area. At the French end of the Tunnel the Coquelles terminal complex occupies four times as much land, allowing space for more artificial landscape features and for commercial development in emptier surroundings.

Besides taking land out of use, a linear development like a railway creates an obstruction between one part of a block of land and another, upsetting its homogeneity. A new line may therefore reduce agricultural or industrial productivity or prevent the easy movement of people or wildlife.

Whilst land itself is a natural resource, much of its surface has already been altered to suit the purposes of man. Wildland—that barely modified by man—is comparatively uncommon globally and rare in populated country. Altered land has acquired characteristics regarded as both good or bad for the environment of a rapidly expanding human population (and mainly bad for other animal species).

Man's landscape-altering activities include the following:

- *Building*: i.e. covering the ground with structures like housing, offices, factories, roads and railways; permanent, often impervious ground cover may be too costly to remove after the structures become obsolete; but on derelict *unpaved* land (such as the earthwork in railway formation) ecological reclamation may be possible even if the topography has been altered.
- *Agriculture*: modern monoculture farming can be as sterile and damaging as buildings except that the soil is more easily restored; but some traditional, usually labour-intensive, agriculture involving the maintenance of hedges, pasture and ponds, has created land now regarded as valuable for both conservation and amenity;
- *Forestry and woodland management*: large-scale coniferous plantation is similar to agricultural monoculture; well-managed broad-leaved woodland is more suited to nature conservation.
- *Mining*: extraction from quarries leaves cliffs on hillsides or lakes in flat country where the water table is high; above deep mines subsidence of buildings, roads or railways can occur; unplanned disposal of waste material can spoil otherwise good quality land.
- *Recreation*: in *natural* surroundings in mountains and forests or on land *created* for particular pastimes (such as golf courses); environmental hazards are erosion, pollution, visually intrusive structures, access roads and parking.
- *Drainage*: extraction of water and control of water levels.

Building, agriculture and mining are commercial activities which compete for land, making it a traded commodity with a very wide range of prices according to the profitability of the use which is permitted. These *commercial land* categories are considered together in Chapter 12.

Conservation land is wetland, woodland, much uncultivated open space and many recreational or agricultural areas which remain the habitat of natural, including threatened, flora and fauna. Effects of railways on terrestrial and aquatic ecology are described in Chapter 13.

Land and features enjoyed by people: recreational land, accessible woodland and open countryside, together with historic man-made structures and some land left free after industrial decline, are dealt with in Chapter 14. Another important and popularly recognised heritage resource is attractive scenery. Chapter 15 describes how railways fit into or intrude upon *scenic landscape*.

Some valuation of the scarcity and sustainability of untraded land has to be made for comparison with that which can be commercially priced. Failure to do so effectively inflates benefits and minimises costs calculated for any scheme which may displace conservation, heritage and scenic land. The issues and options in environmental *evaluation of land resources* are set out in Chapter 16.

All scarce land resources require protection against new development of industry, housing, commerce or transport systems. Land is scarcest in populated areas just where such development is most likely to be promoted. Road and rail connections to carry traffic from the Channel Tunnel are needed in south-east England, not in the empty areas of Wales or Scotland.

A first step in assessing land resources affected by railway development is to record current land utilisation. This may be taken from such land use or ecological maps and geographical information systems (GIS) as are available. Sources of land cover data comprise:

1. mapping undertaken accurately but on a piecemeal basis over a number of years;
2. modern mapping or GIS based on satellite imagery or aerial photography interpretation (API).

Modern land cover mapping has been summarised for land cover by the Institute of Terrestrial Ecology (1993). Some systems (such as the EC's Corine Land Cover Project now complete in western continental Europe) are comprehensive but record no land elements smaller than 25 ha; others are used mainly to record changes, i.e. to assess global environmental degradation or recovery rather than to define the detail necessary for planning a 15-m-wide transport route. In the UK the Countryside Survey and the associated Land Cover GIS/mapping (based on 25 m cells) is probably the most comprehensive source for planning railway routes. Topographical and land cover mapping can be updated by ground inspection and adapted as necessary to highlight the sensitive features described in Chapters 12 to 15.

When land resources have been identified, the next step is to identify their physical characteristics; besides ecological data these include the subsurface water table and geological conditions that explain current land use and determine the potential and hazards of future exploitation or conservation.

11.3 WATER RESOURCES, RIVERINE AND COASTAL LAND

11.3.1 Water Availability

In most inhabited parts of the world, fresh water is scarce. Man builds dams and controls watercourses to maximise what he can extract; where conditions favour underground sources these are also exploited. Use or overuse of surface and ground water causes a reduction in stream flow and lowers water levels; these effects restrict supplies to downstream users and threaten the existence of aquatic creatures and plants.

Reservoir storage and river diversion projects have more significant effects on the quantity of water resources than do railways or roads. Nevertheless, if railway tracks create barriers across watercourses, this can affect the nature of wet places.

11.3.2 Watercourses and Wetland

Streams and rivers occupy a small part of total land areas; but the narrow riverine strips or marshes beside them and the wider estuarial flats where they approach the sea are often areas of particular ecological interest. Unless they are already canalised, the courses of streams frequently intersect the straighter lines of any railways built along their valleys.

Railways affect watercourses and wetland directly where they cross them, i.e. by the footings of bridges over rivers and by earthworks interrupting minor drainage. The crossings have to be planned and constructed to accord with the strategies of the authorities responsible for the watercourses. Usually this requires building bridges or culverts in such a way as to interfere least with flows and floods, movements of fish and riparian conditions. More is said about crossings of aquatic conservation land in Chapter 13.

Some modern railways are planned to run below average ground level, sacrificing the economics of equal cut-and-fill construction to meet 'environmental' objections to obtrusive trains. This practice has occurred in Germany and local pressure is seeking it in England and Switzerland. In undulating country this results in some minor valleys and watercourses being crossed *below* their bed level, i.e. by cuttings. Hydraulic and aquatic problems, related to siphoning or diverting of watercourses, arise where drainage channels cross the railway at ground levels between 2 m *below* track level (below which a culvert can be provided beneath the track) and 8 m *above* track level (where an aqueduct can pass above the trains).

These problems seldom occur in traditional practice where railways are generally aligned above flood levels. Modification of surface drainage, for example at a long embankment, is then the main land resource issue.

Assessment of railway impacts on inland water resources can be based on:

- a record of the extent and nature of watercourse or wetland crossed and a description of the type and hydraulic characteristics of the crossing structure envisaged;
- a longitudinal section of the proposed line's vertical alignment showing the anticipated surface and ground water (phreatic) levels.

In populated country many watercourses are already subject to artificial constraints, whether from works to accommodate nineteenth-century railways or from more modern drainage or flood protection measures. Some railway embankments now form the sides of flood storage reservoirs; the largest dam in England also serves as an embankment for a motorway across the Pennines. Several early railway embankments double as sea walls or river guide banks.

The earth's surface is not entirely stable. Volcanoes and earthquakes still result from orogenic stresses and glaciers continue to modify high level cliffs and valleys. In Britain some rivers still meander naturally but the main active land-forms are river estuaries and coasts.

11.3.3 Estuaries and Coasts—Active Land-forms

Estuary sediments are subject to interactions between regular tidal and more variable river flows. Man-made embankments—whether to carry railways or to constrain river courses—make considerable impact on these interactions and on the geomorphological, ecological and scenic characteristics of estuarial wetlands and mudflats. Britain's estuaries are exceptional global resources; whilst many have been extensively altered by reclamation, flood control and transport works in the past, any future changes are intended to be strictly limited in accordance with sustainable land resource and coastal zone management strategies. The Department of the Environment (1992) has issued stringent guidelines to restrict developments in areas subject to flooding solely to those which cannot be located elsewhere.

Coastal land is periodically under very evident attack from the sea. From earlier times steps have been taken to protect natural and man-made structures from erosion and undermining by means of seawalls, breakwaters and groynes. More recent science throws this philosophy into doubt since it is realised that the effects of the sea are not felt simply as a direct perpendicular attack on the shore. Awareness of littoral drift whereby some beaches are starved of sand or sediment but accretion occurs elsewhere has led to the re-examination of the effectiveness of seawalls on their own. Broader strategies such as managed retreat are being devised in which coast lines may be sacrificed where their loss is inevitable. Any controls of erosion or accretion are managed for the optimum benefit of all land and marine resources (English Nature 1992).

Whether for railways or for other coastal protection, therefore, *sea defen-*

ces have to be planned as part of comprehensive coastal strategies. The *scenic* impacts of railways across estuaries or along coasts are discussed in Chapter 15.

11.4 OPEN SPACE AND 'UNSPOILT' COUNTRYSIDE

This heading covers a very broad concept of relatively undeveloped land and water, available for conservation or amenity now or in the future. Many of the land categories described in Chapters 12 to 15 could be involved.

The barrier created by a new transport route may impinge on the character of the space or wooded landscape which it crosses. Railways tend to avoid hill country, where the least developed space exists, by following valleys or tunnelling under high ground. But if a line has to be built across populated countryside between two cities it is most unlikely that *any* route can avoid all areas regarded as unspoilt.

It is therefore necessary to measure the extent of such country which is affected. Definitions of quiet or open country which could be applied include

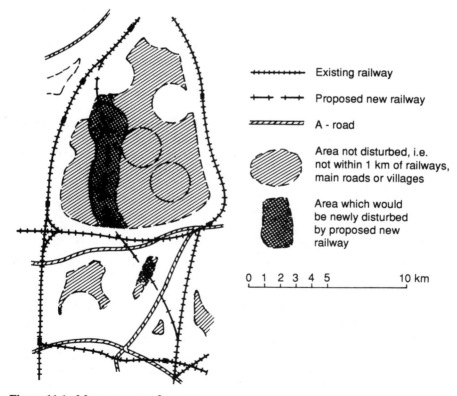

Figure 11.1. Measurement of remoteness

the following:

1. Areas designated for protection from development in county structure or independent plans for rural preservation; these areas could be classified in various categories from relatively featureless but fertile farm land up to designated areas of outstanding beauty.
2. Those areas which currently attain certain more uniform but arbitrary 'quietness' or 'remoteness' criteria, e.g. land areas
   - with no groups of over, say, 20 buildings or 5 hectares of private residential or industrial space, and
   - more than, say, 1 kilometre from an A- or M-classification road or a railway.

In either case the reduction in open space can be measured, compared with the reductions on alternative routes and assessed against the proportion of open space still remaining in the region. As an example one can measure the impact on the lightly inhabited country crossed by the 1982 East Coast Main Line Selby Diversion (Figure 11.1). In an area of $300 \, \text{km}^2$ in mainly rural land south of York, 33 per cent was originally quiet country, using the second definition. This was reduced to 28 per cent by construction of the 24-km-long new railway line.

Complications arise since 2 per cent of the quiet land was *replaced* as a result of abandonment of the original main line through land which was no longer disturbed by trains. There would also be further *loss* of quiet countryside to the mining activity which required the railway diversion in the first place.

Other resource impacts could be assessed in terms of river and stream crossings and areas of different land uses affected.

ROUTE SELECTION

11.5 HISTORIC DEVELOPMENT OF TRANSPORT ROUTES

This section describes the issues in transport route selection up to the end of the nineteenth century. One purpose of examining such distant history is to explain how space, topography and land use were perennial issues in planning the British railway network. That network forms the bulk of today's system. It was influenced by environmental issues of the day and determined much of the pattern of urbanisation and land use that took place thereafter.

The other purpose is to describe how the last main line, completed in 1899 and now mostly abandoned, was designed and constructed so well that, *except in the London area*, it could have been adequate for any of today's high speed trains.

Roads in flat country were laid out in straight lines by the Romans. They were often raised above ground level and surfaced with hard pavement materials. Bridges were built at river crossings. Because of their direct alignment and substantial construction, Roman roads are often still in use today.

Road-making was virtually discontinued for over 1000 years after the Romans left. Maintenance was regarded as a charitable activity in early medieval times, then almost totally neglected. Improvements started from about 1550 with allocation of responsibilities to parishes and, from the late seventeenth century, by improved 'turnpike' roads financed by tolls. During the long period of neglect there were nevertheless certain developments which affected subsequent road alignments:

- Even after Roman times, trunk roads continued to serve as highways for armies; unfed and often hostile soldiers preyed on the people unfortunate enough to live nearby; through roads were therefore unwelcome near most settlements and their routes tended to avoid all except the cities which were already connected.
- To make conditions more difficult for highwaymen, the 1285 Statute of Westminster prescribed 'that highways leading from one market town to another shall be enlarged . . . so that there be neither dyke, tree nor bush, whereby a man may lurk to do hurt, within 200 ft of the one side and 200 ft of the other side of the way' (Chambers's Encyclopedia 1955).

The first of these developments determined the alignment of early bypasses, which in more peaceful times then provided focuses for new villages. The second provided a 120-m-wide transport right-of-way; with sufficient foresight this could have provided space for at least some lengths of modern transport systems. Occasionally the space was indeed used for road widening; more often field enclosure or building development took advantage of eventual relaxations in the Statute. In some rural areas broad strips of that 120 m reservation have survived, protected by 'common land' or 'village green' status.

In the first half of the eighteenth century *new* military road building was undertaken by General Wade in the highlands of Scotland. These roads were in sparsely inhabited, mountain country; they followed valleys and passes, some of which were subsequently used by modern roads or the Highland Railway.

By the beginning of the nineteenth century road improvements were in hand using techniques developed by Macadam and Telford. Generally these made use of existing road alignments some of which were by now more divergent from the Roman straight line model or the graded routes of the mountain military roads. Eccentricities such as abrupt right angled bends had resulted where earlier byways and some highways had been diverted around the buildings and estates of influential landowners.

Early railways used human or animal traction. Distances were comparatively short, for instance from mine entrances to riverside wharves. Owners often controlled the intervening land so there were few property disputes. Gradients had to be modest for horse haulage; but the advent of water power and then stationary steam engines enabled steep cabled-haul inclines to be incorporated.

Narrow water channels had been constructed as headraces for mills, leading from stream offtakes at gentle gradients along valley sides. But *navigation canals*, mostly built between 1760 and 1830, had to be comparatively wide (generally about 15 m including the towing-path) and absolutely level between locks. In broad valleys canals ran within a short distance of the main rivers. In undulating country canal routes could shorten river meanders and follow a slightly steeper route by the use of locks. At the head of steep valleys canals pierced the watershed by tunnels or, by following more circuitous routes, crossed intermediate valleys on aqueducts.

Conventional railways, for locomotive-hauled trains, replaced road coaches for passenger traffic; rail freight services displaced those on rivers and canals as well as some coastal shipping. Because of the gradient limitations, the alignment of railways was not dissimilar to that of canals. In flat or gently undulating terrain, many canal routes had been selected along alignments subsequently suited also to railways. Indeed railway entrepreneurs bought out the canal companies to kill the competition. In doing so they acquired the land which could then be used as a corridor both for delivery of construction materials and for the track itself.

Thus the Great Western Railway shares a corridor with the Kennet and Avon Canal and the River Kennet itself for 40 km from Reading to Hungerford (Figure 11.2). This railway then follows the canal for a further 20 km up a tributary valley to Pewsey. The London and Birmingham Railway ran near the Grand Union Canal for much of the 40 km route from Watford Tunnel to Bletchley; the Great Western Birmingham line follows the Oxford Canal route and the Midland Railway the Trent and Mersey Canal for similar distances.

Railway construction must also have profited from canal experience in the siting and design of structures and tunnels in high country. Telford's 1805 Pontcysyllte aqueduct, by which the Llangollen Canal crosses 120 ft (37 m) above the River Dee, set a magnificent precedent for many fine stone or brick railway viaducts which followed. Long railway tunnels under the Pennines and through other uplands were adjacent or close to the existing canal tunnels.

The Cromford and High Peak Railway in Derbyshire was even built to link two existing *canal* systems. Much of the railway crossed a plateau with a number of intervening inclines too steep for canal locks to be practicable. Two sections were so steep that cable haulage had to be used instead of locomotives; that above the Derwent valley at Cromford was still in use in

Figure 11.2. The Great Western Railway was aligned close to the Kennet and Avon Canal

the 1950s. Elsewhere in Britain acceptable gradients were usually steep enough to obviate any need for abrupt changes in level, although cable haulage was provided initially for gradients out of certain city terminals—Liverpool (Crown Street), London (Euston) and Glasgow (Queen Street). Abroad, switch-back reversals were adopted to negotiate particularly steep mountainsides; later more satisfactory but expensive solutions for gaining height were achieved by spiral tunnelling.

In selecting longer distance routes across rural Britain, opportunities for level or moderately graded track were first identified. There then remained a choice among a number of options with different environmental consequences. Except at the Euston incline, Robert Stephenson used no gradient steeper than 1 in 330 for the London and Birmingham Railway. Brunel's Great Western Railway was able to incorporate both easy gradients and long curves where it ran up the Thames valley, but resorted much more to embankments, cuttings, tunnels and steeper gradients in the hillier country further west.

However, in the English countryside property rights were obstacles often as powerful as topography in the way of ideal railway alignments. Some large landowners sold their land readily to railway promoters if the traffic generated was likely to be to their commercial advantage. Others insisted on substantial realignments around their estates; 'Lord Harborough's curve' at Saxby in Leicestershire is an example of an otherwise unnecessary diversion in open country (Turnock 1990: 33). Watford Tunnel was created only to avoid a line across the *outskirts* of the estates of the Earls of Essex and Clarendon (LMS 1947: 3). Eton College objected to Brunel routeing the Great Western Railway close to its premises, so the main line ran through Slough instead. However, when Queen Victoria needed access to nearby Windsor, a branch line was built, proudly on continous viaduct and leading to a grand terminus. The line passed close to Eton but provided no station in that town.

Elsewhere tunnelling beneath property was considered as a means of overcoming objections but could rarely be justified. An excusable exception was the shallow tunnel provided to preserve the pastoral calm of historic Haddon Hall in Derbyshire. The Midland Railway provided ample visual compensation to its passengers on other sections of this scenic route. The railway has closed but the tranquility of the Hall is now mildly disturbed by traffic on the adjacent A6 road.

Most railways in Britain had been constructed by 1880. The alignments of the few new lines that were constructed at the turn of the century were able to take advantage of the operational lessons gained from the established steam railway system. Some relaxation in gradient restrictions was permissible for the powerful and well-proven locomotives by then in use. Relatively simple missing links could therefore be made for the Great Western Railway from Swindon to the Severn Tunnel (via Badminton), from Princes Risborough to Banbury (via Bicester) and from Westbury to Taunton (via Castle Cary).

More ambitious was the West Highland line from Glasgow to Fort William (1984) and thence to Mallaig (1901) and the Highland Railway extension from Strome Ferry to Kyle of Lochalsh. These lines, offering fast ferry connections to the Hebridean ferry ports, are commonly recognised as classic scenic routes.

The Great Central Railway's London extension, completed in 1899, was conceived as a fast rival to the existing lines from Manchester and South Yorkshire to the capital. Partly this optimistic idea sprang from the strong interests of the promoter, Sir Edward Watkin, in the South Eastern Railway and in the Channel Tunnel which, in the event, was only implemented 100 years later. The error in investing in the Great Central, which never paid a dividend, lay in assuming that a new line could rival the existing competition and indeed that traffic demand would continue. Be that as it may, an opportunity was taken to plan and construct a fast and easily graded line, mainly along a completely new route, from Nottingham to London.

Environmentally, construction of the Great Central Railway throws interesting light on the relatively mild effect of a railway on the countryside and towns of central England; this is in spite of the amount of earthwork and structures needed for a high-speed alignment and perhaps in contrast with the effect of motorways and related development half a century later. The requirement for competitive operation dictated that as far as possible the line should be on an independent alignment and that new, exclusively Great Central, stations should be built at all the towns served. Thus new stations competed with those of the Midland Railway (at Nottingham, Loughborough and Leicester) and the London and North Western (at Rugby). But at Aylesbury a station was shared with the Metropolitan Railway. For 147 km from Annesley, north of Nottingham, to Quainton Road, near Aylesbury, the new route was selected without great difficulty to run through some of the emptier areas of England. These areas were still rural because they had remained at some distance from the corridor of the Midland and London and North Western main lines built 50 or more years earlier.

Access for the Great Central Railway to stations in the centres of towns had to be achieved by tunnelling (as at Nottingham) or embankments and viaducts (as at Leicester) to negotiate the urban areas which had developed after the first railway boom. At the London end, the obstacles to any totally new route were formidable. As it happened, Sir Edward Watkin had strong interests also in the Metropolitan Railway. This suggested the solution of sharing the Metropolitan route into London with a separate new terminus at Marylebone. The latter was located on the boundary line, from Paddington to Liverpool Street, south of which no main line railways had been permitted since the limit was designated in 1846.

Today much of the Great Central Railway route north of Aylesbury is abandoned. Meanwhile any remaining buildings and numerous fine brick viaducts provide evidence of what was a well-engineered and environmentally sympathetic route through rural England. It is understood that the line was once considered for reinstatement as a possible new high speed route to the North; but further investment in the West Coast Main Line was the option subsequently adopted.

The main obstacle to resurrection of the Great Central route was evidently the perennial difficulty of providing a satisfactory route into or through London. That city's transport problem at the end of the twentieth century is to provide routes for cross-country traffic direct from the northern and western main lines towards the English Channel and Europe. The existing Thameslink and West London 'Extension' lines and the planned Crossrail and Union Railway are no doubt part of the solution although these are intended mainly for passenger traffic.

11.6 ROUTE SELECTION FOR NEW RAILWAYS

11.6.1 Initial Broad-band Route Selection

A new railway is required to meet a specific need or to implement a long-term regional transport plan. A route is selected primarily to join two or more points as economically as possible, i.e. the most direct practicable alignment.

There are three other targets in route selection—to provide the required operating capacity and speed, to minimise construction cost and to protect the environment.

Line operating capacity is determined by the number of tracks and the way in which different types of trains are to be scheduled. Train speed is constrained by severity of gradients and curves, both influenced by topography. Performance standards dictate horizontal and vertical alignment criteria; if these criteria cannot be achieved within cost and environmental constraints, then the standards must be lowered.

Major items of construction cost which influence route selection are tunnels, bridges, deep cuttings and high embankments. A cheap alignment following the ground contours means steeper gradients and sharper curves. Construction costs are estimated by engineering calculation and minimised by comparison of different options for suiting the topography, whilst allowing acceptable operating charateristics.

Environmental objectives in initial routeing are:

1. To avoid valuable features of heritage, habitat or landscape.
2. To avoid centres of population not served by the line.
3. To take account of regional development and land use plans.
4. As far as accords with points 1 to 3, to make use of existing transport corridors (roads, railways, canals and lineside development) or to share new corridors with new roads.

There is a fundamental area of contention between cost minimising and avoiding features of environmental value. This is because the market price of the latter is usually very low. Equally contentious will be any application of shadow prices to enhance the value of environmental resources. However, the possible solutions are examined in Chapter 16 where it is proposed that, where adjusted values can be applied, this should be done first in initial route selection.

Preliminary engineering, costing and environmental planning processes lead to selection of one or more broad route bands. This is usually the first stage in route selection; it precedes detailed alignment which tries to avoid particular obstacles whilst meeting operational requirements.

An essential component of an environmental assessment is a quantitative description of the proposed route, its features and impacts. This is as necessary in proposing an initial corridor as in assessing a final detailed alignment.

British Rail (BR)'s first announcement for the Channel Tunnel Rail Link in July 1988 disclosed that four alternative route bands across Kent were being examined. A small scale map (BR Board 1988) showed these approximate routes together with urban areas and bands of land 'preferred' or 'restricted' on environmental grounds. The operational and environmental issues for each route band were assessed and compared before announcing, in March 1989, the selection of an amalgam of these routes.

A full alignment study, involving engineering and environmental studies and public consultation, was then undertaken along the chosen band. After some adjustments to the tunnel sections and other features of the alignment, the section outside London was fixed and a 240-m-wide strip of land was reserved. This was intended to preclude the possibility of any land outside that strip being affected. Subsequently, however, construction of the rail link was postponed and the Government decreed that completely new routes should be examined. This was a setback for transport and development planning in the region, although it temporarily removed some matters of contention.

However, the planning experience gained was not entirely lost; the eastern end of the route remained similar in many respects and some aspects of the new route into London reflected common issues that had already been aired.

11.6.2 Alignment

Broad-band route selection will have determined the type of engineering features which will be involved and will have confirmed what operational performance can be achieved and what environmental constraints must be accommodated. These will be the paramount criteria in designing the horizontal and vertical alignment—the second and more precise stage of route selection.

Horizontal alignment can avoid sensitive obstacles but high speed train operations preclude tight curvature. Adjustments can be attempted, for instance to miss historic structures or scarce types of habitat, or to run closely parallel to an existing railway or motorway; but where these aims cannot be achieved within the alignment criteria, then either environmental sacrifices must be made or the design speed of the line must be reduced.

Vertical alignment depends upon topography and permissible gradients. With powerful modern locomotives, steep terrain is less of an obstacle than it was 150 years ago. Short steep gradients can be accommodated to the extent that, in normal running, trains can maintain speed to the top; or, in the event of signal or emergency stops, gradients can nevertheless be negotiated, albeit slowly, from a standing start.

Entry routes into cities continue to provide major alignment problems. Large, expensive buildings are serious obstacles to ground level or elevated railway development in cities, whilst deep structural foundations, underground services and existing metro railways complicate bored tunnel alternatives. Opportunities sometimes arise to make use of older but redundant railway viaducts, cuttings or shallow tunnels where their alignment suits new traffic. Examples are the TGV *Atlantique* route into Paris Montparnasse or an abandoned road route now providing metro tracks across the Tyne at

Newcastle. Underutilised routes sometimes provide similar opportunities; the eastern end of the North London Railway was considered for the route of Union Railways' Channel Tunnel trains into St Pancras; although in this case one reason why the option was not adopted was the difficulty in maintaining an essential freight link during new construction.

11.6.3 Use of Existing Transport Corridors

One of the ways of minimising new impacts on unspoilt countryside is to use space within or adjacent to existing transport corridors. This should be a priority consideration in initial broad-band route selection. Close parallel alignment of trunk roads and railways makes efficient use of land space (see Figure 11.3). However, practical difficulties arise where there are obstacles to widening the basic corridor, for instance due to lineside or roadside buildings, and in accommodating different curves or gradients to those of existing railway tracks or roads.

High speed rail alignment is less flexible than that of trunk roads. As a consequence areas of land may be trapped between road and railway. The Cologne to Frankfurt *Neubaustrecke* route has adopted short gradients as steep as 1 in 25 in order to run parallel to motorways. Even so, German planning has found it generally impracticable to align new lines directly beside *existing* autobahns. In France new TGV lines have been constructed

Figure 11.3. A transport corridor. The M1 motorway runs adjacent to the former Midland Railway main line near Mill Hill, north of London

successfully along the corridors of planned *future* autoroutes. The TGV-*Nord* line has also been twinned for 130 km of its length with the existing Al autoroute, although for less than half this length can the two routes run precisely parallel. Elsewhere irregular pieces of land are cut off between road and railway, some of them more than 50 m wide. Such pockets of land are seldom of much commercial use; but they *may* have conservation value.

It may transpire that *maglev* vehicles can operate at high speeds even on the relatively tight curvatures of road routes. In the USA the possibility has been mooted of a very high speed *maglev* line using an elevated guideway above the New York to Albany highway. This would avoid the necessity for very high expenditure in land acquisition—one reason why the concept of elevated suburban railways is being resurrected.

In Britain the long process of planning the Channel Tunnel Rail Link (CTRL) was always firmly directed at sharing transport corridors where possible. One of the original route options shared a corridor with existing BR lines over nearly all its length but was discarded on grounds of a long journey time from London to the Tunnel. Another made much *less* use of existing routes; it was eventually rejected on environmental grounds and because it was not compatible with Kent County Council's requirements for *the use of existing transport corridors*.

The CTRL route announced by Union Railways in 1993 is aligned parallel or near to the M20 motorway for considerable lengths west of Ashford and near to the M2 south of Gravesend; it runs beside existing railway lines both east of Ashford and by the London Tilbury and Southend line north of the Thames between Purfleet and Barking. Indeed, apart from the tunnels, there are few sections on the surface where the route is not located in some form of existing transport corridor.

Where very high speed trains are likely to share new routes with commuter, semi-fast or freight traffic, consideration must be given to provision of four, rather than two, tracks. This could be necessary to provide sufficient capacity at peak periods for trains travelling at different speeds. It is difficult and expensive to widen a transport route subsequently where it was not designed for its ultimate capacity.

When lateral constraints are tight, the width of two-track railway formation can be restricted to that of permanent way, ballast, lineside accessories and drains—say 10 to 12 metres for two tracks—by constructing bridges or retaining walls rather than embankments or cuttings.

Use of valuable land resources is closely related to integrated transport and land use planning, especially in the development or redevelopment of shared transport corridors. Land in such corridors has to accommodate any main line railways, local rapid transit, through trunk roads and motorways, local distribution roads, housing, industrial and commercial development, and whatever amenity and conservation areas are suited to the land and water features and to the built infrastructure.

11.6.4 Flat Country

The highest speeds and least cost of construction and operation are associated with the flattest, emptiest countryside. Such spacious areas are now scarce in lowland Britain. An exception was the 24 km stretch of country crossed by the East Coast Main Line Selby Coalfield diversion, already referred to. Only one village was visible from the line and no serious environmental obstacles were posed.

Practical considerations dictate that the track formation should be a metre or two above level ground to allow for cross-drainage under the route. On soft foundations imported material may have to be used in low embankments to provide adequate bearing capacity. In flat country, local drainage patterns may be ill-defined and easily distorted by new earthworks. Ample culverts need to be provided, in addition to bridges over the larger watercourses. However, drainage should not be 'improved' to such an extent as to damage natural wetland habitat. Where wetland cannot be avoided, carefully sited low viaducts interfere with land or water habitat much less than solid embankments.

In Germany, where recent railway planning tended towards the 'hide it at all costs' approach, one section of the new Mannheim–Stuttgart line has been built in a cut-and-cover trench below the water table. The expensive works involved have been offered as a solution in crossing waterlogged ground. However, one may wonder whether the long-term effects on the excavated surface and intercepted ground water may not be more significant than any visible disturbance from a surface railway.

11.6.5 Hill Country

We have seen in Chapter 4 that modern motive power can haul passenger trains at speed up long gradients of up to 1 in 75 or for short distances on slopes as steep as 1 in 25.

Freight trains cannot normally tackle the latter. Nor is the cant of the track on curves for heavy, slower freight trains the same as that designed for safety and comfort in high speed passenger trains. So the vertical and horizontal alignment parameters for fast passenger lines is not ideal for freight use.

Whilst modern traction has the power to haul trains fast up or down long-established hill routes, it may be limited in doing so by the curvature of track designed for more moderate speeds achieved in the past. In planning high speed routes in mountainous territory in Europe, 'it may well be more economical to build a completely new railway than to try to realign and upgrade the old line' (Green 1991: 146–147). It is difficult to dispute this general conclusion for applications in particularly steep country and where the old lines, built at only moderate capital cost and subject to speed restrictions, followed contours closely. However, before accepting the necessity for a new route over a long distance, consideration should be given to the environmental impacts and land-take of new routes, and to establishing the practicability of retaining certain sections of the old route, especially where only modest speed restrictions would avoid disturbance to a particularly sensitive environment.

In hilly or undulating country, the direction in which a railway line crosses or runs along the macro- or micro-topography influences its impact on the landscape. If the line clashes seriously with the topography or the vegetation, hard or soft landscaping measures—earthwork or planting—can be introduced, as described in Chapter 15.

11.6.6 Access and Infrastructure

Where a railway route severs existing highways, any necessary alterations to the latter should be examined as an opportunity for planned improvement rather than as an obstruction. The same applies to realignment of other railway lines, footpaths, power lines or underground cables. In each case planners should ask the following questions:

- What is the most appropriate solution for *all* interests?
- Can new access be provided (e.g. a new road or bridge)?
- Can the new access be better than the old in meeting current requirements?
- How much will the new access cost?
- What will be the increased or reduced value of properties as a result of the revised access?
- If replacement access cannot be provided, what effect does this have on the value or utility of land cut off?

Level crossings are an economical and environmentally satisfactory way of allowing footpaths or minor roads to cross railways. For operational and safety reasons, they are often not permitted where traffic is heavy or speeds in excess of 160 km/h are required. However, BR (Prideaux 1991) has found that properly protected and observed full barriers can allow higher speeds across level crossings.

Public utilities are buried, like gas pipelines, or carried above the surface, like power cables. Subsurface lines may be buried under embankments, diverted along cuttings or occasionally taken over them. For minimum land disturbance, diversions and crossings should be shared by roads and utilities wherever practicable.

11.7 ROUTE IMPROVEMENT FOR EXISTING RAILWAYS

New alignment or widening of existing lines may be required for operational reasons:

- to straighten them and permit higher speed;
- to provide additional tracks to increase capacity;
- for connections between old and new routes.

High speed alignments may involve extensive earthworks or even new tunnel bores to accommodate larger radius curves; where they divert from existing transport corridors, these works have the same environmental implications as have been discussed for completely new railways.

One feature of higher speed railways is the need for increased safety precautions; this includes elimination of major level crossings and all sections of 'tramway' shared with road traffic. Another consideration in upgrading a line is whether the clearances need to be increased—to accommodate electrification equipment or a larger loading gauge. The viability of making these improvements may depend largely on the practicability, cost and environmental effects of raising bridges on the unaltered sections of the line; or the limited size of a tunnel may be an insuperable obstacle to continuing use of a section of line.

Other reasons for realignment could be new high priority development on the land concerned or because of civil engineering or geotechnical problems threatening track stability. The Selby diversion was needed because of the risk of subsidence under the old line when a large new coalfield was brought into use. A railway in Africa needed relocation because of the high occurrence of land slips blocking the route in certain sections. Many former colonial railways, built at minimum cost by following surface contours, can easily be realigned using modern earthmoving equipment. However, the same environmental constraints about using up land resources apply and in tropical conditions the problems of erosion caused by topsoil stripping and excavation can be serious.

Provision of additional tracks is efficient in terms of land use and involves less earthwork than a completely new route. In undulating country with cuttings and embankments the average width of railway land between fences will be at least 20 metres. Adding tracks adjacent to an existing line will therefore be economical in land-take. Whatever the height of bank or cutting, each extra track still adds only 3.5 m to the land requirement; even part of this land-take can be avoided where the expense of retaining walls can be justified instead of extending sloping earthworks.

Connections between old and new routes are difficult to align where they occur outside railway land. Especially at junctions, the problems of 'trapped' land are involved. On the other hand, if the old route is to be abandoned, this provides opportunity for the old track land to be reinstated for amenity, conservation or some utilitarian purpose.

One of the key issues in the operational planning and indeed financing of the Channel Tunnel Rail Link has been the connections which should be made between the new systems and the old. These can allow local traffic from Kent or Essex to use the new fast lines into London. The options have not been resolved at the time of writing. However, it is interesting to note that one possibility mentioned (Union Railways 1993: 99, D.5.2) is a connection that might use the 'Gravesend West' line, a long-abandoned branch railway. The necessary track widening and reinstatement would involve clearance of at least one side of what has become a linear nature conservation area.

12 Residential, Commercial and Productive Land

12.1 IMPACTS ON CURRENT LAND USE

Land acquisition for any type of new development causes:

- loss of the land covered and of the utility of the residential or commercial facilities on it;
- a reduction in the total land space occupied by such facilities in the region.

Land covered by a railway is permanently lost to its original owners and function. So are any buildings that had been constructed on that land. Direct monetary compensation and an obligation to move out must be the outcome.

In the case of a long continuous structure like a railway the utility of adjacent land may also be reduced, at least temporarily. The line may create a barrier between farm buildings and outlying fields or between houses and shops, schools or social centres.

There are degrees to which severance of access within areas like farms or golf courses affects the utility of land. At one extreme it may be worth while to provide a bridge, at the railway's cost, so that the isolated fields can be reached as easily as before. At the other, pockets of land may become so isolated, for instance between a new railway and an existing road, that further working is not economically practicable; the land can be purchased and turned over to railway use or to wilderness.

For each type of property affected, environmental assessment has:

- to determine the nature of the land taken and classify it as to its actual or potential utility;
- to quantify the direct local land-take and numbers of buildings lost;
- to identify any severance or similar effects on the utility of adjacent land or buildings;
- to recognise any measures whereby wings of buildings or small parcels of land can be saved or the integrity of groups of buildings or small communities can be preserved;
- to examine the impact on total available resources of each type of land.

The nature of losses incurred relates to the function of the land taken. These functions are:

1. Housing—residential property from high density apartment blocks to isolated buildings in spacious grounds.
2. Commerce and industry—offices, shops, factories, foundries, mines or quarries.
3. Agriculture including horticulture, commercial forestry and associated agro-industries.
4. Access roads, footpaths and public utilities which are integral with all these categories.
5. Educational, cultural and recreational buildings, playing fields or parks associated with housing and commercial areas.
6. Conservation and heritage land or watercourses.

Items 1, 2 and 3 are marketable property for which a price of acquisition or compensation can be estimated. Except where elements of category 6 are involved, any resource value to posterity is not likely to exceed that market price. Category 4 items are not always saleable; but the construction cost of any replacement can be estimated.

Buildings in category 5 can be valued but the market value of recreational or amenity land is not fairly representative of its ultimate social or environmental value. Nor is that of category 6; the assessment of impacts on these types of land resources is discussed in subsequent chapters.

12.2 RESIDENTIAL PROPERTY

12.2.1 Impacts on Property

The Stockton and Darlington Railway, opened in 1825, was the first public railway for which people's land had to be acquired. The landowners objected and were circumvented, placated or overcome by parliamentary process (Westwood 1977: 9). *This is what still happens.*

Routeing a new railway through residential property causes:

- an obligation on the railway developer to pay compensation for the land required—this is a significant financial planning factor;
- a reduction in housing stocks and perhaps a need for new residential land resources—this is an environmental impact;
- concern and disruption to individual property owners.

Railway development costs can be minimised by seeking to avoid all residential areas. This means that:

- cheaper land will be sought first—this may threaten land of high environmental value unless other constraints against damaging such land are enforced;

- where some buildings obstructing the route are of special heritage value, these may have to be specially avoided, protected or moved;
- in passing through city areas of very high land prices and expensive buildings, tunnelling may be cheaper than demolition.

Initial route planning for new railways is undertaken along broad bands which avoid as far as possible the main villages. More precise alignment then attempts to avoid specific housing clusters and isolated, particularly historic, buildings.

The long radius of curves demanded by high speed trains reduces the possibility of avoiding every obstacle even in rural areas. A number of houses may remain, particularly in roadside rows, which are still directly in the way. Besides having to suit operational requirements, equitable alignment faces competing environmental and property objections.

As a mitigating measure, retaining walls can be built to preserve particularly valuable land features or buildings which would otherwise be destroyed by cutting or embankment slopes.

The *reduction in housing stock* is the number of housing units to be lost; this is a readily understood and therefore politically significant statistic.

Substantial housing loss is possible where a railway has to be widened to accommodate extra tracks in urban or suburban areas and where properties are close to the original boundary line. In redressing the balance elsewhere, some land adjacent to railways may be well suited to new housing development, for example:

- in space available at obsolete sidings or marshalling yards;
- along existing or new lines.

The last does not imply a resumption of closely-packed ribbon development but spacious housing in country where planning policies allow new building *only* near to transport infrastructure. The Antwerp to Rotterdam railway, south of the Dutch frontier, passes high quality housing parallel to the railway, leaving unspoilt the large areas of open country not near any road, railway or settlements.

Individual property owners are concerned first about the actual location of the route. Are *their* properties to be lost or will their environment suffer from close proximity? Any environmental statement must include a map of the route, indicating the properties to be demolished and those within a certain distance of the project boundary. Owner's other concerns relate to the compensation they will get for property affected and to blight and uncertainty.

12.2.2 Compensation—Property Lost

Property has to be acquired where it obstructs the route. Compensation to the owner is subject to compulsory purchase regulations unless some prior direct arrangement has been made.

Beyond the legal or market minimum value, a more generous compensation formula can be fixed if it is judged necessary. Indeed this may be needed to create a situation in which the project itself receives any statutory local approval; or it may strengthen the, probably fragile, image of the railway promoter for subsequent planning applications or land acquisition.

A typical compensation formula might be composed of (a) the current market price of a similar property in a similar (pre-railway) situation, *plus* (b) a fee representing the real costs of acquiring replacement property and as disturbance compensation.

Argument is inevitable over the actual level of these elements. But, for (a), prices of all but exceptional or isolated properties are available through fair examination of published data, giving careful attention to the date of transactions in rising or falling markets. The level of any disturbance fee (b) may be more contentious. The UK Department of Transport normally pays only for (a), as assessed in England and Wales by the District Valuer. But London Docklands Development Corporation has added payments of type (b) in order to speed the public planning process.

Since 1991 compensation of up to 10 per cent above the market value *may* be payable, according to circumstances, for house purchase for road and rail links.

12.2.3 Compensation—Property Affected

If it can be shown that there are impacts, other than land-take, which will reduce the value of *off-line* property—such as noise of trains—then there can be recourse to statutory compensation (in the UK under the Land Compensation Act 1973). In practice, for Department of Transport projects, the only compensation usually given to occupiers who do not lose their property is provision of insulation where certain noise limits are exceeded.

Nor is any compensation usually given for severance within communities beyond ensuring some sort of access to every property. However, in environmental assessment some quantified assessment should be made of the impact of severed access of routes from houses to shops, schools, etc. The numbers of people and journeys affected and the distances concerned are suitable units for measurement.

12.2.4 Blight

In the context of transport development, 'blight' is adopted as a term for adverse effects on nearby properties. Effects on historic buildings, for instance, can be physical deterioration arising from vibration or, less directly, from severed access or spoilt environment; the latter can lead to reduced numbers of visitors, less income, attention and maintenance, and perhaps vandalism. More commonly blight is synonymous with a reduction in house prices; this results from more sellers than buyers coming forward once any suspicion of environmental change arises.

Because main line railway development is new to modern society, facts about related housing blight and popular concern are only now being revived. However, the following observations can be made:

1. People who actually live near railways are not commonly disturbed by trains. Such people are not representative of those who feel threatened by new lines but their experience is relevant.
2. The external impacts of new *roads*, mainly noise and air pollution, are well known; similar data is available for railways. But there is also general awareness of the usefulness of roads, even to people who have to live near them. Motorists like fast through roads and town or village centre environments benefit from bypasses. Railways seldom provide such benefits.
3. Where there is a station, there will be public transport benefits and other impacts on the community. Local house prices may rise as more people are attracted to the area.
4. Early motorway schemes encountered strong objections. Today people displaced or closely affected by new road works continue to object. But their more fortunate neighbours will opt for the driver and bypass benefits and will not support the objectors' indignation. However, where a railway passes a village without any station or travel benefit, many inhabitants may oppose the scheme. Few will actively support it.
5. As a result, the prospect of a new railway in the neighbourhood generates vociferous and widely publicised indignation. The paraphernalia of protest bedecking a village—an attempt to prevent the line being built—will undoubtedly cause prospective house buyers to avoid the area.

Therefore opposition to a projected railway, expressing local fear of its environmental impact, can affect the housing market and reduce prices.

Owners of houses near the Channel Tunnel Terminal at Cheriton or close to the 1989 route of the Channel Tunnel Rail Link were given opportunity, by Eurotunnel or British Rail respectively, to sell their properties if they felt that they were blighted by the proposed developments. At Cheriton, Eurotunnel adopted resale strategies intended to preserve the communities. However the compulsory purchase of the properties which had to be demolished may have resulted in apprehension and antagonism in Kent against the Tunnel and against the related new rail link in subsequent transactions.

On the various rail link route options, market opinion was that blight on home sales would remain until a final decision on a route was made. However, even at a time when the housing market was entering a slump, demand for houses in Ashford was claimed to be good. This was attributed to the chances that the rail link *would* be built and that the town would benefit.

12.2.5 Uncertainty

Undetermined routes are the main source of uncertainty. Needs for public consultation and for advance publicity cause preliminary information about any road or rail scheme to be widely circulated. But the final route cannot be certain until all the interests and issues have been resolved. Meanwhile a question mark hovers over people's plans, particularly to move into the area; naturally this affects house values.

On the Channel Tunnel Rail Link, British Rail recognised this problem in 1989 and defined a broad corridor, 120 metres either side of an initial centre-line. This corridor was mapped and made public with an offer to buy any property which lay within it. Many home owners took up the offer which was to remain open until either the railway was complete or, as actually occurred, a decision not to proceed with construction on most of that corridor was taken.

An understanding of the uncertainties and blight which complicate property acquisition for railway development is useful planning background. However, these very uncertainties preclude blight as such being taken into account in environmental assessment; instead the effects of noise and other forms of specific disturbance with which blight is associated are directly assessed.

12.3 OFFICES AND INDUSTRIAL PROPERTY

Because commercial development tends to take place in dense urban areas or specially planned out-of-town complexes it is seldom an obstacle to new inter-city transport routes.

Any offices or factories that do have to be demolished are a clear loss to their owners who will have to be compensated in the same way as householders. Industry and commerce are economic activities. The costs of disruption may be measured either in financial terms, such as loss of profitability to owners or the cost of replacement, or in economic terms, such as a reduction in regional production. Where the industry disrupted is a polluting or a cleansing one, then environmental gain or loss has to be evaluated.

Environmental assessment of railway routes through industrial land should comprise description of the extent and type of industrial or commercial activity affected, indication of the options available for continued production and recognition of any land use revisions.

Whilst loss of land for a new railway might cause a reduction in industrial activity, this could sometimes lead to an improvement in the existing environment. There may be a need or opportunity for new planning in the industrial land remaining; or revived or alternative industries may benefit from new rail sidings.

Transport and economic changes have led to manufacturing decline in city centres and to new types of industrial or service activity elsewhere. Land space was left where railway yards, sheds or workshops and docks or shipyards were no longer needed. Some of this space is still available, if not for new transport infrastructure then for industry, warehouses, offices, housing, recreation or wildlife habitat. Railway planning requirements should be determined before other development takes up any remaining space on land accessible by rail.

Where new railway construction takes place on derelict industrial sites there may be danger in disturbing contaminated ground. Environmental risks must be assessed as well as the means of rendering the ground safe. Technology is available for investigating and dealing with contaminated land. Chemically polluted land can be treated, hazardous materials like asbestos located and removed, and methane gas released. If the costs of such measures are excessive, total avoidance by realigning the line may have to be considered.

Certain industrial activities, such as quarrying and gravel extraction, create scenery like cliffs or water surfaces which many find attractive. Disused quarries may provide suitable entry points for tunnels into hillsides. Similarly, where new railway routes are to cross areas suitable for gravel extraction—a subject of common local contention—there is an important opportunity for integrated and environmentally sensitive land use planning.

12.4 AGRICULTURE AND FORESTRY

Where a new railway crosses developed countryside, land to be acquired is predominantly agricultural. In England this could be everywhere except near towns, steep escarpments or on northern and western mountains and moors.

Particularly in the South-East, rural areas can be a patchwork of arable land interspersed by villages, pasture, woodland and public open space. This section is concerned with fields and buildings used permanently for crops or animal husbandry—the commercial elements of farms—rather than with their less productive fringes more suitable as ecological habitat.

The area of agricultural land that is affected by railway development can be measured in hectares. Its quality can be gauged by its potential productivity in terms of topography, drainage characteristics and soil chemistry. In Britain such assessments are already available on maps showing the Ministry of Agriculture, Fisheries and Food (MAFF)'s Agricultural Land Classification (ALC). Affected land areas of each class should be reported in Environmental Statements:

1. These should differentiate between land which is taken out of use entirely and that subject to various degrees of severance—from areas which can be reached by new 'accommodation' bridges or underpasses to those which become totally isolated and can no longer be commercially viable.
2. They should identify the number and size of farm holdings which are affected, including any whose size will be rendered unviable; indicate any scope for land rationalisation and impacts on farm economics and employment and on farm-based non-agricultural activity.
3. They should indicate the dominant types of agriculture in the affected holdings and in the region as a whole; note that broader categories of farming land than ALC classifications, such as Areas of Special Significance for Agriculture, may be designated locally to reflect regional economic and environmental goals; these regional planning factors should be noted as should any local drainage or micro-climatic changes which rail construction may cause.
4. They should identify land which will be required temporarily for construction operations, any interruption to farming activity which will be caused, any hazards to high value crops (e.g. dust creation during earthmoving), temporary access and service arrangements needed, and the requirements for subsequent land reinstatement.

ALC classifications are a ready measure of the comparative *long-term* value of land for crop production; as such these classifications are well suited to *environmental* assessment of agricultural resources. In monetary valuations, for instance for compensation assessments, the value of the land, buildings and equipment is related more directly to current productivity and current market demands.

ALC land categories vary from Class 1, which is excellent productive land, to Class 5, which is unattractive for almost all forms of agriculture. Very roughly, the worse the agricultural rating of land the better is its value for conservation or amenity. 'Many areas of semi-natural vegetation of high conservation interest, such as floristically-rich grasslands and heathlands, are associated with infertile soils' (Marrs and Gough 1989: 30). 'Improvements' to land by application of chemicals or drainage enable the ALC rating to be raised where in its 'unimproved' semi-natural state it was more suitable for flora and fauna. Exceptions to the trend occur where good quality land is still divided by long-established hedgerows, interspersed by woodland or drained by sensitively managed streams.

The better class agricultural land, for instance in alluvial valley bottoms, is often that offering the most direct, well-graded alignment for railways. It will not therefore be surprising if the ALC class of land crossed by a railway is above the average in the region concerned. This is no disaster if agricultural land is abundant and if other types of valuable land resources are thereby spared.

There are a number of impacts of agricultural activity which can themselves be environmentally negative—overuse of chemicals, erosion, drainage, intensive factory farming. Recently there has been a surplus of land under the plough in Europe. There is therefore a case for railway alignment to take agricultural rather than scarcer conservation land. However, the farming industry is not necessarily at odds with conservation; in particular, MAFF designates Environmentally Sensitive Areas (ESAs) which retain special landscape or conservation interest where the quality as wildlife habitat can be enhanced by encouraging specific forms of agricultural management. ESAs therefore fall into a category of conservation land, the subject of the next chapter, which transport routes avoid where possible.

Meanwhile agricultural land resources are more common than conservation land and therefore less sensitive to development. Admittedly, built-up land or that covered by railway permanent way formation is not easily restored to agriculture. But, whereas scarce conservation land is easily ploughed up, the converse is not true; for instance it takes 100 years to turn a field into botanically diverse woodland.

Commercial forestry involves long-term cropping. In Britain it is undertaken mainly as coniferous plantations which, unlike agriculture, are suited to poor quality (ALC Class 5) land. Forest production takes place in areas too remote for competing development and too spacious to be affected by railways. Any interaction between railways and forestry is more related to conservation (Chapter 13) or to scenery (Chapter 15).

In Siberia or North America forestry has been practised as systematic felling of natural timber resources. In some cases, unsustainable extraction has been materially assisted by rail transportation.

13 Nature Conservation

13.1 ECOLOGY AND HABITAT

Ecology is the science of relations between organisms and their environment. Their environment includes the surroundings in which they breed and feed, i.e. their habitat. Some wild flowers and most animals and birds are protected. It is destruction or degradation of their habitat which threatens valued species of flora (plants) and fauna (animals, etc.).

A common result of land development, for transport or anything else, is the depletion of wildlife habitat. Wetlands are threatened throughout the world; typical British resources at risk include ancient woodland, heath and natural grassland.

Long-established parcels of land are host to many inter-dependent species in biologically diverse ecosystems. Any description under the broad headings given here is therefore necessarily general; its purpose is to relate recognised types of conservation land to the particular problems of route selection for the long narrow strips which constitute railways.

13.1.1 Conservation

A key world environmental objective is the preservation of land most likely to sustain growth of species of flora and fauna otherwise threatened by extinction. Globally, most of this land is wild. 'Wildland' is defined, in the World Bank's conservation programme, as 'territories of water and land which have been only slightly or not at all modified by modern man, or have been abandoned and have reverted to an almost natural state'. However, by this definition, there is little true wildland left in the more densely inhabited countries. In England human activity has been altering the landscape for over 6000 years. But modern man's influence might practicably be taken as that resulting in 'planned', as opposed to 'ancient' countryside, and generally consequent upon developments taking place from about 1700 (Rackham 1986: 5).

Recognised as the most valuable natural or semi-natural habitat areas in Great Britain are the 5400 Sites of Special Scientific Interest (SSSIs). These have been identified by the former Nature Conservancy Council (NCC)—now English Nature and its Scottish and Welsh counterparts. SSSIs are designated mainly for biological but some for geological reasons. Together SSSIs cover over 17 000 km². This 7 per cent of the nation's land area includes National Nature Reserves (NNRs) as well as international categories such as Ramsar sites (wetland) and the European Community's Special Protection Areas (SPAs, for wetland birds) and—in the future—Special Areas for Conservation (SACs, for habitat).

SSSIs represent the best authenticated, officially recognised areas of valuable habitat in Britain. Even so a number in the more populated areas of England have been damaged by government-approved road schemes; and new railway schemes are unlikely to leave them all totally unaffected.

National Trust property includes recreational and scenic land and historic buildings as well as some SSSIs, NNRs and other conservation areas. Some National Trust land is 'inalienable' and can only be acquired through a full parliamentary process—a greater deterrent against transport developments than other categories of protection. Indeed the foundation of the National Trust arose partly as a result of opposition to railway development (Farrington 1992: 51).

Besides nationally designated conservation land there are numerous tracts of soil, water and vegetation constituting wildlife habitat. These vary greatly in extent and character. Some are of unestablished or arguable conservation value; some have been recognised at local level as Local Nature Reserves or other designations. For instance, in Kent SNCIs (Sites of Nature Conservation Interest), assessed by the county wildlife trust but recognised in local authority plans, cover a number of ancient woodlands, grassland, quarries and *carrs* (wet woodland). These sites are not usually of SSSI quality but are so numerous that every projected Channel Tunnel Rail Link route through the county would have affected some of them.

Because of the complexity and variability of conservation resources, their sensitivity to interference by transport infrastructure and the significance of any loss can only be determined by expert assessment.

13.1.2 Ecological Assessment

Expertise provided should cover:

- the characteristics of railway routeing and construction, such as penetrating or slicing off the edges of woodlands and creating barriers to movement or drainage;
- the particular type of habitat, ecosystem and species at risk.

Occasionally, reconnaissance reveals an important habitat that has never been previously identified. More frequently local naturalists are already aware of the characteristics of places of interest in their area. But often no detailed investigation will have been performed before a development proposal requires it.

The first step in ecological assessment for a new railway should be recognition of designated sites. This should be followed by a survey to identify other areas of potential interest. The significance of these resources is then evaluated; the likelihood of their being affected and the scope of further studies can then be decided with the railway planners.

An ecological assessment should provide the following:

- A description of the area, the ecosystem or habitat it provides (with any local wildlife corridors or buffer zones) and the species it sustains (with observations on their rarity and fragility); climatic, geological and drainage characteristics of the site; history and influence of human activities.
- Prediction of the impacts (damage or alteration to habitat), their magnitude and spatial extent, how and when they will occur and how long they will last (temporarily, short or long term, or permanently).
- The ecological significance of these impacts (in terms of national or local resources, diversity and scarcity); viability of preservation of habitat and species; priorities for route avoidance or special protection.
- Proposals for minimising damage or, in rarer cases, for compensating for it.

13.1.3 Flora and Fauna

Flora is seasonal. Orchids, for instance, are invisible over much of the year. Even expert botanists cannot detect all species if their investigations have to be undertaken at the wrong season. A number of visits over a year are ideal for a comprehensive assessment.

Most of the vegetation with which people are familiar is planted by man, for instance trees, shrubs and hedges in villages, suburbs and towns. Much recreational parkland is totally man-made and in the Dutch lowlands even includes artificial wetland areas. Thus the simpler types of ecosystem can be created as well as destroyed; but newly created habitat is less likely to encourage the rarer species requiring particular micro-climates and soil conditions or seeds no longer generally available. Woodland or flood pasture may take a century or more to mature whilst natural conditions existing in very old forests may, if destroyed, never be replaced because the climatic conditions in which they were originally created no longer exist. More commonly some species can only be reintroduced where they are protected against human interference or where special measures are taken to maintain the condition of their habitat.

Many apparently natural woodlands or meadows have in fact been used by man for generations, at least until the present century. They are only semi-natural as they have been *managed* to promote both high yields and long-term conservation. Biological management could be equally effective when applied to the landscaped features of engineering structures created for the convenience and rapid transport of man; whether such management could be guaranteed for the centuries necessary to create complex ecosystems is questionable.

Any need for protection of *animals* depends upon their scarcity and the fragility of their habitat. Man has not only destroyed much of the old order but has instituted major changes in the dominant species. For instance, planting of beech woods or reservation of moorland for hunting has increased the habitat of introduced species like grey squirrels or deer. The latter are more abundant and in more varieties in Britain than they have been for 1000 years (Rackham 1986: 49). On the other hand, if a rare species is threatened, it is usually because of the small size and fragility of the eco-system in which it lives. Obviously, new railway routes should avoid these sensitive locations where possible. Where they cannot, the practicability of providing new habitat for endangered species can be investigated—whether high roosts for bats, underground 'sets' for badgers or ponds for crested newts. It has to be admitted, however, that the costs of creation are high, the intricacies of habitat dependency and replaceability not always fully understood, and the chances of success low.

13.2 TERRESTRIAL HABITAT

Land areas of importance to flora and fauna in Britain include:

- woodland of different sorts;
- hedgerows;
- various categories of uncultivated open country;
- inaccessible or unoccupied pockets, including fenced or disused railway land.

Where railway construction is likely to affect any of these, the consequences for habitat have to be assessed and ways of mitigating damage have to be planned. Finally, opportunities for conservation within railway boundaries have to be recognised.

13.2.1 Woodland

The term 'woodland' covers a wide range of trees from commercial 'needle-leaved' conifers to native 'broad-leaved' species in parks, managed copses or relatively wild forest. Actual species of naturally occurring trees are related to local geology, soil chemistry and hydrological conditions as well as to human influences.

Birds, other fauna and low level vegetation for which the woods provide habitat depend in turn on the condition of the trees and of the ground. There must be light for new growth, whether it is let in by natural demise of trees, by timber felling or by pollarding and pruning. Woodland management involves selective cutting of tree trunks and branches to maximise timber and wood production or to enhance and diversify other plant growth. Management techniques can be varied to suit particular commercial requirements or conservation strategies.

13.2.1.1 Broad-leaved woodlands

Broad-leaved woodlands in Britain have been exploited and managed by man throughout most of their history until comparatively recently. Exceptions are found only in the most inaccessible places. Untended trees will fall down and regenerate themselves from time to time according to a natural cycle. If timber is felled, trees of most species spring up from the severed stumps or from buried suckers. Most broad-leaved woods will survive unless every tree is grubbed out and the land is converted to another use. However, the time intervals and manner in which trees are felled (for trunk timber) or coppiced (for poles) will determine the growth conditions for ground vegetation and fauna. Woodland which is methodically harvested may well survive longer than that which is left to natural growth and competition.

When traditional management is abandoned, the period during which vegetation reverts to a wild state is at first likely to be regarded as one of neglect. Natural regeneration of woodland then follows slowly; but biodiversity, one goal of conservation, is achieved as much by appropriate management as by time.

13.2.1.2 Ancient woodlands

Ancient woodlands have been in existence for over 400 years. Those in seminatural condition, managed but without any substantial new plantation, reshaping or revised drainage, are a still common but much threatened feature of south-east England—so common that motorway construction through Kent cut through several of them; so threatened that Union Railways' 1993 maps of environmental effects list 'cumulative effect on various ancient woodlands along the route' as the *only* major impact of the proposed route from London to the Channel Tunnel. One reason for the sudden decision by the British Government in July 1993 to abandon a plan to construct a road through Oxleas Wood may have been the fact that it was the only ancient woodland remaining in London.

The value of such woods can be assessed ecologically, as a collection of species or an ecosystem which could not be re-created, *or* as historic landscape, in terms of the layout and ancient features particularly near present or former boundaries.

New railways should seek to avoid the most valuable sections of ancient woodland. This can be difficult for high speed lines which permit only large radius curves. Shaving off the edge of an old wood may cause as much or more long-term loss as driving right through it; this is because of particular ecosystems at the edges of woods and because of historic boundary features such as lines of trees or earthworks. It is important in detailed route alignment for woodland experts to establish the relative merits of different woods in respect of size, shape, situation, historic features and diversity of species. Some old woodlands have been ill-managed or inadequately felled—creating poor conditions for renewal—and these may have less justification for preservation.

Biggins Wood, a 5 ha fragment of ancient woodland, was removed by recent railway development—the Channel Tunnel Terminal at Cheriton near Folkestone. Part of the wood was relocated elsewhere after considerable research as to soil transfer and ways of promoting and managing suitable ground cover before the new tree canopy is available (Buckley and Knight 1989).

13.2.1.3 Commercial conifer plantations

Forestry can be regarded as a form of long-cycle agriculture. Trees are planted at a density to preclude any low level vegetation and are felled in blocks, allowing no variation in age. As a contribution to conservation, some modern forests are 'mixed' and include managed stands of traditional deciduous trees. Meanwhile, most hill forests still include remnants of older growth and wilder features in broken ground or near watercourses.

Coniferous forest has been planted beside the West Coast Main Line to Glasgow where it climbs Beattock Bank; many similar situations arise in the more forested areas of Europe. Were any such lines to be widened, there would be no great loss to natural resources where a strip is cut through short-lived conifers to accommodate extra tracks. Indeed it may provide the opportunity to reshape adjoining areas of the forest to accommodate expansion in broad-leaved trees. This expansion may most practicably take place in 'strip banks', ribbons of broad-leaved woodland along the lines already created by streams, forest boundaries, footpaths or railways. This form of habitat reconstruction can take place by sustaining such natural broad-leaved communities as still exist, by encouraging their return or by expanding the species along colonising fronts (see Anderson 1989: 129).

13.2.2 Hedgerows

Hedges differ from woods in that:

- they constitute boundaries rather than incorporate them;
- they are more exposed; the large-crowned, short-stemmed trees in hedges

offer habitat and appearance different to that of the same species in thicker woodland;
- their length provides connections between other areas of nature conservation.

Hedgerows may be as diverse as woodland in their ecological significance. Generally their age can be related to the number of tree and shrub species they support. Their significance as historic boundaries is also an element in their assessment. There is a great deal of 'difference between an ancient saxon field boundary and a quickset Parliamentary Enclosure' (Fiona Reynolds 1990: 4).

Destruction of hedgerows may be inevitable where they cross new railway routes. Across ancient countryside it may be impossible to avoid all the numerous historic hedges. Fragmentation of the more sensitive or naturally diverse lengths should be minimised. The siting of any incidental features like bridge approaches or construction haul roads should avoid damaging hedges and ditches along country lanes or should limit such damage to one side only.

13.2.3 Uncultivated Open Countryside

Grassland, heath or moorland may be extensive and unobstructed; or they may be divided into fields by hedges, walls or copses. In the past some may have been drained or ploughed and cultivated. Unenclosed land, generally less suitable for agriculture, may be that most suited to wildlife. Even arable land, such as downland recently cultivated but of marginal agricultural value, might in time be restored to the more valuable habitat it supported before first being ploughed. Meanwhile, railway planners need to recognise in route selection the comparative scarcity and vulnerability of uncultivated land.

Some *grassland* is meadow, mowed for cattle feed, often near streams and periodically flooded; it might be threatened by railway land-take along broad valley bottoms. Most grassland is pasture, i.e. grazed by sheep or other animals sufficiently to prevent higher growth; upland pasture is extensive and remote and therefore less susceptible to development.

Chalk downland turf, unrepresented in national parks and hardly found outside Britain, is a much depleted, botanically rich ecosystem. Its survival depends on grazing as a management method. It has been reduced by lack of grazing, ploughing and use of herbicides and fertilisers. On the other hand, where railway lines cut through downland in cuttings, the turf may have been in an undisturbed condition for so long that it gains vegetation that only mature chalk downland can support. It is surmised by ecologists (Palmer 1993) that *Cerastium brachypetalum* (grey mouse-ear) is found in railway cuttings because these are the *only* areas where suitable, open chalk grassland now remains.

Lowland heath could be vulnerable to any type of development. It provides rare habitat suitable for some of the scarcest reptiles and birds.

Dune heath is a stage in the stabilisation of sand dunes. In temperate countries it is one of the seaside features recognised in 'heritage coastline'. In more arid climates moving sand dunes are a less welcome feature of desert conditions; attempts to stabilise them are more directed at keeping railway lines clear of sand than of preserving any habitat the dunes may threaten.

Moorland is usually extensive, peaty and remote. Its cover can be kept stable by grazing or burning but it may turn into woodland if left totally undisturbed. Moors are common in Britain where they support a distinctive and extensive type of vegetation. Scottish railways cross Rannoch Moor, the 'forest' of Badenoch, Culloden Moor and the remote expanses of inland Caithness. It is most unlikely that any new railways would be needed in such remote regions but, if they were, the ecological disturbance caused would probably be modest compared with the effects of commercial afforestation and land drainage.

13.2.4 Inaccessible Land

Land which is wild because it is difficult to reach includes cliffs, both natural on the coast and man-made at disused quarries inland. Other relatively inaccessible locations are found in dense, fenced off woodland, railway or road reservations, or on islands. As well as providing undisturbed habitat, these are important features of scenery.

Pockets of unused land occur in both rural and urban situations where buildings, quarries or railways are abandoned. Those which are left wild will grow into scrub—long coarse grass, bushes and small trees. On fertile ground scrub is a stage in eventual development of new natural woodland. However, this new woodland may be more valuable in the medium term if some form of management—selective cutting or extra planting—is undertaken to produce growth most beneficial to particular wildlife species.

13.2.5 Assessment of Terrestrial Habitat

Ecological reconnaissance along the route corridor of a projected railway should cover all land not paved or built on, particularly where it has not been 'improved' by drainage or chemicals or extensively ploughed. There will be a number of land categories, including much arable and some industrial waste land, which are of low ecological significance; but there may be some isolated ecosystems of exceptional scarcity or scientific interest.

Description of habitat must include soils, drainage, significant species and their situation, extent and sensitivity to disturbance. Combinations of species and the overall number of different species in the habitat are also relevant.

Classification of species and ecosystems by ecologists obviously follows appropriate botanical or zoological systems. But it is useful for non-biologist planners if the significance of habitat is interpreted in more familiar terms. Physical attributes of the habitat itself warrant explanation as do the various layers of vegetation, from ground cover through field flora and low-growing undershrubs up to shrubs and trees.

Situation features relevant to impact prediction concern:

- whether effects of development such as railway construction can be reversible;
- whether any wildlife corridors will be fragmented or any critical boundary or buffer zone conditions will be affected;
- the viability as ecological units of any radically reduced areas;
- how access to the area, for viewing or management, may be gained;
- how protection of habitat, from people and other hazards, is to be effected.

Quantitative data is important in monitoring changes which occur to habitat—the numbers of specimens, physical measurements, areas of land or water and lengths of hedges or streams, all explained by notes on location, situation and type of habitat. Data of this sort can then be used to determine the fragility and continuing viability of each affected habitat. Essential elements in railway route selection are the size, quality and comparative conservation value of similar features in the same area.

13.2.6 Mitigation Measures

Terrestrial habitat can be preserved or maintained—to a greater or lesser extent—by sensitive route alignment, design for damage limitation, provision of wildlife corridors across the line and, occasionally, reconstruction of alternative habitat elsewhere.

13.2.6.1 Route selection

Careful investigation of alternative alignments can minimise destruction. For instance, heath or ancient woodland could be given a high avoidance priority. Priorities have to be allocated where there is any choice available, taking into account:

- expert opinion as to the value of one conservation area compared with another;
- judgement on the importance of protecting wildlife habitat compared with that of preserving people's property;
- legal and political realities.

13.2.6.2 Damage limitation

Damage can be restricted by fine adjustments to horizontal or vertical alignment, e.g. to spare certain critical vegetation, drainage or boundary features, or by protecting habitat adjacent to the railway. Old woods cannot be re-created by new planting; but the extent of damage can be restricted by limiting land-take. Raising rather than lowering the alignment can be less destructive in some circumstances; shallower cuttings achieve a narrower swathe through woodland; placing track on viaduct rather than level ground or embankment reduces the loss of sensitive ground like wetland or wildlife corridors. The CTRL route proposed in 1990 was to have crossed scenic parkland at Boxley, near Maidstone, by a viaduct rather than embankment where the alignment crossed a valley. This would have permitted movement of amphibians, preserved the landscape viewed through the viaduct and maintained an economic vertical alignment. The route proposed by Union Railways (1993: 97) is on a similar alignment but the level is not defined. However, it appears that the viaduct solution may be rejected for a ground level option, partly to meet local authority wishes to reduce the visual impact of trains.

Sections of recent broad-leaved woodland or coniferous plantations which have to be taken can be replaced or extended in another direction. Planting of new trees is also a feature of 'soft' landscaping. But in addition to any visual improvements that landscaping provides, there are important ecological factors to be considered in its concept—for example, the type of trees or ground cover, the layout and spacing of introduced plants, the role of natural regeneration and any requirements for subsequent management.

13.2.6.3 Wildlife corridors

Animal crossings are already provided by bridges or tunnels across some railways and roads. Experience with these can no doubt show how such facilities can be used more widely. British Rail has installed 24 badger crossings on the Southampton to Portsmouth line (Guild, 1990), the M40 motorway provides both bridges and a tunnel for deer (Moore 1990), and small tunnels beneath TGV lines and British roads assist the annual migration of amphibians to breed in water. The narrow width of double track permanent way means that the actual tunnel length can be as short as 10 or 12 metres.

For animals that are cautious of using tunnels, practical encouragement is needed by:

- suitable soft floor material, rather than hard paving, at crossings;
- barriers to prevent animals using old, severed routes; badgers are not deterred by railway lines and quickly learn to cross them; where third rail

electrification is installed badgers will continue to attempt the crossing until those who escape electrocution understand the track's lethal nature.

In more remote territory the provision of corridors for wildlife to cross obstructions may be urgent. Early railroads in the American West, being unfenced, were much less destructive to wildlife than the hunters who followed. But, more recently, the Alaska oil pipeline has been cited as a serious obstruction to wildlife migration. Modern railways should make ample provisions for whatever wildlife they may obstruct.

13.2.6.4 Creation of habitat

Planting appropriate flora may attract targeted species of displaced fauna. For instance, the M40 in Oxfordshire had to cut through blackthorn bushes where colonies of black hairstreak butterflies breed. Transported mature and new blackthorn trees were planted between the new road and nearby undisturbed woodland to interest these butterflies before their original habitat was destroyed (Moore 1990).

New planting cannot replace all the characteristics of ancient, long-established features; it is intended to enhance and extend existing natural resources as much as to replace anything that has been destroyed. The layout should accord with that of existing woods, hedgerows and other man-made features (including any new railway). Some natural protection against the prevailing wind may be necessary. More observations about planting are included in Chapter 15 where it is described as a means for enhancing the lineside landscape.

Physical transplanting of mature trees can be attempted; this was undertaken experimentally and with at least partial success in Kent. Trees up to 8 m high were moved whole, larger trees as coppiced stools. Ground flora was also transferred, with extra management measures as the shade regime required from the canopy of new and transferred trees took five years to mature (Down and Morton 1989).

Woodland, parkland or wetland can be created within a variety of spaces including in transport corridors. New broad-leaved or mixed 'community' forests, to be planted in populated areas of Britain of the size of that planned in the Midlands ($384 \, km^2$), are as fresh a concept as are new railways. There are great opportunities for integrated land use planning; it can allocate controlled access, recreational amenity, wood production and wildland development whilst taking into account the transport infrastructure, buildings and farm land which are already there.

13.2.7 Conservation on Railway Land

Railway lineside land in Britain is believed to provide 30 000 hectares (Guild 1990) and *all* railway land 54 400 hectares (Bunce and Jenkins 1989) of relatively undisturbed habitat; presumably the latter figure includes the considerable areas of disused railway lines and sidings. Beyond a narrow trackside strip on which vegetation is purposely suppressed on working railways, encroachment by wildflowers can be encouraged. Still further back on modest slopes and still within the fenced area, tree growth can be permitted. Flowers like cowslips, which have become comparatively rare in accessible places, have flourished in railway cuttings. Vegetation can be planted on appropriate types and thickness of topsoil—perhaps 30 cm depth for trees or shrubs, but little or none if wildflowers are to thrive.

Grass cutting and shrub thinning of slopes are occasional but essential activities which have to follow construction. Near the track itself, clearance with suitable herbicides is routinely practised. The practicability and cost of routine maintenance are factors in the choice of species for lineside vegetation. Throwing out of old ballast on to the grass verge is a practice to be avoided.

On disused railway land there is opportunity either to plant vegetation and create conditions appropriate to any redevelopment or else to leave it to nature. The latter course initially encourages flowers common in derelict areas such as rosebay willow herb or Oxford ragwort—now a typical railway plant but imported originally from Mount Etna via Oxford Botanical Gardens. Ashes, laid instead of stone ballast beneath some sidings, are much more penetrable by weeds and wildflowers.

If ground conditions are suitable, wildflowers give way in a few decades to secondary woodland of pioneering species such as oak, birch, hawthorn or ash (Rackham 1986: 68). Other trees cited as common in old railway situations include hazel, willow, holly and alder (White 1986: 218). If access for people to walk along the former track bed is required, then there may be problems in keeping the way clear. Plants like blackthorn, which can penetrate railway ballast, are so tough that they dominate growth on the track and prevent passage unless they are frequently cut. So nature reserves on old lines may need management, even selective weedkillers, in the same way as the trackside strip of working railways has to be kept clear.

13.3 WATERCOURSES AND WETLAND

13.3.1 The Scarce Resources

Rainfall feeds watercourses and sub-surface aquifers. There is heavy demand for abstraction from both these sources—for domestic, industrial and, in drier climates, agricultural water supply. Higher living standards and population growth increase demand and reduce the quantity of river flow. Impacts on the ecological quality of riverine areas result from polluted effluents and from civil engineering. Banks are destroyed or courses altered by construction of river training walls, bridge piers or culverts through

embankments. Railways along river valleys require these engineering structures.

Special, often fragile, conditions for plants and wildlife exist in the beds,
banks and borders of watercourses (streams, rivers, lakes and ponds) or on
'wet' land, flooded permanently (swamps), seasonally (like water meadows)
or tidally (salt-marsh).

Wetlands referred to as *marshes* are generally those in which the waterlogged soils are inorganic. *Fens* in eastern England are lime-rich peats and
silt soils fed by rivers rising in chalk districts. When drained they are very
fertile for agriculture—but are no longer wetland! *Bog* is a term for lime-
deficient peats, common in wet upland tracts.

A quarter of the British Isles is or was some kind of wetland (Rackham
1986: 375). Most of this is upland bog in areas where transport routes are
neither common nor seriously destructive. Much more vulnerable are wild
river banks and marshes; 1 million hectares have been lost in The Netherlands; in other developed parts of Britain or Europe lowland marshes and
any surviving undrained fens are fast diminishing and continue to be threatened by drainage improvements and sinking water tables.

Flora and fauna which can be affected by construction across wetland are:

- animals and birds which feed in wet areas or streams, or vegetation which
  flourishes in damp conditions;
- freshwater plants and fishes;
- marine creatures able to withstand salinity changes in estuaries and the
  birds which prey on them at low tide.

13.3.2 Assessment and Classification

In many important wetland areas an expert assessment will have been made
previously by a national or local environmental body or by river management authorities. Where this assessment has resulted in designation as an
important conservation area, it places a strong constraint on any development.

Classifications for statutory protection of habitat start at international
level and extend down to local areas of interest. Britain ratified the 1971
Ramsar Convention on 'Wetlands of International Importance'. Under the
convention governments undertake to conserve both wetland generally and
sites designated in particular. All wetland International Nature Reserves
(INRs) or EC Special Protection Areas (SPAs) designated in Britain are also
Sites of Special Scientific Interest (SSSIs). At county level and locally there
are other categories of nature reserves and conservation areas.

Many of the local and all the previously undesignated sites affected by a
new railway development will have to be freshly assessed. The aquatic ecologist should describe any threatened wetland feature in such terms that it can

be compared with similar features elsewhere and to report any unique or fragile features. Ideally, for a watercourse, its dimensions and flow (with seasonal variations) should be measured, the water's visible quality, the nature of the stream bed, and growth on the banks noted. Signs of aquatic or terrestrial life should be described, as should any evidence of pollution. Freshwater wetland, watercourses and lakes can be classified as organic or inorganic, lime-bearing or acid, and in terms of seasonal water flows and levels. The chemical and biological condition of inland lakes and tidal flats alike is critical to bird life and the viability of their major migration routes.

Classification and evaluation of wetland starts from its very definition. In the USA there is a statutory process for 'wetlands verification', to determine the extent of land affected by any potential development and whether it should thereby become 'jurisdictional wetland'. The criteria for this determination are the hydraulic properties of soils, the type of hydrophytic vegetation and indicators of wetlands hydrology. However Kusler *et al* (1994) have pointed out the dangers of using the absence or presence of 'hydric' vegetation; that vegetation may be missing for several years where there are long cycles of water level variation.

Scientific studies needed in actual route planning for a railway may be less legally-slanted, at least initially. But the same criteria should be used to define the functional values of different pieces of potentially affected wetland as a guide to determining the least-disturbing alignment and type of crossing structures.

In Britain, the statutory water pollution watchdog is the National Rivers Authority. However, nature conservation is only one of its duties which include drinking water quality monitoring, flood defence and land drainage. The interests of conservation seldom complement those of drainage.

13.3.3 Railways, Development and Wetlands

As long ago as 1829 the Liverpool and Manchester Railway was constructed across Chat Moss whilst in the 1890s the West Highland Railway had to be founded on the bogs of Rannoch Moor. The lines were eventually floated on hurdle structures or dumped material. But the techniques followed precedents set in constructing tracks across the fens 5000 years earlier (Rackham 1986: 382). Such man-made works are hardly more than incidental, however, in the diverse natural drainage pattern and landscape of extensive moorland.

It is the remnants of wetland in less remote country which are more seriously threatened by new routes or by other forms of land development. Most arable crops originated in middle eastern semi-arid climates and cannot thrive on wetter, undrained land. As a result, wetland resources in the inhabited parts of Britain and Europe have been reduced by drainage for arable agriculture more than by construction of transport infrastructure. Some government initiatives have been taken to discourage further drainage and destruction of wetland but there are inconsistencies in their application.

The Thames Estuary encompasses important areas of remaining coastal grazing grass, such land having apparently been reduced from around 14 000 hectares in the 1930s to only 4600 hectares less than 60 years later. Much of the Estuary, from London to Tilbury on the north bank and further to the Isle of Sheppey on the south, comprises the East Thames Corridor development area. The area is evidently judged to be sufficiently extensive to allow both residential and economic development, for which there is little space elsewhere in south-east England, and preservation and protection of the natural wetland resources.

Before a Channel Tunnel Rail Link route through the East Thames Corridor was seriously considered, the British Government was accused of acting inconsistently in decisions concerning development in the Estuary (*New Scientist* 21 April 1990). The controversy arose when an application for film studios and 'theme park' development on Rainham Marshes, including 450 hectares of the largest SSSI in Greater London, was favourably received by the Department of the Environment. Elsewhere in the Estuary, on the Isle of Sheppey, the Department was less receptive to a proposal for port, industrial and residential development also threatening SSSIs and marshland. Perhaps a deciding factor was that the Sheppey marshes were designated for special protection under the Ramsar Convention whilst the Rainham site is only one SSSI among several.

The 1993 Union Railway route proposes to avoid causing any critical damage to Thames wetlands by:

- adopting a route precisely adjacent to the existing Tilbury line for 3.5 km, taking only a narrow strip of land at the edge of the the Inner Thames Marshes *SSSI* along the north side of Rainham and Wennington marshes (Union Railways 1993: 100, D.5.8);
- entering tunnel beneath the proposed Ramsar and SPA lagoon and marshes at West Thurrock (Union Railways 1993: 100, D.5.6);
- suitable design of a crossing 'considered to be at the least sensitive part' of Ebbsfleet Marshes *SNCI* (Union Railways 1993: 99, D.5.5).

Various routes for the Union Railway crossing of the estuarial River Medway were evidently investigated. These involved searching for the most sensitive crossings of historic grazing marshes and rough grassland. 'Experience in engineering the alignment in this area indicates that the [ecological] effects can be significantly reduced by careful location of the viaduct piers' (Union Railways 1993: 97, D.4.4). It also seems evident that much more expensive options suggested for hiding the line by sinking it into or below the river bed could have seriously upset the aquatic environment.

British *river estuaries* are recognised as areas of critically important wetland habitat; they are more numerous and varied in character than those anywhere else in Europe. Britain has therefore a global responsibility for maintaining the integrity of its estuaries. Some examples of existing railways which run beside estuaries with inter-tidal mudflats and of major importance to large numbers of wildfowl and waders are described in the last section of this chapter. The impact of railways on estuary scenery, which is closely related to ecology as well as to geological and morphological conditions, is reviewed in Chapter 15.

13.3.4 Impacts and their Mitigation

Possible impacts of railways on aquatic habitat relate to:

- effects on river regime of engineering structures such as training walls, embankments or bridge piers and abutments;
- loss or relocation of sections of river bed or banks, ponds or marshland, where covered by railway formation;
- changes in surface or ground water level (raised upstream, lowered downstream) because of constrictions on flow of water;
- effects on water quality;
- disturbance of flora, fauna and habitat during construction.

13.3.4.1 Engineering structures

Any structure built in the bed of a watercourse is liable to interfere with the pattern of stream flow and the stability of banks. Walls supporting a railway along the side of a river act as training walls; their location, length and layouts must be planned as part of river engineering to suit other uses, structures and conservation interests along the stretch of watercourse concerned. Embankments across low-lying ground can either destroy or create wet conditions according to how they modify natural or semi-natural land drainage.

In wide alluvial channels of some of the larger rivers of Asia or Africa, bridge piers form obstacles to flow which, during floods, can result in scour around the foundations or changes in the course of the channel. In these conditions bridges need deep piers and possibly also major river training works. Even on narrower, more stable river courses in Britain, scour in high floods in the early 1990s has been responsible for collapse of railway bridges—across the River Ness at Inverness and the Towy on the Central Wales line.

Bridges over rivers or streams should leave sufficient clearance, not only to accommodate flood flows but to retain much of the original vegetation and micro-climate and, on unconstrained streams, to interfere as little as possible with channel conditions.

13.3.4.2 Relocation of sections of river

River channel relocation may be necessary where a railway alignment follows a meandering watercourse. Relocation of habitat can be arranged through the following steps:

1. Construction of the new river section in the dry, preserving as far as is practicable the gradient, width, depth and line characteristics; provision of berms at the banks for waterside plants and to aid the subsequent establishment of suitable river's edge conditions.
2. Collection of plants from the condemned section of the river (or elsewhere); temporary placement of these in wet conditions, such as behind gabions in a nearby section of the river.
3. Replanting in the new section when water is diverted into it.

Any meanders or elbows of abandoned river which remain may provide scope for habitat creation. Creation of new ponds can be undertaken to allow transfer of aquatic flora and fauna to any site of the right shape and soil conditions that can hold water and provide habitat. Ponds are no longer in general use for watering cattle and are becoming scarcer. So the wildlife, especially amphibians, which live in them suffer accordingly.

Urban developments and new railways in cities frequently involve relocation of watercourses or minor river works. Any development in the King's Cross and St Pancras area for the Union Railway terminus will create opportunities for preservation or extension of wetland reconstruction to follow those already taken—such as the ponds created at the Camley Street site near the Regent's Canal.

13.3.4.3 Changes in water level

The effects of constrictions in watercourses, such as at railway crossings, are difficult to predict because of the variation in stream flow and lack of hydrological data for small catchments. However, data can be synthesised by statistical methods calibrated for the local climate and catchment characteristics. These, together with hydraulic calculations, can be used to predict water levels and to plan that rivers flood sufficiently to ensure seasonal wet conditions but without excessive risk to structures and property.

Maximum interference with natural drainage results from diverting the water through a totally enclosed culvert; the least disturbance occurs beneath a wide bridge. Where a railway route crosses rivers or cannot avoid marshland, construction of viaducts, with carefully located piers, is greatly preferable to broad embankments covering a wide band of land and pierced only by narrow culverts.

However, new routes for surface drainage may also provide opportunities to create aquatic habitat, for instance in 'balancing ponds'. These are constructed where drainage systems collect water before crossing under a railway. Instead of making the ponds sufficient merely to contain short-term overflows, they can be excavated to greater depth and lined with clay or fibre mats so as to retain water throughout the year; and sides can be shaped to provide a suitable bank profile. The main problem with roadside balancing ponds is maintaining water quality; railway conditions may prove more amenable.

Construction of bridges or culverts on the larger streams may affect navigation of boats. On the smaller watercourses the structures may create high water velocity and physical obstacles, affecting aquatic life.

13.3.4.4 Effects on water quality

Water can be contaminated by herbicides or other spilled materials arising from railway maintenance. Watercourses may also be polluted by erosion of soft materials, especially when building bridges or culverts, or by oils or chemicals from trains or during construction. Contaminated effluent must be collected in separate drainage sumps where, if there is sufficient space, it may be cleansed by natural biological means; otherwise the effluent may be piped elsewhere for conventional treatment.

13.3.4.5 Construction

Impacts of construction of railways on watercourses have been described in Chapter 10.

The physical effects of different construction operations can be determined by engineers and their impacts on habitat can be assessed by aquatic biologists. The most environmentally acceptable engineering solutions can then be devised taking into account:

- the hydraulic performance of environmentally acceptable channels, a subject on which the results of comprehensive research is available (HR Wallingford 1988);
- the alternative methods that can be used for bank protection of waterways and drains (Hemphill and Bramley 1989).

Other expert advice which is available for both aquatic and terrestrial nature conservation relates to the use of vegetation in civil engineering (Coppin and Richards 1990) and to techniques for biological habitat reconstruction (Buckley 1989).

13.4 RAILWAYS AND NATURE RESERVES

There are in Britain many stretches of *disused* railway track, particularly in cuttings, where wild flora and fauna have taken an unhampered hold of the long, tree-lined and often inaccessible strips of land. Numerous nature reserves have been designated to incorporate them among a large number of similar sites arising from termination of human activities at quarries, sand or clay pits, reservoirs, canals, brickworks and factories. The exposure of geological strata at rock faces in quarries and railway cuttings, similar to the erosion of cliffs caused by sea action, is another prime example of the benefits of railways to conservation.

The relevance of railways to the existence or creation of semi-natural habitat arises because:

- the land is often protected from interference or development;
- seeds of other species may be brought in on trains;
- different light or micro-climatic conditions exist on opposite sides of cuttings and embankments;
- wetland features can be created (or damaged) where railway works interrupt cross-drainage;
- coarse track-supporting ballast is often laid on finer, cindery material—infertile layers which, in time and where railways are abandoned, nevertheless encourage a high diversity of wild plants.

Figure 13.1. Floral habitat on railway banks, near Churston, Devon

LIVERPOOL JOHN MOORES UNIVERSITY
LEARNING SERVICES

Figures 13.1 and 13.2 show how habitat is created on cutting slopes and around ponds beside working railways.

The conservation value of railway land has been highlighted by Sargent (1984). A study of species-rich vegetation in London (Smart 1989) looked at three areas of well-developed botanical diversity on abandoned railway sidings of from two to thirty hectares and of ways of creating similar habitat on another wasteland by importing substrate material such as chalk, limestone, sand and clay to encourage growth typical of several different seminatural habitats.

There are a number of conservation areas which abut on to or are traversed by railways *still operating*. An example of the latter is Marazion Marsh in Cornwall, the remnant of a lagoon separated from the sea by a shingle ridge. The lagoon has been firmly separated from the sea for several centuries; a map of the Exeter to Penzance road dated 1675 already showed a causeway along the shingle nearly 200 years before the railway track was laid. Today the main line crosses right through the 85 ha SSSI where reedbeds support wildfowl, waders and passerine migrants; even so, the railway itself is not generally noticeable except when a train passes. The western

Figure 13.2. Ponds and surrounding wetland habitat, near Henley-on-Thames, Oxfordshire. Lineside ponds were created as 'borrow' pits during railway construction, resulted from interruption of drainage or were deliberately provided as 'balancing ponds'

side of Insh marshes on the River Spey in Scotland is a more extensive example of a railway crossing a valuable stretch of aquatic and wildfowl habitat.

In other locations railways, like major roads or coastal defence works, are the man-made barriers which form the boundaries between different land uses or land-forms—drained or built-up land on one side, semi-natural habitat on the other.

The St Erth to St Ives branch line, also in Cornwall, forms the bank of the Hayle estuary at Lelant Water SSSI, an area of inter-tidal mudflats on which birds depend for their invertebrate and algal food supplies. The simple railway platform for park-and-ride travellers to St Ives provides an ample viewpoint.

Many much larger estuaries provide similar mudflat habitat. Among the most important for tens of thousands of birds are the estuaries of Cromarty Firth in Scotland and the River Exe in Devon. The railway near the northern shore of the former is merely incidental. However, the Exe estuary is defined by railway embankments for 7 km on each side; some sections of these embankments serve also as flood protection bunds. The south-western corner of the Exe estuary is a 1200 ha SSSI/SPA/Ramsar site of international importance for wildfowl and waders, with mudbanks, mussel banks, salt marshes and reeds. It may be observed that passing trains, even when they whistle, are totally ignored by the feeding wildlife. It has been noted that certain species no longer visit, possibly because of disturbance (Burrows 1971: 118) but there is widespread evidence that trains (and other machines) do not disturb birds as do people and dogs.

The mouth of the Exe estuary is almost blocked by the degenerate sand spit of Dawlish Warren which extends from the high ground and railway station for four-fifths of the distance across the estuary to the opposite bank at Exmouth. The Warren provides its own sand dune habitat (as well as a golf course, beach facilities and car park) but serves mainly to protect the estuary mudflats from the open sea. The spit is unstable, as are many estuary mouths, and has been in danger of complete destruction (Burrows 1971: 179) so protection has to be continually in hand. It is possible that the fragility of the spit is partly due to the lack of littoral drift material from the coast further west where any erosion of the cliffs has been prevented for 150 years by a railway sea wall.

Another coastal, but cliff rather than estuarial, SSSI traversed by a railway is Folkestone Warren (140 ha), a landslip area comprising chalk cliffs, chalk grassland and scrub (Figure 13.3). Ardill (1987: 179) explains that 'erosion and seaspray give rise to five different communities of plants, from the most salt-tolerant near high-water mark to chalk grassland species at the cliff top'.

This steep reserve is also noted for butterflies and moths; in addition it is an international reference site for geological stratigraphy. A sea wall was

Figure 13.3. Folkestone Warren, Kent. The instability of this chalk landslip area for long discouraged human interference. The railway to Dover crossed the terrain with difficulty and required construction of a sea wall to reduce erosion at the cliff foot. The improved stability, coupled with continued remoteness outside the railway land and the beach access track, encourages a rich diversity of growth including orchids, chalk grassland, sea cliff flora and ungrazed scrub

built primarily to protect the railway which had been damaged by slips earlier in its history.

On the other side of Folkestone and slightly further inland is yet another SSSI—that of the Folkestone to Ettinghill Escarpment. The physical nature of the reserve is perhaps only marginally affected by the railway works of the Channel Tunnel terminal built up to the foot of the escarpment. However, the scenery above the terminal station makes a spectacular arrival point for visitors from the Continent.

In the same area the smaller Holycombe SSSI combines unusual calcareous fenland supporting rare flora and fauna with both archaeological finds and another geological reference site. The Channel Tunnel entrance alignment was adjusted and natural drainage preserved to minimise destruction of the habitat. Archaeological and geological evidence has either been preserved or fully investigated in advance of disturbance.

Other nature reserves have been created in the form of man-made reservoirs or in marshes created between rivers and railway embankments. Walthamstow Marshes in north-east London is such a reserve, reputedly the last ancient grassland in the Lea Valley, bounded on the west by the river, on the north by a reservoir and on the east and south by railway lines. Essentially these barriers have prevented encroachment by roads and buildings.

Thus railways have, for over a century, acted as defensive barriers against development on land which they have severed.

14 Heritage and Amenity

14.1 INHERITANCE AND ENJOYMENT

This chapter is concerned with the impact of railways on heritage and amenity resources. These resources are buildings or land which are valued for human activities other than purely utilitarian.

Heritage is what may be inherited—natural or man-made features of lasting value. Valuable features which we have inherited include diverse wildlife habitat, attractive scenery and historic structures. Chapter 13 has assessed the ways in which railway development can affect habitat. Chapter 15 will examine how railways relate to scenery. This chapter is concerned firstly with man-made structures.

Amenity is pleasant surroundings, natural or constructed—some place or facility which can be enjoyed. *Culture and recreation* are forms of enjoyment, the first with intellectual connotations, the second more physical ones.

Railway construction may damage heritage features or spoil an amenity. The noise or sight of trains might affect people's enjoyment.

14.1.1 Conservation of Heritage

Historic structures cover a range of antiquities:

- archaeological remains: Roman brick foundations and mosaic floors; prehistoric megaliths or earthworks (Section 14.2 below);
- medieval or later buildings of particular structural interest (see Section 14.3); stately homes of high architectural value, richly furnished or in historic landscapes;
- railway structures themselves, some of which have already achieved heritage status (Section 14.4).

Land resources of amenity value include:

- recreational areas where people go to enjoy themselves, whether informally or in organised sport (Section 14.5);
- disused railway land suitable for amenity or conservation use (Section 14.6).

14.1.2 Impacts of Trains on Amenities

The sight of trains is not relevant to theatrical, orchestral and similar audio-visual cultural activities taking place *indoors*. These may be affected by noise or vibration, although the latter can be repelled by suitable design of modern auditoriums.

There are two types of *outdoor* activity which could be affected by passing trains. One is serious sport or cultural activity which demands all the attention of performers and most of that of spectators. The other encompasses less formal activity like walking, riding, fishing, playing golf and enjoyment of private or public gardens.

On most lines the passing of trains is so occasional that their impact on people's enjoyment of historic buildings or landscapes is likely to be neutral. At castles open to the public and close to railway lines, like Conway in Clwyd or Stokesay in Shropshire, it would be unusual to hear of annoyance at passing trains.

There are many examples of sports venues overlooked and passed by railways, for example racing at Aintree, golf at Wentworth or cricket at Old Trafford. The disturbance is no more noticeable than that caused by incoming flights to London's Heathrow Airport passing over Twickenham and Wimbledon, global heritage centres of rugby football and lawn tennis. There seem to be few precedents that railways or even aerial activities intrude on cultural or sporting enjoyment.

14.1.3 Physical Damage Caused by Railway Construction

The significance of physical disturbance by railway development of heritage and amenity resources can be assessed as follows:

- first, by determining what cannot practically be reconstructed and may be lost or damaged when a railway is built;
- next, by identifying what can, perhaps with difficulty, be protected or resited;
- last, by assessing the regional significance of any losses or gains.

14.2 ARCHAEOLOGY

Necessary steps in assessment of archaeological impacts resulting from new development are:

- recognition of features of possible significance;
- investigation and description;
- advice on what action should be taken to preserve or record significant features at sites affected by construction and on any opportunities to enhance their situation.

14.2.1 Recognition

New evidence about man's oldest structures has usually to be sought below ground level. Most ancient works which show above the surface have long been recognised. Deeper structures and buried treasure are only likely to be unveiled by excavation—by machines constructing foundations, wells or railway cuttings, or through deliberate archaeological exploration.

The possible presence of hidden features can be deduced from the location and configuration of known structures in an area. Other superficial clues, in favourable geological conditions or where land has not been disturbed by deep ploughing, are given by variations in topography or vegetation. These clues may be in lines of 'crop marks' or 'soil marks', ancient boundaries or spatial relationships seen most clearly from the air. Aerial and satellite photography is in any case a common source of much other investigative data including physical, geological, botanical and aquatic information. There are close connections between geology or topography and ancient landscape features, e.g. settlements along spring lines or fortifications on high ground. In separating prehistoric human activities from geological events, remote sensing can help to differentiate, for instance, between man-made earthworks and glacial features like pingos and eskers.

14.2.2 Investigation

If superficial or aerial evidence indicates archaeological features of potential importance then more thorough exploration may be justified. In planning such exploration, expert knowledge and established archaeological practice have to be applied:

- To determine as much about the site as possible without breaking the ground.
- To decide whether and how to undertake some form of remote sensing, geophysical investigations, physical exploration such as digging trial trenches across sites or more thorough examination of single features.
- To assess the significance of all the features revealed and to recognise any necessity to afford special measures when railway construction takes place.

In any region there may be a wealth of information about local discoveries or those made elsewhere in similar circumstances. But sites where actual developments are proposed have seldom been subject to thorough expert examination until detailed planning requires it. In the Cheriton (Folkestone) Channel Tunnel railway terminal area, 80 per cent of the many archaeological sites now known were revealed during detailed field assessment undertaken for the Tunnel project. On road schemes in Britain many unknown sites are discovered even in areas which have already been investigated intensely (Friell 1990).

14.2.3 Preservation and Record

Massive monuments like city walls or Stonehenge are both national heritage and historic evidence. Lesser archaeological features are widespread and only occasionally of remarkable individual quality but contribute to evidence about how and where our distant predecessors lived. Recorded information about such features may be more important than their preservation *in situ*.

Individual removable artefacts—articles of ancient workmanship varying from chips of pottery to buried treasure—can be revealed by deliberate exploration or by chance excavation such as during railway construction.

Preliminary archaeological assessment of each section of a railway route should determine the likely existence of artefacts or significant situations. Archaeologists can recommend what further investigation and data collection is justified *before* construction commences.

Following investigation, recommendations should be made:

- in exceptional circumstances, that the railway alignment should be adjusted to avoid a particular feature;
- occasionally, that engineering measures be taken to protect an ancient structure or an archaeological setting;
- usually, concerning measures that should be taken during construction.

Isolated monuments or stone patterns lying directly in the line of construction may be judged suitable for straightforward removal to an adjacent site or into a similar historic environment elsewhere. For those elements which cannot practicably be moved there may still be a possibility of protection as has been provided, for instance, for Roman mosaics in the foundations of certain modern office buildings.

Precautions during railway construction concern the way in which working site layouts and access routes are prepared and how earthwork, particularly excavation, is to be undertaken. Action to be taken when items of archaeological interest are revealed should be defined in construction specifications and as a result of monitoring and advice by archaeologists on site. Generally ground should be left as undisturbed as possible; even if sites are buried under railway track or an embankment, future generations can always re-excavate them.

Full records—photographs, descriptions, locations and measurements—should be kept during both investigations and construction. The objective is to add to archaeological knowledge about the region or historic period concerned.

14.2.4 Opportunities for Enhancement

Construction of a railway or other land development may provide the occasion to enhance the situation of ancient historic features. Apart from the valuable knowledge gained in investigation, it may be possible to improve access, presentation and preservation both *in situ* and by collection of artefacts for a museum.

For instance, a new station might be planned at a 'green-field' site where a large space is required for car parking. This could be in an area with megalithic tombs, isolated historic buildings and an ancient track system or among a profusion of hidden buried features. Such a case actually occurred at a 'parkway' station site in Kent on the 1989 Channel Tunnel Rail Link route. The fields at such a site might be of little interest to most members of the public, compared with the attractive scenery nearby. Design of a station and car park could, however, be suitably combined with preservation of historic structures or with exhibition and explanation of artefacts. The site would then be of much greater general attraction. Access to the points of interest could be by a system of paths suited both to the ancient track system and to the layout of the station and car parks.

14.3 HISTORIC BUILDINGS

There are many thousands of buildings in Britain which are recognised as heritage structures and listed as worthy of some sort of protection. It is difficult to build a new railway in populated areas without affecting some of them. A key requirement in environmental impact assessment is therefore recognition of the relative importance of individual structures or groups of buildings. A route may then be chosen which future generations may agree has caused the least loss of regional heritage wealth.

After expert assessment, four courses of action are possible to deal with buildings which lie within initial route bands. These are avoidance, protection, relocation and demolition.

14.3.1 Expert Assessment

Many historic buildings are remarkable for the nature and quality of their construction. They cannot be easily reproduced by modern building techniques. It is for experts to assess their characteristics and advise on their significance. Transport planners and engineers encountering historic structures on a route require advice regarding the following points:

- The *magnitude* of a structure: whilst skilled masons can repair or even extend old buildings of cathedral-like proportions, they cannot reconstruct the originals, if only for economic reasons.

Figure 14.1. The town walls of Conway were adapted, nearly 500 years after their construction, to accommodate the North Wales main line

- *Structural type* and building materials: the relations between, for instance, available timber, methods of sawing beams and assembling them in frameworks, and the structural characteristics of the resulting building affect both its significance as worthy for preservation and the practicality of reproduction.
- *Skills* and unique or rare qualities in construction and decoration: the carved ornamentation in old buildings or ironwork in railway station columns are as important as the massive construction of Conway Castle and city walls (Figure 14.1) or the Forth railway bridge.
- The *setting* of buildings in groups and amidst other heritage features: this affects the feasibility of removal of a building to another situation or of demolition of part, rather than most, of a group of related buildings barring the route.
- Factors regarding *use and upkeep* of historic structures which may determine whether protection or preservation is practicable.

Official and expert recognition of the significance of historic buildings and their preservation is entrusted to bodies like English Heritage, the National Trust and county councils. In the same way as conservation land is recognised by *SSSI* status, buildings are designated as heritage monuments or listed as Grade I, II and II* buildings. But, as with nature reserves, some are not yet recognised; assessments during investigation of railway, road and similar projects sometimes reveal other buildings that warrant safeguarding.

14.3.2 Preservation by Complete Avoidance

In route planning, prime concern is to avoid those historic features whose loss might be most regretted. Railways have occasionally to run adjacent to heritage property or even through part of extensive parkland estates; but—apart from some Victorian masterpieces sacrificed in station redevelopment—rarely has railway construction involved destruction of a major building of recognised national heritage status. However, buildings of undoubted historic value but less established repute are as numerous in well-inhabited country as are small areas of wildlife habitat or scenic interest. Priorities among all these categories have to be established in equitable route selection.

14.3.3 Protection of Adjacent Features

The setting of historic groups of buildings is bound to be upset if a railway has to pass right through them. The least destructive solutions involve, firstly, precise alignment to spare the most significant structures of the group; then, if the costs can be justified, any structures that can be moved

Figure 14.2. The main line railway through Princes Street Gardens, Edinburgh is one of the minor features of man-made embellishments of the natural landscape

can be suitably rearranged; finally consideration can be given to bridging over or under the tracks to connect the severed pieces of the group.

Preservation *in situ* of adjacent buildings can be aided by structural support to protect walls or foundations and by measures to minimise subsidence or vibration. Drainage design can control ground water drawdown.

Strips of land or ancillary features which are important in the setting (surroundings) of historic structures can be saved by retaining walls at cuttings or embankments. Such measures, increasingly evident in new road construction, are not new. As long ago as 1855 the Great North of Scotland Railway built retaining walls to preserve the edge of a country graveyard and to solve other land acquisition problems.

On a grander scale it has been possible in the past to fit, rather than to impose, well-aligned railways into city architecture. Princes Street Gardens in Edinburgh are the very heart of the city, lying as they do in the valley between the castle and tall buildings of the Old Town on the south side and Princes Street and the Georgian New Town to the north. Yet trains pass right along the bottom of the gardens (Figure 14.2). The railway is an inherent part of city centre layout. Its place in city scenery is described in the next chapter.

14.3.4 Relocation of Historic Buildings

If buildings cannot be avoided, perhaps they can be be bodily moved or completely dismantled and re-erected elsewhere. Examination of foundation and structural conditions should reveal whether relocation is possible; a detailed plan and cost estimate will indicate whether it is feasible.

Removal as a single unit—by jacking and wheeled transport—may be practicable for modest structures with timber frames and only light, replaceable foundations.

The viability of piece-by-piece reconstruction depends on the practicability and cost of dismantling and re-erection, and on the value to posterity—perhaps in a different setting—of the reconstructed building. One practical consideration is the capability of the reconstruction team to record, dismantle and reassemble the structure and to perform all the skills, some probably ancient, in erection, jointing, roofing, mortar, wall surfacing and detail finishes. Another key factor is replacement of broken or decayed elements.

Construction of the M40 motorway extension, completed early in 1991, included the dismantling and relocation of a seventeenth-century Grade II building near Warwick (Moore 1990). Union Railways' (1993) environmental proposals include dismantling of a listed building in Kent; two other Grade II buildings were dismantled rather than demolished when clearing space for the Cheriton Channel Tunnel Terminal—in each case presumably with a view to re-erection elsewhere. However, the value of any relocated edifice may be reduced if it is moved out of its original environment, e.g. away from surrounding buildings in a homogeneous group.

14.3.5 Demolition and Historical Record

In choosing which of a number of buildings should be sacrificed for railway alignment, expert guidance is again needed. Among second order old buildings it is not only the more substantial, built by relatively prosperous merchants or farmers, which warrant investigation. Much less common survivors are labourers' dwellings, many of which were pulled down during nineteenth-century sanitary reform. Smith (1985:12) notes 'Any surviving dwellings of the very lowest orders of society . . . will merit the investigator's time and attention before many 'better' structures of a century or two of greater antiquity.'

The key requirement when a structure is demolished is for a record to be made. Expertise in the historic period concerned and in vernacular (regionally characteristic) architecture should be applied—first in overall assessment of the building as to its proportion and its setting. This preliminary overview will indicate which particular features should then be investigated and recorded—photographically, on drawings or by description. Of particular interest to future historians may be dates of ownership, various uses and stages of extension or rebuilding; structural framework and materials; and techniques of roofing, walls, finishing and decoration.

14.4 OLD AND NEW RAILWAY ARCHITECTURE

14.4.1 Historic Achievement and Future Opportunity

There is a huge heritage resource in Victorian railways. Some of this is seen as industrial archaeology, some as railway architecture. A definitive British work on the latter is *Railway Architecture* (Binney and Pearce 1985) written by members and associates of SAVE Britain's Heritage. This stresses the value of a wide variety of railway features, decries the destruction of some of these in post-war modernisation and offers suggestions about the modern reuse of what may otherwise become derelict structures.

A first opportunity for railway planners is in continued use of existing railway features—in new railway development or, where elements are redundant, in conversion to other activities. However, preservation should be undertaken for the best effect on the total environment, which might sometimes be better *without* old buildings of unexceptional character.

A second opportunity lies in developing new architecture appropriate to twenty-first-century railway development and for posterity.

Railway architecture is evident in the following:
- *Structures* carrying the line—bridges, viaducts and tunnel portals often

exhibit outstanding features of Victorian engineering with both functional and decorative architecture; modern railway systems should incorporate the best of these suitably contrasted with appropriate contemporary equivalents.

- *Major city terminals*—monumental stations are now city heritage; this justifies preservation of the best features such as stone masonry, brick-work, steel columns and roof frames.
- *Country railway stations*—some of the few still in use do not require per-manent station staff; buildings can be preserved as dwelling houses, in isolated locations, or as small offices or shops in more built-up surround-ings; often stations are fine examples of local styles and building materials (see Figure 14.3).
- *Locomotive and carriage works and sheds, offices*—redevelopment can be similar to that of disused mills and warehouses, preserving the best exter-nal features in reconstruction; new uses can be related to development of adjacent land formerly used for sidings.

The possibilities for preservation of old railway structures and the sensi-tive construction of new ones are best illustrated with reference to viaducts. The incorporation of heritage structures in modern development can be examined with reference to stations.

Figure 14.3. Stamford. A country town station in vernacular architecture

14.4.2 Viaducts

Great Victorian bridges have only recently been recognised as part of our national heritage. Fifty British railway viaducts are already listed for preservation; many of those still carrying trains are not. The need to appraise their heritage value arises:

- on lines of only marginal economic value, like that over the Pennines from Settle to Carlisle, where continued train operation depends on expensive repairs to key structures;
- on closed lines when safety considerations or development plans require their repair or demolition.

Among older structures still in use Ribblehead viaduct is in the first category. Public appreciation of its heritage significance was made evident over many years during which closure of the line was mooted. Its present reprieve and recognition of the line's heritage and social value are only possible as long as subsidies are found to pay for maintenance and repair of the viaducts. Meanwhile, plans backed by English Heritage seek to designate as a conservation area the whole Settle and Carlisle line on which the viaducts stand (*New Civil Engineer* 7 March 1991).

Britain is rich in brick structures—buildings, bridges and retaining walls. Construction of the last railway built into London, the Great Central in 1899, involved a number of fine viaducts, notable for their graceful design and the quality and complexity of their brickwork. One of these viaducts (shown in Figure 9.3) still carries trains across the Misbourne valley in Buckinghamshire. Without any structural alteration it now allows the M25 motorway to run through its arches.

Another prominent multi-arch brick viaduct carries the line from Swanley to Chatham across the Darent valley in Kent (Figure 14.4). The viaduct overlooks the villages of South Darenth and Sutton-at-Hone on the northern side and is seen in a rural but not outstanding setting from the south. The 1989 Channel Tunnel Rail Link (CTRL) route was to have passed over a parallel viaduct on the southern side facing *away* from the villages.

Nevertheless, a great deal of concern was expressed in the villages about the noise of the new trains that would be generated on the new parallel structure. Fears were also expressed about the effect of a new structure on the landscape. The latter is already dominated by the existing viaduct; the visual impact with an extra structure would have been unchanged on the village side and very similar from the south if a sympathetically spaced second structure blended with the brick arches of the first one.

The confused attitude to the appearance and architecture of this and other railway viaducts can be illustrated by a series of media events as follows:

1. In 1989 a very young lady appeared on television in Channel 4's 'Com-

Figure 14.4. This viaduct at South Darenth in Kent was to have been duplicated by a similar structure for the route proposed in 1990 for the Channel Tunnel Rail Link

ment' slot and pleaded that the 'lovely viaduct' at South Darenth should not be spoilt by building the new (CTRL) railway in front of it.

2. In 1991 the Government announced that, because of the opportunity to regenerate the East Thames Corridor, the CTRL route would now come into London from the east and that therefore there would be no new construction in the Darent valley.

3. A painting in the Royal Academy 1992 Summer Exhibition entitled 'Celebration' depicted a group of people overjoyed at good news (presumably 2 above) and celebrating beside a house in front of a viaduct with a (conventional) train going over the top of it.

4. A letter to *The Times* (Edward-Collins 1992) referred to another, closed but listed, railway viaduct and asked 'by what logic a structure which is of no further use has to be retained when, if it were not there already, its construction today would be opposed by . . . the same people who now support its preservation'.

Railway viaducts are recognised widely as important visual amenities. But in some quarters anything new is bad. Older structures may be regarded neutrally or occasionally as eyesores, but when they are threatened they become industrial archaeology in need of restoration. There may be unworthy candidates for preservation, especially where structural improvements have been insensitive to the original designs. Ugly strengthening measures, in the form of massive blocks placed (many years ago) around the piers of Big Water of Fleet viaduct in south-west Scotland were apparently the reason for the structure being removed from a protective listing by Historic Scotland (Gill 1992).

To achieve a positive approach to railway viaducts, engineers and architects need only take advantage of recent or imminent advances in technology and materials for all types of bridges. Cable-supported bridges can already give the rigidity required for long-span railway crossings; and traffic-activated control systems will be able to ensure the necessary aerodynamic stability. Development of new plastic and fibre-reinforced materials, advanced adhesives and composite designs will be used both in new bridges and in the extensive work which will be required for rehabilitation of older structures.

New opportunities therefore lie in the design, materials and methods of construction of viaducts; but little is new in their siting. Twinning of viaducts, for instance, first occurred more than 150 years ago when the Great Western Railway duplicated Telford's Llangollen Canal crossing at Chirk on the Welsh borders (Figure 14.5). Classic examples involving *different* types of

Figure 14.5. The railway at Chirk on the Welsh border crosses a viaduct twinning that of the canal constructed a few decades earlier

bridges which have occurred since are the rail and road crossings over the Forth and the Tamar and *three* bridges (and now a submerged tube) across the Conway. The Union Railway will probably duplicate the M2 motorway crossing of the River Medway in Kent. Use of common sites for a number of bridges increases the historic and structural interest at those sites whilst preserving other stretches of the valley in their semi-natural state. As for popular opposition to new structures, this was certainly aired, with less effect but more justification, in some of the early urban viaducts built above houses in cities like Manchester in the 1840s.

Disused viaducts are not attractive just as impressive civil engineering monuments. They actually *lead* somewhere. Their role as spectacular footpaths is nothing new. The sight of the Pont du Gard, a Roman aqueduct, has been an attraction to visitors for centuries; to walk across it has always been exhilarating. But, particularly where, in some railway age viaducts, the solid masonry around the piers hides softer bulk material inside, a century of wear and wet weather calls for repairs which pedestrians alone are unlikely to finance. Organisations like the Northern Viaduct Trust have managed to attract the funds and to satisfy planning requirements about the acceptable appearance of the renovated structures.

Expenditure to ensure stability of a structure, rather than to carry trains, is not necessarily excessive. Smardale Gill viaduct, on the long disused Barnard Castle to Tebay line, was repaired only as a monument. This contribution to grand landscape was achieved both at modest cost and at minimum disturbance to valuable ecological habitat by a combination of carefully planned approach, traditional (steeplejack) access for repairs and effective treatment methods (Hayward 1990). The viaduct's status was raised to that of a Grade II* building only when it began to deteriorate visibly, more than 20 years after the tracks had been removed.

There is no reason why modern bridges should not eventually attain a heritage status similar to those on disused lines. But nor is there any obligation for totally new railway structures to follow nineteenth-century fashions. Modern equivalents may be more appropriate. Since completely new railway bridges have been needed only rarely in the recent past, there may be lessons to be learnt from the design of modern road bridges. Some of these are magnificent, some mundane and unnecessarily utilitarian. Otherwise acceptable railway bridges may be spoilt by visually obstructive or intrusive additions such as opaque noise barriers or unnecessarily massive overhead electrical equipment. Where it is necessary to impose major new structures into the landscape, open competition can be encouraged for their conceptual design among the best architects and urban and rural landscape designers. Review by bodies such as the Landscape Advisory Committee is a frequent requirement. Themes can be adopted for new railway infrastructure to provide modern equivalents to castellated tunnel portals and Victorian station façades.

14.4.3 City Stations

Stations have two parts—the platforms and tracks (which when roofed over were referred to as 'train sheds' by the pioneers) and the ticket halls and passenger and staff accommodation in adjacent buildings. Early emphasis was on spaciousness. Opportunities were taken to locate accommodation buildings so as to present a grand façade facing the city streets—at the buffer end of the train shed if the station was a terminus or at the side where through trains were accommodated.

It is sadly too late for preservation of the Doric Arch at the entrance to Euston, demolished as an obstacle to post-war station modernisation, or of the arched train shed roof at Manchester Central, not needed when the whole station became redundant. However, a superb example of station reuse is the Musée d'Orsay in Paris where magnificent architecture and steelwork now encase a national art collection. St Pancras's Victorian Gothic frontage, for most of its life an office building, may revert to its original use as a hotel whilst the grandly roofed train shed will one day provide spacious accommodation for the Union Railway.

City station redevelopment since the war has gone through stages of total clearance (as at Euston), rearranged platforms and replacement of roof structures, and above-train development of new commercial property.

Arched station roofs—hammer-beamed as at Bristol, huge spans as at St Pancras or curving as at York or Brighton—are a fine example of the best Victorian engineering. The roof of Paddington Station is a Grade I listed structure and English Heritage supervised its renovation. Glazed arched roofs on the grand scale followed a style initiated by the famous greenhouse designer, Joseph Paxton. He influenced such roofs for country stations in his native Derbyshire, after he had created his glasshouse at Chatsworth but before he planned the Crystal Palace. Such magnificent structures are not always feasible for modern railway stations, which are often underground, beneath existing stations or airports. But the Channel Tunnel rail terminal at Waterloo sits beneath an asymmetric and tapered transparent roof of modern design, different but equal in grandeur to that of the best wrought-iron structures in Victorian train sheds. In Europe, new stations at Seville, Lyon and Charles de Gaulle airport outside Paris have taken a prominent place in modern architecture and one planned for Utrecht could be equally impressive.

Some city stations have been submerged by overhead property development. London's Liverpool Street preserves its best roof features with new concourses and high level walkways, financed as part of the surrounding office blocks of the Broadgate development. But New Street station in Birmingham is lost under a major shopping complex whilst the airline terminal fitted into the roof at London Victoria sharply reduces the light and spaciousness of the platforms. Fortunately, the mid-1990s are seeing the emergence of a new generation of stations in Western Europe in which light, space and colour are given much more emphasis and in which roof spans and architecture incorporate new materials and techniques in the same way as do modern bridges.

14.4.4 Contry Stations

New stations outside towns are usually associated with car parks. Green landscaping could reduce the intrusion of rows of cars into rural views. In the twenty-first century public transport is likely to be dominant in the cities, private cars in the country. Parkway stations should balance, with an appropriate local but unashamed style, the grander façades and wide train sheds of sensitively modernised city stations. Some twentieth-century stations (such as Bristol Parkway) were built with narrow platforms and short cantilevered roof sheets; as a result, they are as starkly utilitarian as the car parks surrounding them. Use of locally traditional building materials and more spacious passenger accommodation would be better suited to the long-term use envisaged for station buildings.

14.5 AMENITY LAND

14.5.1 Popular or Remote Recreation

Land on which people can enjoy themselves varies from that specifically set aside as public playgrounds or parks, through sports grounds and golf courses to any more remote areas in which individuals have reason and opportunity to wander.

In *sports arenas* people enjoy watching the actions of skilful players. The spectators are the 'crowd'; they are seeking communal pleasure, not quietness or solitude which passing trains might disturb. The facilities provided are constructed both to permit high standard sport and to accommodate a high density of people. If railway or any other type of development should destroy or spoil these facilities, then the damage, compensation and necessary mitigation can be assessed in the same way as for any commercial property.

Remote countryside is the opposite type of land. Whilst not all walkers are seeking solitude, they are certainly avoiding crowds and traffic noise. They are attracted by scenery and deterred by unexpected blemishes on it. The impact of new railways in these circumstances may be judged in terms of nature conservation, described in the last chapter, or of scenery which is the subject of Chapter 15. Any *diminution of remoteness* should be assessed for the region concerned on the basis proposed for quiet countryside in Section 11.4.

14.5.2 Access to Amenity Land

In a well-populated region like south-east England, most land which is not built-up or under arable agriculture is likely to be hilly, wooded or with water features and attractive as an amenity. Such land may be:

- generally available to access by pedestrians, horses or mountain bikes;
- reserved for particular open air sports such as golf, gliding, shooting, moto-cross, war games or water sports;
- private estates—some of which are open to public access whilst others are near to or traversed by public footpaths.

The amount of land actually lost to railway development is restricted to its narrow width. However, a railway can also act as a barrier in recreational countryside. It can sever roads leading to entrances or car parks and can cut off footpaths or bridle-paths on which walking and recreation is enjoyed; but loss of amenity value can be limited or even reversed by suitable provision of access across the line.

The route and standard of *roads* across new railways should be decided in conjunction with the owner's or countryside agency's plans for encouraging or restraining people's movements and parking in particular areas.

In replacing or rerouting *paths* across railways, designers should:

- maintain or enhance ecological and landscape features;
- acquire wayleaves or purchase land where necessary to enhance the value of the footpath network in the vicinity of the railway;
- maximise viewpoint potential, including from the top of railway earthworks or overbridges;
- provide sympathetic, rather than visually obstructive, railway fencing.

14.6 USE OF DERELICT RAILWAY LAND

Disused railway land is a valuable resource. Its disposal and reuse is an important development opportunity with considerable environmental implications. In heavily populated areas amenity use may be an attractive long-term opportunity. Alternatively, planned building on spare railway land may relieve from development other land in the vicinity which would be better conserved.

14.6.1 Old Railway Lines

The pre-1960 British railway system resembled what engineers call a 'redundant' structure. It had more lines than were necessary to join up the key centres. Subsequent line closures included most of these redundant routes. The longest main lines closed were the Great Central, from London to Sheffield, and the 'Waverley' route from Carlisle to Edinburgh—both over 150 km—but there were also many country lines, for instance in Wales, north-east Scotland, Norfolk or Cornwall, which lost their train services as traffic between smaller towns turned to the roads.

New uses for disused lines should first include possible revived railway use. Otherwise any long strip of spare land is an opportunity for improved communications in another mode. In hilly or attractive rural surroundings this will be footpaths. In flatter country routes may become farm tracks and occasionally cycleways. In urban areas occasional opportunities arise for using old railway routes as sections of road bypass, such as at East Grinstead in Sussex. There are numerous examples of new trunk road sections built along disused railway routes but these almost invariably require wider cross-sections and additional earthwork. Conversion to tracks or roads may be straightforward or may require expert landscape architecture. Where gaps have been left at bridge sites, a suitable new structure has to be provided.

Non-transport infrastructure installed on disused lines has included a radio telescope near Cambridge and electric power transmission lines carried discreetly through the Old Woodhead Tunnel instead of in the open across the Pennines.

Commercial uses include acquisition by adjoining property owners to acquire extensions to their land on which to erect new buildings or to expand farming operations. But these block access along the original line which probably has a much greater long-term value as a light rail route, pathway or wildlife corridor.

Environmental uses include returning land to nature, touched on in the last chapter. Success of terrestrial habitat may depend upon protection or management for particular species. But provided the land retained is still a continuous strip, it is analogous to long hedgerows and provides scope for biological migration. Because of the hard ballast surface and freely drained formation, old tracks provide aquatic habitat only at the bottom of embankments or, rarely, in wet patches in cuttings or tunnels.

14.6.2 Disused Railway Yards

Numerous empty land areas, up to a hectare in extent, were made available by the discontinuance of freight services at town and country stations. Disused goods depots and coal yards are common locations of empty railway-owned land. Less common but much more extensive are marshalling yards, engine and coach sheds, and locomotive, carriage and wagon construction and maintenence works. About 10 such areas were located around the outskirts of London alone. Typically these were around 20 hectares in extent. But the former Great Eastern Railway depot at Stratford covered over 50 hectares.

Wide areas of derelict railway land can be put to many uses including the following:

1. Provision of new railway facilities; on the Channel Tunnel Rail Link, for instance, new stations and depots might use at least part of the disused space formerly devoted to Ashford and Stratford railway works; for freight transfer operations, 10 hectares is enough space for transfer of at least 250 000 tonnes of freight per year in a road/rail freight depot with storage facilities, or 1 million tonnes or more at an intensively used container terminal.
2. Recreational land, for which there may be a high demand in urban areas; this use was included as an adjunct of a 1989 proposed King's Cross/St Pancras station development plan.
3. Car parks; the whole of the former Brighton locomotive works is now a station car park.
4. Commercial and residential development, if items 1 and 2 do not claim higher priority; such development is a popular source of short-term finance for railways.
5. Continued disuse, taking appropriate measures for protection or management, e.g. as a nature reserve—at least until a future generation justifies a better purpose for it.

Planning authorities and railway landowners have the opportunity to balance the demand of commercial, transport and environmental interests. They should regard disused railway land as an asset of particular value and carefully control how flat surfaces are used or built on, how cuttings are filled and bridges and embankments removed or whether these should be retained or managed in their disused condition.

15 Railways in Scenic Landscape

15.1 PAST PRECEDENTS AND FUTURE PLANS

Scenery is enjoyed by people but exists as a resource. Most scenery has been altered by man and it is man's use and misuse of land resources which determine whether attractive scenery will remain so for future generations. Grand scenery dwarfs man's activity; subtler scenery is more easily modified but may in itself constitute historic landscape.

This chapter describes railway situations in different types of scenery. Sections 15.2 to 15.5 look at how railways affect scenic views in a variety of conditions from grand mountain landscapes to flat country and from rugged coastlines to desolate upland.

Section 15.6 examines, with particular reference to south-east England, the effects of railways on views of attractive but populated countryside; the impacts are probably typical of those often encountered in planning new European high speed lines. Section 15.7 deals with circumstances which arise where railways enter any cities with particularly scenic features.

In each situation it is necessary:

1. to describe the scenic features which are attractive and potentially long-lasting;
2. to identify scenic impacts of railways;
3. to determine, by examining established examples, whether these impacts are significant and what will be the implications of railway development on regional scenery.

In this way the lessons about scenic impact of existing railways—where indeed there is *any* impact—can be used as precedents to predict what may be done to avoid unnecessary blemishes to views by future railways.

Section 15.8 summarises the risks of scenic intrusion and the measures that can be taken to best preserve the landscape.

Scenic features which have to be assessed include, firstly, visually significant land-forms, physical geography, vegetation and ecology; next, the semi-natural but man-managed fields, woods, wetland, streams and drainage; and then the man-made features such as buildings, quarries or roads.

Scenic impacts on each type of view can be examined:

• first, in terms of men's activities, from ancient industry and agriculture to modern roads and structures;

- next, of transportation routes generally and *linear* developments;
- and then, of railways with their characteristic straight alignment, viaducts and tunnels.

Adverse impacts comprise loss or severance of specific scenic features, and creation of harsh discontinuities or discordant structures. Railway construction may also interfere with physical processes—such as the flow and morphology of rivers and estuaries—which determine the nature of scenery over a wide area. Concordant effects can be enhancement of man-made landscapes by introduction of sympathetically aligned or architecturally welcome features.

Scenic implications of new railways in *particular situations* can be assessed in comparison with similar but existing situations. The *regional* significance of new impacts on scenery can be gauged in terms of the extent of similar scenery, spoilt and unspoilt, elsewhere in the locality.

15.2 MOUNTAINS AND UPLAND

15.2.1 Mountain Scenery

The most scenic features of mountain country—listed in order of their permanence—are cliffs, glaciers, waterfalls or streams and trees. People's preference for these depends upon their particular interests.

If mountain topography is steep it provides fine views, both looking up at slopes and cliffs and looking down from the top. Branch line railways bring people to admire this scenery and to explore it. In less steep but still elevated upland or moorland the views are less spectacular, certainly in photographic terms; the often empty and uniform nature of high plateaux attracts less popular interest but is still valued for its contrast to inhabited country. Some main line railways have to cross such territory.

15.2.2 Human Intrusion into Mountain and Upland Scenery

Mountain scenery is among the most natural of landscapes. Nevertheless, it can be spoilt, or at least rendered less natural, by human development.

The impacts of man-made activities in upland areas were first occupational, related to agriculture, livestock, forestry and mining. These have been followed for a century by hydro-electricity generation and especially by tourism, sport and recreation. A demand for transport to scenic mountain resorts ensued, first by railways, later by roads and cableways.

Three types of railway situation arise in mountain scenery:

- very steep rack-and-pinion or funicular railways which are built to reach mountain peaks or summit ridges;

- moderately steep railways, usually narrow gauge branch lines with rack sections, traversing mountain sides or narrow valleys to reach mountain resorts (in Switzerland) or former mining communities (as in North Wales);
- main line railways, for faster and heavier trains, which have to cross broad mountain ranges through tunnels or by the lowest passes in generally less steep upland.

Rather than seriously intruding into mountain landscape, railways usually *give access* to places in or near to it. An intrusive exception is any railway taking a direct skyline route to a summit.

15.2.3 Mountain-Top Railways

Sharp mountain peaks are outstanding elements of scenery. Less spectacular summits are still good viewpoints.

Railways make no positive contribution to scenic resources. They can provide easy transport for people who are prepared to pay to reach a summit—whether to see the view or to descend on foot or skis, at leisure or at speed.

The sharpest peaks are generally inaccessible to conventional railways. But suspended cable *téléférique* systems can go anywhere sufficient finance and helicopters can effect their construction. *Téléfériques* and their high terminal stations are an undoubted scenic intrusion; but they are not railways and are therefore of only marginal relevance to this book. They are an important part of the infrastructure which attracts people to the resorts and, together with chair lifts, are the most intrusive element in the scenery of high mountain slopes—much more discordant in their abundance than any mountain railways.

The Eiger–Jungfrau range in the Swiss Alps is no dull round-topped hill; it has a sharp serrated summit ridge, permanent snowfields and notorious cliffs. Even so the *Jungfraujoch Railway* was tunnelled inside it, not to an actual summit but to a col at a height of 3454 metres. Because the upper part is entirely in tunnel, the railway itself cannot be seen in the views of the Eiger North Face or the adjoining snow slopes. It is the associated restaurants, hotel and viewpoints which impinge on mountain-top scenery; and these visual blemishes would not exist without the railway.

The highest open air railway in Europe is that from Zermatt to the Gornergrat (3089 m). Although this viewpoint is well below the attitude of Monte Rosa, the Matterhorn and the other peaks nearby, the line is certainly in the *mountain-top railway* category. However, even the open air Gornergrat Railway can hardly be judged a scenic intrusion in an area from which cable cars and ski-lifts extend prominently across snowfields to even greater altitudes.

In Britain most mountain ranges have been eroded for rather more millions of years than have the Alps. As a result the summits and ridges are usually less high, sharp and steep, with some cliff-lined north-facing corries resulting from the last ice age but generally with more accessible shoulders. Railways could be built up almost any of these hills but commercial demand has been such that only the highest summits in Wales and the Isle of Man actually succumbed.

The *Snowdon Mountain Railway* ascends 955 metres along 7.5 km of steep mountainside route emerging for the final 500 m length along the top of a ridge. The summit of the mountain is a popular and therefore usually crowded place; by no means the majority of people go there by train.

The great popularity of hill walking is itself a problem to mountain scenery in the more accessible and easily erodible hills of Snowdonia or the Lake District. Construction of a new railway would not be allowed to any summit today. But the railway from Llanberis to the top of Snowdon was built nearly 100 years ago. Even then it was seen by some as a sinister commercial intrusion into nature, by others as a welcome attraction. A contemporary description (Ward Lock 1897:177) asserts that 'the whole route has been selected so as to cause as little disfigurement to the mountain as possible'. The same range of opinions still prevails today. But, since it is part of our Victorian heritage, probably a majority of people who notice the railway accept it. Seen from neighbouring hills, the sight of steam trains struggling

Figure 15.1. Intrusion by the Snowdon Mountain Railway arises more from its contrast with the background scenery than with the comparatively dull foreground of the gentle Llanberis ridge up which the train climbs

up the north-west ridge of Snowdon is surprising but not uninteresting; seekers of true solitude have in any case to seek less popular hills. If a future generation wishes to remove the railway, the track could easily be dismantled leaving less trace of its existence at higher levels than that of the substantial footpaths.

The impact on the environment of mountain-top railways has to be assessed more in terms of the buildings which serve the train passengers than of the track.

15.2.4 Steep Mountainside Railways

Where they lack cliffs or rock features, the *scenic nature* of mountain slopes depends mainly on whether they are forested. Trees may be natural, as in much of the Alps or in the few remaining areas of Scots pine in the Highlands. But most of the pines were felled or their natural regeneration prevented by grazing animals, leaving open deer forest, grouse moor or sheep pasture. Much British mountain woodland is commercial plantation.

Where a railway traverses a mountainside, trees may well make it invisible except for glimpses of a passing train. So railway lines cutting through the trees will seldom be any more discordant than the straight edges of forests or the sharp changes in tone between trees of different types or maturity.

Where they are not hidden by trees, railway earthworks, structures and track are much more visible. The significance of these railway features in the landscape depends upon how many man-made elements already exist. Such elements are few on remote Scottish hillsides; they are more numerous in North Wales where there are stone walls, remains of quarries and mining cableways, hydro-electric pipelines and electricity transmission lines.

Situated in the Snowdonia National Park, the *Ffestiniog Railway* is now a key feature of tourism, the industry which replaced slate quarrying during the twentieth century. Tourists travel to Snowdonia primarily for the scenery; but many also visit its enthusiast-run railways to combine sightseeing with industrial archaeology and the mechanical fascination of Fairlie patent steam locomotives.

The Ffestiniog Railway's lower route is through thick woodland with occasional lakes; in middle distance views, the trains are unobtrusive, often invisible among the trees, noticed only by occasional whistle notes.

After climbing through the broad-leaved woodland section the railway emerges in higher open ground where both modern and early industry are far more prominent in the surrounding landscape than is the track itself. Modern development is represented by commercial forestry and the Ffestiniog hydro-electric pumped storage scheme; the latter has introduced a lake (lower reservoir) and power station beside the line and, higher up below the col of the Moelwyns, the Stwlan dam and its small deep upper reservoir. The railway itself, now realigned to avoid the lower reservoir, incorporates the

Dduallt loop, a feature more characteristic of Alpine or Himalayan railways.

The Ffestiniog Railway's upper terminus is at Blaenau Ffestiniog, a town at the centre of the most visibly intrusive, perhaps stark, evidence of the earlier and very extensive slate industry. So prominent are the waste heaps and so ordinary the slate roofed terrace houses of the town itself that the area has been excluded from the Snowdonia National Park. Other signs of old slate workings are common within the Park itself—unroofed cottages, mine adit entrances and often steeply inclined formation of horse or rope-hauled 'plateways'. All these weathered remnants, blended now with dry stone walls and natural rock and scree scenery, have become permanent features of British slate or mineral bearing hillsides. The less easily weathered and more prominent high waste slate dumps are aspects of industrial archaeology much more intrusive on landscape than narrow gauge railways.

15.2.5 Main Line Railway across Upland

Uplands are the higher parts of a country—wide enough to permit any surface transport routes which need to cross them but not necessarily steep enough to be scenically outstanding.

In the Alps, the main north–south routes have to cross mountain ranges so high that long (14 to 20 km) tunnels are inevitable. But east–west routes, such as that from the upper Rhine valley to that of the Rhone, run parallel to the main ridges; the *Furka–Oberalp Railway* runs across upland at an altitude of over 1000 m for more than 90 km between the Rhine and Rhone headwaters. This is more than twice the equivalent length on any of the major north–south routes. Although there is one very long tunnel, completed as a short cut only in 1982, the great majority of the line is in the open.

To travel on the Furka–Oberalp route is a popular scenic experience; but the country traversed lacks outstanding mountain scenery except in occasional distant glimpses. The few resorts passed, such as Andermatt, exist for the ski slopes rather than for summer pleasures. When roads and railways do not tunnel beneath mountain ranges they seek the lowest, most practicable passes available. These tend to be in the more barren and unremarkable watershed terrain.

The scale of Alpine mountain ranges is such that several main lines require long or spiral tunnels to negotiate them. The same is true in mountainous areas of Canada, New Zealand or Iran or the fiords of Norway. Scottish uplands are less formidable. Despite its name and although it crosses the highest railway pass in Britain, the 820 km long former Highland Railway system included only three tunnels with a modest total length of less than 720 metres!

In England the *West Coast Main Line* climbs 280 m in 50 km up Shap Fell, between the Pennines and the Cumbrian mountains, and then Beattock

Bank, of similar height, in southern Scotland. These are all comparatively gentle uplands too wide to tunnel through but suited to moderately steep surface routes.

Railways across Scottish moors are usually single track which minimises any impact on the scenery at ground level. Some sections, like the *West Highland Railway* across the eastern side of Rannoch Moor, take routes where no access is available to car-borne visitors. But solitude, with fine distant views, is assured for walkers alighting from trains at roadless stations like Corrour.

In wide open upland spaces, cropped by sheep or deer, managed for game birds or commercially forested, conditions are less truly natural than they are in many fenced railway enclosures. The cuttings and embankments of railway lines, such as the line over Shap Fell, provide linear strips along the lineside which are greater in extent as habitat for flora and fauna than they are as ballasted permanent way.

Scenery near passes is often less spectacular than in the valleys by which they are approached; the upland is grand mainly in its remoteness, wide extent and occasional views of mountains in the background. Often the terrestrial and aquatic habitat, and therefore the nature conservation potential, is rich because of that remoteness. Human development comprises forestry, livestock ranching, limited recreation and occasional roads. The visible impact of railways in such scenery is even less significant.

15.3 VALLEYS AND GORGES

River valleys are natural transport routes. They are also important elements in scenery; and a greater length of main line railway runs along broad valleys in gently undulating country than in the steeper mountain scenery described in the last section.

Scenic damage can result when road or railway construction disturbs the green features of softer landscape or the rockier ones in harder geology.

Valleys are associated with views of slopes, often wooded, and with pastoral or riverside features at the valley bottom. The cliff scenery of *gorges* in hard crystalline limestone contrasts sharply with the often austere plateaux they intersect. But some of the wider gorges have provided relatively easy routes for main line railways through upland areas which elsewhere pose formidable obstacles.

15.3.1 Railways in Valley Landscapes

Striking valley scenery is associated with geology (rock and softer landforms) especially in gorges and upland valleys; with vegetation of all types; and with water features. The most extensive views usually relate to hillsides and valley slopes, the more intimate ones to valley bottoms and river banks.

Railway routes *along* valleys keep as close to the surface contours as is practicable and affect mainly the scenic pattern of features on the slopes. Routes *across* valleys have to be elevated above the valley bottom levels on more visually intrusive viaducts or embankments.

The perceived beauty on *valley slopes* is most commonly related to their vegetation. The sight of natural or mixed woodland, scrub or moorland contrasts favourably with that of more familiar and utilitarian growth such as arable fields, regimented vineyards or horticulture. Assuming that the climate and steepness of slopes are not such as to make erosion a serious danger, temporary displacement of vegetation need not be an everlasting loss of scenic resource. Most grassland can be reinstated in a few seasons, scrub woodland in a decade or two. However, it takes a century to re-create tall mature woodland or chalk downland whilst thick road metal and railway ballast seal the ground indefinitely unless they can be removed or covered.

Valley bottom wetland and river features are more sensitive and scarce; so damage can be permanent. Protection of such terrain is a matter more of nature conservation than of landscape preservation; ecology and scenery are in any case closely related. Provisions to protect wetland and watercourses, discussed in Chapter 13, can be of benefit also to valley bottom scenery.

15.3.2 Routes along River Valleys

Railway construction *along* valleys involves alterations to natural topography and drainage, according to the limitations on track curvature which operational requirements, particularly speed, dictate.

If an acceptable alignment can hug the hillside, only modest or one-sided cuttings or embankments are needed; but for faster services in steeper terrain deeper cuttings, higher embankments and occasional tunnels or side valley crossings become essential. The visual significance of contour-following alignments is a continuous but not easily discernible railway track; the straighter track is more occasionally but prominently seen.

The slopes of cuttings are only prominent where high one-sided cut along steep hill sides reveals the underlying formation. Railway cuttings need to cut through both hard and soft rocks. In softer formations they provide valuable protected habitat for flora and fauna; in hard rock they offer interesting evidence of subsurface geology not normally seen. At the end of Chapter 13 we saw examples of geological as well as nature conservation associated with railways. Like quarries, railway cuttings carry both biological and earth science interest. *Earth science conservation in Great Britain* (NCC 1991) describes a strategy for the conservation of geological SSSIs and less formally recognised rock formations, fossils and land-forms. Many of the latter are relatively erodible formations such as glacial relicts—features of the scenery to be avoided by new roads or railways but which could be mistaken for artificial earthworks or ready sources of construction materials.

In the softer non-glacial geology of south-east England four rivers—Wey, Mole, Darent and Medway—flow north towards the Thames, cutting through the North Downs between Guildford and Maidstone. All accommodate active railway routes.

Typical of these valley routes are those railways which follow the modest Darent and the more substantial Medway rivers in Kent (see Figure 15.2). These rivers have cut gaps 10 km in length through the chalk of the North Downs. Railways run along both valleys connecting the towns below the steep southern side of the Downs with those on the more gradual northern slopes of the London Basin or the Thames Estuary.

Lines from Sevenoaks and Maidstone join to run northwards down the Darent valley then cross towards Swanley and London—one of the 'boat train' routes used by passenger traffic from the Channel Tunnel until the high speed rail link is built. The present Darent valley line from Otford to Eynsford fits unobtrusively into the landscape below the wooded escarpment on the eastern side. Its route is alongside a road; half is in cutting. Although trains can be heard, they can be seen only on certain sections. The track is not visible across the valley.

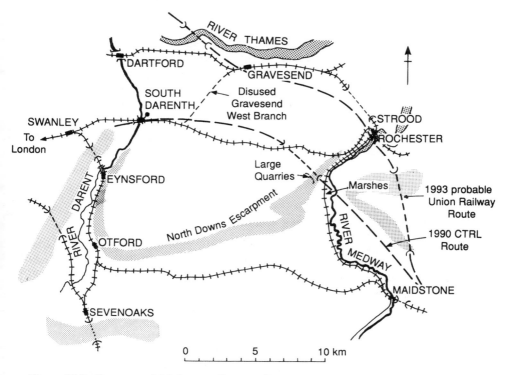

Figure 15.2. Darent and Medway valleys—railway routes

Along the left bank of the much wider Medway valley runs a line from Maidstone to Strood (Rochester). The width of the valley enables the line to follow a low level course, unobtrusive in comparison with the chalk quarries, a cement works and ribbon development along a main road. At the northern end of the valley another railway, that from Swanley to Rochester, joins the west bank at a higher level. From the eastern side of the valley neither line can easily be detected unless a train is sighted.

Except where viaducts cross the valleys, neither railways nor single carriageway roads are easily discernible among the woods or hedges of any unspoilt countryside. The existing railway routes along these river valleys are no more visually unattractive than any of the other man-made structures along the transport corridors.

15.3.3 Routes across Valleys

A main scenic impact is the sight of high viaducts or long embankments where a railway crosses from one side of a valley to another. The appearance of viaducts as heritage structures has been mentioned in Chapter 14; their impact on total valley scenery is one which is inescapable but certainly, for nineteenth-century brick arches, one that is aesthetically acceptable.

High and prominent viaducts carry two railway lines across the River Darent's valley through the North Downs. Arguably any long viaduct solution is visually preferable to any high embankment. Such an embankment obscures most of the Darent valley's profile where the M20 bypasses the otherwise attractive village of Farningham. The M20's visual impact is certainly more obstructive than the railway viaducts at Eynsford and South Darenth.

Railways cross the River Medway by lower level bridges at Rochester and Maidstone—respectively north and south of the steep-sided part of the valley through the Downs; so there is no greater visual intrusion at the crossings than that of any ordinary urban bridges.

15.3.4 Routes through Undulating Country

An example of a railway network in equally complex but less populated country is that of the former Great North of Scotland Railway in the north-eastern shoulder of Grampian region. The line from Aberdeen to Keith (53 km) and thence to Inverness is the remnant of a system north-west of Aberdeen, once four times as long (see Figure 4.3); the surviving route passes through major and minor valleys in softly attractive, well-wooded or pastoral sub-montane countryside.

Evidence of the lines which were closed in the 1960s is not easy to detect except at structures like the viaduct at Cullen. Elsewhere there are many locations where the original earthworks have slightly modified the local drainage pattern; but most of these modifications are no more substantial than similar works related to roads, structures or even farming. The line still operating is an integral part of a mature (century-old) transport pattern which is as natural in scenic character as most of the other local landscape features. Neither the single track railways nor the single carriageway roads can be said to detract from this sort of scenery.

These subjective assertions about the acceptability of century-old railway lines are bound to be challenged if applied to new routes—with some justification for high speed alignments less suited to the natural land contours. However, with the great variety of railway lines that have been built through valleys of different character, it is likely that some reasonable comparison can be made with a mature example of a line of similar character in similar surroundings. This is not to assert that *all* new high speed alignments are scenically acceptable; but that fair comparison will show that cases where damage to valley scenery is a sole reason for rejecting an alignment are infrequent.

15.3.5 Railways through Gorges

The Royal Gorge on the Arkansas River in Colorado, USA, is over 1000 ft (300 m) deep with precipitous granite walls and a very narrow boulder-strewn bed. Under the current philosophy of the Preservation of Wilderness, construction of a railway through the gorge would be unthinkable. However, a century ago the Frontier Tradition of the time saw the gorge only as an outstanding opportunity; for four years from 1876 a 'Great Railroad War' was waged between competing railway constructors for possession of the gorge (Strahler and Strahler 1977:392). The shooting war was settled by a Supreme Court ruling and the winning company built the railway—taking the most practicable, economic but environmentally damaging solution to a mountain crossing problem. Note, however, that the scenic value of the route was not ignored. By 1916 open-top observation cars were attached to all daylight trains passing through the Royal Gorge (*Railway Magazine* 1966:507).

Most examples of railway routes through gorges in Europe are much less blatantly damaging, although it is only in the wider gorges where other man-made development has already been intensive that new routes might be permitted today.

Eastern France is a region where a number of wide gorges have been used by main line railways. Typical examples are the pre-TGV route near Dijon and the line through the Jura to Geneva near Bellegarde. Secondary railway lines through narrower constrictions are common alpine features.

In Britain most lines through gorges have now been closed but their place in the scenery can still be studied. North of the present terminus at Matlock, the former Midland Railway main line cut through the Derbyshire limestone gorges of Monsal, Miller's and connected dales. John Ruskin was one of the lovers of unspoilt Monsal Dale. He led heated opposition to the construction of the railway, regarding it as typical of the commercialism of the industrial revolution with its lack of concern for nature (Leleux 1984:183). Now the permanent way has been removed and the track bed and viaduct are a country walk. There were riverside walks when the line was still operating but the disused railway track probably gives better views of the valley as a whole.

The Pickering to Whitby railway, winding for 10 km through the glacial gorge of Newton Dale in the North Yorkshire Moors still operates under private preservation. No less a lover of solitude and hill scenery than the Lakeland mountain guide writer Wainwright, tramping his Coast-to-Coast Walk for television, pronounced this active railway a welcome feature of his trek.

Some railways have been abandoned for 50 years or more. The Welsh Highland Railway, which penetrated the very heart of Snowdonia, has become an attractive footpath beside the steeply flowing Glaslyn river. It leads through a long, dark but accessible tunnel behind the most spectacular part of the Aberglaslyn gorge. The gorge has been an attraction to tourists for more than two centuries, even before the narrow gauge railway was laid. Nevertheless, the track of the latter remains a popular feature (Condry 1987:156). The intrusion on the landscape is no stronger than that of any ancient road whilst its utility for pleasure is considerable.

In these particular cases there is evidence that active or disused railways are perceived—currently and by some people—as welcome elements of valley scenery. Nevertheless, a steep valley or gorge is undoubtedly more scenic if its natural beauty is preserved and it carries no transport route greater than a footpath. Where routes have been long established, a railway track is less visually intrusive than a heavily trafficked road. A single corridor of transport development may be more acceptable than a network of routes over the valley floor. However, any remaining unspoilt narrow valley scenery should invariably be spared from encroachment by modern transport.

15.4 FLAT COUNTRY

The main visual feature of flat country is its *lack* of natural scenery. As a result, the scenic impact of railways in such country is rarely significant. However, environmental assessment for a new line requires a statement of any visual obstruction that will occur and of views that will be afforded to passengers. If there is any high ground from which the railway can be seen then its *linearity* may be one of the significant visual features.

If there *are* any views, then the intrusion of railways into them should be examined in the light of other imposed influences. For example, where there are no man-made structures or evidence of human activity, the land traversed can be described as open wilderness; its nature is determined by the vegetation which the climate and geology encourage. If the land supports cultivation or buildings, these dominate the landscape.

15.4.1 Open Wilderness

Scenically the Nullabor Plain of Western Australia, the American prairies, the steppes of central Asia or the forests of Canada have nothing to offer but a surfeit of solitude. The railways across them do no more than add a feature of slight interest. Yet a high proportion of the world's railway mileage is across such empty territory.

Upland plateaux are usually less uniformly dull except in their foreground. Railways cross Arizona through wide, gently sloping desert backed by steep mountain ranges and occasionally incised by deep canyons. Some Scottish highland lines offer similar views on a more modest scale but amid less arid foreground. In 'big country' tracts like these early railway lines introduced limited access and some settlements when they were built. But most later development has resulted from road access. Solar power farms, airstrips and missile bases appear occasionally in the American West and more conifers have been planted in Scotland. Whilst much of the the original wildness and solitude remains, there is undoubted intrusion along narrow but continuous and conspicuous transport corridors and in more local but extensive man-made developments for which there is no space in populated areas.

15.4.2 Cultivated and Populated Flatland

If there is any aesthetic value in flatland scenery, it is likely to be recognised by painters or photographers. Dutch landscape painters recorded such natural features as existed in the flat Netherlands—meadows, streams, ponds and trees; they had to couple these with man-made houses, windmills or steeples to make a satisfactory background for their main subjects of people or cattle.

Today the attraction of such rural scenery is still related to trees and vegetation but is diluted by intensive agriculture, canalised rivers and villages of uniform dwellings for more concentrated populations. Railway *track*—without earthworks in flat country—is itself dull although *trains* may be of visual interest.

The pattern of developed flatland is linear in nature—boundary features (walls, hedges, fences), water navigation, irrigation and drainage (canals and ditches), roads and agro-industrial or commercial buildings. Railway lines are also straight and fit relatively unobtrusively into the limited perspective possible at ground level.

In really flat country, linear and other patterns are unlikely to be clearly revealed except from the air. The town of Spalding lies on a railway line (once it lay on three) which runs in a series of straight lines between Peterborough and Lincoln. Around Spalding is about $1000\,km^2$ of very flat fertile and reclaimed fenland devoted to arable agriculture. Apart from electricity transmission lines and occasional villages, trees and church spires, there are no features except the large open fields and an intensive rectangular system of drainage ditches. The railway is the only irregular surface feature because it runs diagonally across the drainage pattern. The line is inconspicuous in the landscape although the occasional train may break the remoteness (or monotony) of the view.

The narrower clay plain of the Vale of Kent is a different case. Here not only can the plain be seen from nearby higher viewpoints, but such irregular topography as exists, combined with the equally random siting of woodlands, farmsteads and hedgerows, ensures a variety of landscape.

Under these circumstances the linearity of the straight railway extending for 40 km with only two very slight bends between Ashford and Tonbridge could be said to have created a harsh boundary in a much less regular patchwork of mainly man-made or managed scenery. However, it is hard to find a view in which this railway line through the Vale of Kent can readily be distinguished from hedgerows.

15.5 COAST AND ESTUARIES

There are numerous examples of coastline railways worldwide. Notable cases are on the Spanish, French and Italian Rivieras and in Sri Lanka—where a railway runs on or near the shoreline for 170 km around the south-east of the island. However, there are sufficient examples around the very long coastline of Britain to illustrate most of the ways in which railways can affect coastal scenery.

15.5.1 Coastal Scenery

Scenery is enhanced where high rock formations are cut back by the sea to form cliffs. Less spectacular but distinctive landscape characteristics are associated with beaches and coastal flatland or with wide, tidally flooded river estuaries.

Railway intrusion into this scenery arises where the topography provides a coastal alignment which is a serious alternative to steeper inland routes. Railways can take advantage of flat coastal lands, cross or circumvent any estuaries and make whatever is the most practicable way along cliff sections. Embankments and the need for sea protection works extend the influence of railway building to physical characteristics and scenery at considerable distances. Cutting into cliffs is another element of railway intrusion on scenic coasts.

15.5.2 Railways along the foot of Cliffs

Cliffs form the seaward end of hilly topography. Where rock formations are thick and uniform, as in chalk on the Kent coast, cliffs are high and straight. Where strata are less homogeneous, igneous rocks may be irregularly exposed or metamorphic and sedimentary formations intensely folded. The harder rocks resist the sea's attacks, forming prominent headlands and the softer layers are eroded leaving coves, often sandy and only accessible if valleys lead inland along these softer bands. These headlands and coves are among the finest features of coastal scenery.

Construction of a railway along a line of cliffs affects the natural scenery in that it:

- cuts back into cliffs to make room for the track;
- sometimes requires structural treatment to the upper slopes of cliffs to ensure their stability and to prevent rock falls on to the line;
- provides a sea-wall to carry the track and to protect it and the cliffs from the force of waves.

Sea-walls are built also as protection for paths, roads or structures located at the bottom or top of susceptible cliffs. Whether provided for railways or other purposes, sea-walls are themselves an intrusion on scenery; they are also responsible for changes further along the coastline—notably by preventing positive aspects of erosion such as a supply of sediment to landform/scenery-making processes. Any measures to cut back or to protect cliffs is interference with a natural asset; on the other hand, where rock is relatively soft or uniform, the effect of storm wave erosion can in time be quite as considerable. A positive environmental feature of cliff works to protect rail tracks is the isolation necessary to run trains which often prevents any road or other form of structural development taking place in the space below the cliffs. Apart from trains, access is available only where there is a walkway on the edge of the sea-wall.

Recognition of the scenic value of sea cliffs has resulted in measures to limit any form of commercial development that might spoil them. For instance, lengths of 'Heritage Coastline' have been designated in England and Wales by the Countryside Commission. With one exception, no railways actually run along the shoreline of these designated lengths, although the East Coast Main Line runs near to one in Northumberland. Some other sections of cliff-lined coast are traversed by railways—in South Devon and along the Cambrian coast. Whilst such route selection might not be permitted today, views from these lines and from those on the Mediterranean Riviera were an immediate attraction to tourists.

The 'Heritage Coastline' which *does* carry a railway line is that between Dover and Folkestone. The fame of the White Cliffs of Dover arises because

LIVERPOOL JOHN MOORES UNIVERSIT
Aldham Robarts L.R.C.
TEL. 0151 231 3701/3634

they are the first sight of England as the sea voyager approaches from France. In fact they lack the purity of the taller cliffs at Beachy Head, being altered by various man-made and natural occurrences. In the Dover area the line of cliffs is broken by a natural valley containing the town and the road viaduct to the docks, whilst the cliffs themselves have been quarried in the past and support such major structures as Dover Castle. West of Shakespeare Cliff 3 million tonnes of Channel (railway) Tunnel chalk spoil have been deposited behind a new sea-wall extending several hundred metres out to sea (Figure 15.3). This has created an altered coastline and cliff bottom landscape. Beyond the second (Abbot's Cliff) tunnel the old railway climbs westward for 3 km through the main landslip area, now incorporated in Folkestone Warren SSSI.

The coastal strip between Folkestone and the South Foreland was defined as a Heritage Coastline because of its 'significance in both landscape and recreational terms as one of the finest stretches of undeveloped coast in England and Wales' (Ardill 1987:180). 'Undeveloped' implies that the well-established railway running through the strip is *not* taken as a significant intrusion into the landscape.

The Dover to Folkestone railway is barely noticeable from the sea

Figure 15.3. Creation of new land—disposal of spoil excavated from the Channel Tunnel—at the foot of Shakespeare Cliff. The original railway to Dover was constructed along a seawall below these cliffs, more steep and stable than those which have fallen into Folkestone Warren in Figure 13.3.

although it can be clearly (as in Figure 13.3) seen looking down into Folk-estone Warren. The land created by the Channel Tunnel works substantially alters the shoreline scenery of one section of the coast and covers some of the classic geological evidence near the cliff foot. Ultimately the scenic impact can be contained within the lower levels of the cliffs where it may look more natural in appearance than the Dover harbour breakwaters and buildings 2 km to the east.

We have referred already to the attraction to railway passengers of the lines along the Mediterranean coast. In Monaco, however, the demands for high-priced space for roads and open space along the sea front led to the railway being diverted in 1964 through a 3.5 km long tunnel; this provided 5 ha of former railway space plus additional areas totalling 8 ha on land reclaimed from the sea with material excavated from the new tunnel.

15.5.3 Coastal Flatland

The softer material eroded from cliffs is carried away by littoral currents and deposited on beaches and spits farther along the coast, or joins alluvial material brought by rivers from inland. The behaviour of estuaries is influenced by any flood or coastal protection works or road or railway embankments which may interrupt these inland or offshore deposits.

Figure 15.4. Railways follow the straightest unobstructed line across flat country already lined by roads or hedgerows. This coastal land near Harlech is alluvium which has formed, been drained and cultivated, all within historic times

In historic times the sea has shaped the coastline, claiming the clifftop town of Dunwich in Suffolk and, even recently, buildings in Yorkshire. But the same forces have created new flatter land below Harlech Castle (Figure 15.4) and on Romney Marsh. Both these new coastland areas carry regional railway routes with no more direct impact on the scenery than that of any other man-made lines across the plains. However, low embankments which maintain railway tracks above flood levels are one of the various man-made obstructions to natural drainage which determine the nature of coastal and estuarial land.

15.5.4 Beaches

Many secluded beaches on rugged, rocky coastlines owe their continuing scenic attraction to the fact that they have no direct road access. Nor is it likely that many railway routes would need to cross such territory. However, there are a few situations where it *is* possible to see scenic coves directly from a train. Fortunately the railway itself usually forms a barrier against further commercial development in such scenery.

Beaches near Dawlish in Devon lie directly beside a coastline railway.

Figure 15.5. At Dawlish construction of a coastal railway provided a level route without gradients although this sharp curve restricts speeds to 70 mph (112 km/h). The cliffs have slipped or been cut back or stabilised at various points and are protected by a sea wall. The latter provides access to fine seaside scenery for pedestrians but the accretion of sand and shingle shows the wall's effect on coastal morphology

Below the cliffs, some of the beaches are accessible by walkways beside the railway; few are directly reached by road. Because of its modest width and alignment suited to the line of cliffs the railway is less obtrusive in the beach scenery than might be roads and roadside buildings (Figure 15.5). Note, however, that the curved alignment of the track limits speeds to considerably less than those of which HST 125s are capable; limited speeds are perhaps better suited to a seaside environment but are, nevertheless, faster than other forms of transport can achieve across this sort of landscape. Any acceptable alignment along the shoreline is only possible because the line of cliffs is mainly straight. At gaps between the cliffs, as at Dawlish itself (Figure 15.6), the railway and its sea wall create a partly artificial coastline; the railway then provides a direct seaside view to its passengers but blocks a clear sight of the sea from the town park. Where deviations from the straight lines of cliffs occur there are rocky headlands, scenic pinnacles and inlets. The railway has to tunnel behind the headlands leaving them intact.

Beaches along coastal flatland are of economic, scientific or amenity interest rather than as scenery. Sand dunes or lagoons may arise, depending on the local geology, drainage pattern and off-shore currents; these lagoons provide areas of ecological interest and often fragile habitat. The same applies to the wide alluvial wetlands and mudflats in river estuaries; new railway crossings—particularly if on embankments rather than bridges—will

Figure 15.6. Dawlish sea front at high tide—man-made landscape dominated by a railway

almost inevitably interfere with natural systems and the state of scenery and nature conservation in the wide vicinity.

15.5.5 River Estuaries

The appearance of railways as seen across tidal estuaries is not fundamentally different from that beneath cliffs or across flat land on the coast. The feature of particular visual impact is usually any bridge or embankment whereby the railway crosses the water. Such crossings are longer and lower than most inland valley crossings; it is not so much that these low structures themselves may intrude on the scenery, more the way that they affect the river regime, the active land-forms and the wider landscape which these constitute.

The comparative length of bridge structure and of the embankment sections affects the visual impact of the crossing itself. The long bridge at Barmouth in Wales is attractive rather than obstructive because it allows water level views through it. But long sections of embankment elsewhere have been used as dams to control tidal and flood flows and to create different conditions upstream that are much more destructive of wetland environment and appearance. In particular there are strong conservation objections to

Figure 15.7. The 'Cob', Portmadoc, built as a tidal reclamation dam (in 1811), has since 1836 carried the Ffestiniog Railway as well as a toll road. The strip of wetland is only partially saline as high tides are impounded by a barrage outside this view. The wildlife on the upstream side is therefore different from that on the estuary side

reclaiming and draining estuarial land for agriculture or building development.

From the lower terminus at Portmadoc in North Wales the Ffestiniog Railway crosses the 'Cob', a causeway built in 1811, some 20 years before the railway was added (Figure 15.7). This dam across the Traith Mawr estuary altered the whole river regime and land use pattern upstream, creating 10 000 acres (more than 4000 ha) of new farm land where there had been marsh or tidal wetland. Even in 1811, concern was expressed about the effects on nature of the 'ugly wall' in this early example of land reclamation; although, as Condry (1987: 180) points out, the poet Shelley called it 'one of the noblest works of human power'. The main loss of wetland habitat is now distant history but remains an impressive example of man's influence on extensive but fragile coastal resources. Meanwhile, from the point of view of the modern observer, the railway and road now crossing the Cob afford views of at least narrow strips of tidal and semi-tidal wetland for wading birds, a not unimpressive foreground against a background of mountain landscape.

There is often sufficient space on the alluvium *along the side* of river estuaries to accommodate road or railway routes. Construction of embankments, with sufficiently wide side-drainage crossings, need make relatively little long-term impact on the ecology of estuaries. Whilst estuary fringe railways blend easily into such scenery they also offer fine views to the traveller, for example on active lines on either side of the River Dovey estuary or from the disused route from Dolgelly to Barmouth Junction on the southern side of the Afon Mawddach. However, any tributary estuaries which are inevitably cut off by the straighter railway line remain in danger of losing their tidal character, appearance and function.

15.5.6 Implications of Railways on British Coastal Scenery

Railways *beneath cliffs* on the English and Welsh coastline include:

- 7 km between Dover and Folkestone (chalk), discussed above; this inconspicuous line traverses the whole length of cliff between the two ports; it leaves unaffected a greater length north-east of Dover; the landscape has been altered around the port of Dover and, more recently, where new land had been created by chalk spoil from the Channel Tunnel;
- 3 km on Portland Bill: disused;
- 7 km near Dawlish: this affects the whole cliff line north of Teignmouth but leaves a greater length unaffected further south; the disturbance to natural conditions is modest and access to cliff foot areas is still available only to trains or pedestrians;
- several short (1 to 2 km) sections along the Bay of Cardigan, where a high

proportion of the former Cambrian Railway is along coastline or estuary; little scenic impact;
- about 4 km in total near Penmaenmawr and Conway on the North Wales coastline; less visual impact than that of recent road construction.

The two lines on the west and north coasts of Wales cross a much greater length (100 km) of *flat coastal land* and other examples in Cumbria and Northumbria add another 50 km; the railway infrastructure along these lengths is of little scenic consequence. There are other locations where railways crossing flat coastal land are more environmentally significant, e.g. at Marazion on Mount's Bay in Cornwall where the railway crosses a marshland SSSI mentioned in Chapter 13.

Along *river estuaries*, railways still operating in England and Wales cover a total distance of about 80 km; half as much again is no longer in use. In some localities railways form the dominant boundaries of both sides of estuaries (14 km around the Exe estuary and a similar distance around the Dovey). Here and elsewhere their influence, which is on conservation land rather than on scenery, is seldom significant.

About another 80 km of rail route skirts the sea coasts of Scotland. Some sections on the western side pass very scenic bays and headlands not easily accessible and mainly admired from the train.

15.6 POPULATED COUNTRYSIDE—RAILWAYS IN SOUTH-EAST ENGLAND

Much of the new European network of high speed railways is being built across well-populated countryside—in northern France and Germany and, when the Channel Tunnel Rail Link (CTRL) is finally implemented, in south-eastern England. Railway planning in these regions has to take into account all the land resources described in previous chapters. It also has to consider the impacts on many scenic areas, varied in character but widely recognised as attractive.

Flat country is not associated with such spectacular natural scenery as are areas of higher relief. Mountain country offers much wider views and variety of landscape but it is too steep and barren to support more than a sparse permanent population. The characteristics of *attractive inhabited countryside* lie between these extremes. Typically they feature:

- variable topography—rolling hills interspersed by flatter, alluvial valleys;
- at least a proportion of unchannelled streams and undrained wetland;
- a substantial area (10 per cent or more) of woodland as well as numerous hedgerows and isolated groups of trees, forming a wide and irregular patchwork together with villages, farms and country roads;
- pockets of urbanisation and corridors of transport infrastructure.

Such are the characteristics of *south-east England*. The basic scenery is of three varieties—chalk downland and steep escarpments with lime-loving vegetation and backed by beech woods, contrasting sandy heights and acid soils, and flatter clay alluvial plains. Subsequent landscape modifications depend on the way in which woodland has been cleared whether for pasture, arable farms, orchards, hopfields, water features or historic villages. In the twentieth century this semi-natural landscape has been threatened by property and road development countered by Green Belt and other conservation policies.

Railways from London towards Dover and the Continent have never been wholly satisfactory, either in performance or capacity. This is partly due to the difficult topography associated with the attractive scenery. In topographical terms the first but most indirect route to Dover was the most satisfactory. A line to Brighton had already been built which tunnelled directly through the North Downs, the slight hills in Balcombe Forest and the South Downs. All these hills stretch from east to west so the railway crossed them at right angles, achieving relatively short direct crossings. From Redhill, south of the first of these obstacles on the Brighton line, a very straight, level and potentially fast route was built eastwards to Ashford and

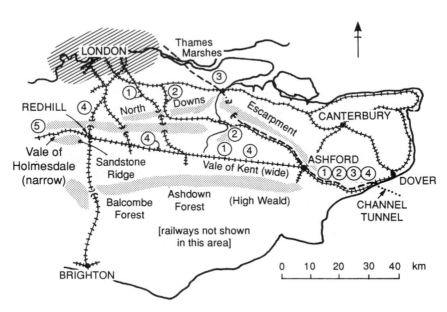

Figure 15.8. Railway routes from London to Dover: (1) Boat Train Route 1; (2) Boat Train Route 2; (3) probable Union Railway route; (4) original route by Brighton line to Redhill; (5) steeply graded, sinuous and picturesque North Downs route to Reading (possible future freight route from Channel Tunnel?)

the channel ports. The total distance from London to Folkestone, near where the Channel Tunnel portal is now located, was 140 km; the current more direct but undulating route is 120 km (see Figure 15.8).

Largely because of the flat terrain, the Redhill to Ashford line remains an outstanding opportunity for a scenically and environmentally sensible route *around* London. However, the great difficulty in increasing the capacity of the existing lines from Redhill through the South London suburbs as well as the distance involved make it as unsuitable as the other existing routes for a high speed line to the Channel Tunnel.

The various projected routes for CTRL as well as those lines which already exist illustrate a number of aspects of railways in the landscape. Scenic impacts which have to be considered arise in routeing lines through chalk escarpments, along the foot of these escarpments, at river valley crossings and through undulating rural areas not covered by these categories.

Finding *routes through escarpments* was always a key problems in planning railways north or south from London. Across the top of the Chiltern Hills or North Downs, they must either find gaps or tunnel beneath the hills. Several gaps through the North Downs now accommodate railways; the scenic

Figure 15.9. View from above Oxted Tunnel, Surrey, one of the routes through the North Downs south of London. Whereas the railway pierces the escarpment directly, the M25 motorway climbs its slopes or, as in this view, runs along the bottom. Note the woodland which either existed before the railway was built or has grown up beside it in the subsequent 100 years.

impact of those along the Darent and Medway valleys has been examined in Section 15.3. But other lines from London to Sevenoaks, the two routes past Redhill and the Oxted line have to pierce the escarpment by tunnels (see Figure 15.9). The Union Railway route to the Continent will do the same further east where it will run under the North Downs after crossing the River Medway.

Because of gradient and curve limitations no conventional railway can

Figure 15.10. Cheriton Channel Tunnel terminal, looking west. At the bottom of the picture is Castle Hill, a point on the chalk escarpment SSSI considerably altered by ancient earthworks. The dominant features of the scenery below the escarpment are the 140 hectare road/rail transfer station, the M20 motorway and the outskirts of Folkestone. Near the top centre of the picture the comparatively modest (9 hectare) Dollands Moor freight sidings are clearly seen but the existing railway from Folkestone to Ashford on its left can be distinguished only by its linearity (QA Photos Ltd)

cross over or through *the top* of an escarpment. Any disfigurement to the land surface is confined to that of *cuttings* leading to tunnels or through low points between the main heights of the escarpment.

Motorways are wider, more sinuous and less sensitive in their higher routes across escarpments. Prominent skyline blemishes have been carved by motorways—the M23 through the sandstone ridge in Surrey or the M3 through the chalk at Twyford Down near Winchester. This is not to assert that permanent alterations to the skyline necessarily constitute blemishes. It can no doubt be debated whether the ancient British earthworks at Castle Hill spoil the Outstanding Natural Beauty of the top of the Downs above Cheriton as much as the Channel Tunnel terminal has intruded on the flatter countryside below (Figure 15.10).

Other dramatic man-made incursions into North Downs scenery are chalk quarries; these leave a major scar on natural landscape or attractive white cliffs, according to one's taste.

In summary, railways rarely harm escarpment scenery because they cannot climb over it. Quarries and roads cause considerable disfigurement; but if the first are accepted, so may be the second in due course. Future generations will have an opinion (but no choice). In well-populated country man-made developments are inevitable. Perhaps loss of a scarce type of conservation land, to agriculture as much as for 'hard' development, is more serious than scenic blemish.

At the *foot of escarpments* is the gentler but complex topography where spurs from the escarpment meet the lower, softer formations. Typical of railways skirting the foot of the North Downs are the cross-country line from Guildford to Redhill and, further east, that from Maidstone to Ashford. The first is steeply graded and sinuous through the narrow gap between the chalk downs escarpment and the Leith Hill greensand ridge, a quiet, wooded or grassland area of great beauty, barely 20 miles (32 km) from the centre of London. The second follows a still undulating but more level line, generally close to the proposed route for the Union Railway to the Channel Tunnel.

The line from Guildford to Redhill provides views of attractive scenery for train passengers; *because* of its sinuous course (unsuitable for high speed traffic), it has little impact on that scenery. Alignments for faster trains inevitably cut below or rise above ground level across variations in the intervening mini-topography (the spurs and vales which run at right angles to the chalk escarpment itself). Where these variations require that railway routes must cross *over* attractive but relatively low vales below the escarpment, then scenic disturbance results. This is often more scenically intrusive than where a *cutting* hides the track through an adjoining spur of often less attractive arable land.

The Union Railway route from London is planned to pass under the North Downs through Blue Bell Tunnel north of Maidstone and then to continue within 1 to 2½ km of the bottom of the escarpment for 27 km towards Ashford. Some sections of the existing Maidstone to Ashford railway, generally nearer to the Downs, can be seen from the top of the escarpment but only when a keen eye spots a passing train. For this reason, and particularly because of its proximity to the much wider and similarly graded M20 motorway, the impact of the new railway on views from the Downs is likely to be minimal.

Wide river valley crossings usually require long viaducts. Existing railway bridges over the River Medway are in the flatter country north and south of the Downs. Two alternative routes for Channel Tunnel Rail Links would have to cross the river from higher ground. The 1993 projected Union Railway route will cross the Medway towards the northen end of the gap, probably near the existing major bridge carrying the M2 motorway. The road bridge is a modern, slim long-span prestressed concrete structure which sets both a precedent and a challenge for any similar crossing proposed for a railway.

BR's 1989 CTRL route was to have crossed the Medway on a long viaduct further south at Halling. This could have entered a tunnel under the Downs through a disused part of the extensive quarry workings, already a prominent feature of the local landscape. A viaduct across the river would be a new and contrasting visual feature but its impact would have been more relevant to disturbance of the Holborough Marshes SSSI which it crossed than to scenic considerations.

Railway viaducts are a blatant but not uncommon or unacceptable feature of valley scenery in populated countryside. Their acceptability depends upon how they blend with other man-made structures in the vicinity, on what can be seen through them and how they affect the land resources around their footings.

Undulating rural landscape is scenic in two respects: firstly, its topography provides numerous middle distance views; secondly, it combines many different features making up overall vistas of varied scenery.

In south-east England these features include numerous woodlands, attractive villages and farms and a complex network of lanes and hedgerows. These are combined in an irregular patchwork, related both to random historical events and to the vagaries of local physical geography. Railways introduce a sharply linear feature into this irregularity.

In the route selection for new railways it is difficult to specify solutions to scenic problems without reference to the other prime environmental requirements. The latter include:

- avoidance of any multiplicity of new routes; these should follow existing transport corridors wherever possible;
- avoidance of areas of nature conservation, heritage and amenity as well as of scenic beauty;
- avoidance of the knock-on environmental consequences of interfering with natural systems.

Transport corridors are often associated with residential and industrial properties. So widening the corridors for *further* transport development will inevitably require the acquisition and destruction of some of these properties; these are of higher commercial worth than conservation land but of little or no heritage value. There is no way in which route selection in populated scenic countryside can avoid this clash of interests. Environmental interests cannot be served by least cost solutions in such circumstances.

15.7 URBAN LANDSCAPE

15.7.1 What is Urban Landscape and Are Railways Relevant in it?

In most cities landscape is composed of buildings, formal parks, people and traffic. Town centre scenery which is commonly admired comprises wide or picturesque streets and squares, fine buildings, sometimes city walls and monuments.

Stations like York were planned to complement rather than to intrude upon the aspect of the city walls nearby or the minster in the middle distance. Continental city stations are often among the most prestigious buildings.

Except perhaps in Lowry paintings, *railway viaducts* may not be commonly regarded as elegant assets of cities. But massive brick arches (as at Stockport) or lighter structures (such as carry the Docklands Light Railway) are not inappropriate to the 1840 or 1990 built landscapes around them.

These aspects of city railways are relevant in total urban transport and environmental planning (see Chapter 4) and to heritage structures (Chapter 14). This section is concerned with the few cities where impressive *natural* scenery—steep hillsides, gorges or water features—occurs as an integral part of the urban landscape.

15.7.2 Railway Involvement in Scenic Cities

To accommodate their population, large towns have grown most successfully in wide flat areas. A few, like Rio de Janeiro or Kowloon, are hemmed in between the sea and steep hillsides; Bogotá, Tehran, Almaty and Athens spread out below a backdrop of steep mountains. In none of these cities are railway lines dominant features in the landscape. But railways enter the heart of Edinburgh and Luxembourg, cities where natural rough and rocky topography complement grand architecture.

We have seen in Chapter 14 how a railway fits into the *architectural* layout of *Edinburgh*. This northern capital is itself situated in a particularly *scenic* position on a slope stretching from the high igneous mini-peak of Arthur's Seat to the Firth of Forth. The narrow valley between the cliffs of the castle of the Old Town and Princes Street, boundary of the New Town, was formerly a lake (the 'Nor Loch'); its reclamation as Princes Street Gardens formed the very core of the city as it is known today.

One hundred and fifty years ago it was proposed that the Edinburgh and Glasgow Railway be extended from Haymarket eastwards through Princes Street Gardens to the site of Waverley station. This extension was at first rigorously opposed by the Gardens' proprietors. They insisted that a 6-ft (1.8 m)-high stone retaining wall be built on either side, a modest measure which has proved perfectly acceptable in visual terms. Steel footbridges were provided for people's access across the railway and to see the trains. These bridges were removed in more recent years but there was such a public outcry that they were soon replaced (Glasgow Museum of Transport 1992).

The railway now occupies about one of the 13 to 14 hectares of the main ('Western') Princes Street Gardens between the formal buildings (Scottish National Gallery and Royal Academy) and the tunnel leading to Haymarket station (Figure 14.2). The railway is thus a feature of interest in the formal part of the city layout rather than an intrusion into it. A landscape painting (Hill 1848) indicates early acceptance of trains as a minor element among the steep slopes, fine buildings and gardens of the city.

Travelling east from central Edinburgh, the East Coast Main Line emerges from Calton Hill Tunnel and passes along a corridor between London Road and the broad parkland below Arthur's Seat. Some of the space in this corridor is filled by sturdy housing blocks and some is redeveloped (Meadowbank Stadium); some is derelict and offers great scope for sensitive corridor redevelopment with the railway embankment as a modest but unhidden feature against a highly scenic backdrop.

The centre of the city of *Luxembourg*, less than a square kilometre in extent, fits into a different but equally scenic situation. Quoting from Baedeker (1897);

> The situation of the town is peculiar and picturesque. The *Oberstadt*, or upper part, is perched upon a rocky table-land, which is bounded on three sides by abrupt precipices, 200 ft high. At the foot of these flow the *Petrusse* and the *Alzette*, which are bounded by equally precipitous rocks on the opposite bank. In this narrow ravine lie the busy *Unterstadte* . . . The view of the town, with its variety of mountain and valley, gardens and rocks, groups of trees and huge viaducts, is singularly attractive.

The viaducts referred to were those of the railway; these remain important and attractive features in the gorges, together with ruins of older structures, lengths of city wall, ancient lower level bridges, riverside paths and a variety of buildings, some notable for fine roofs or façades. The railway structures are now historic features, certainly no more visually intrusive than the very modern European Court of Justice on the far side of the gorge or the long-span modern road bridge joining the top of its highest cliffs.

15.7.3 Future Railway Development in Urban Scenery

Many urban railway routes and some roads have become redundant as the pattern of transport services has been rationalised. Together with nearby redevelopment of outdated building or industrial sites, there is often ample scope for scenically attractive development in conjunction with such new railway or light rail services as are required in cities.

Environmental aspects of city rail facilities have been identified in Section 4.5. Most urban railways are underground. Where they are on or near to the surface it is their noise and vibration rather than any visual impact which is most commonly objected to by town dwellers today. Modern trams are similar in appearance to short trains and have apparently been accepted favourably on the Continent for many generations. City rapid transit systems offer welcome opportunities for environmentally-sensitive planning of green, car-free public transport routes and cycle and pedestrian lanes.

15.8 SCENIC COMPATIBILITY AND MITIGATION

15.8.1 Suiting Railways to Scenic Patterns

Most of the examples we have examined show that railways are reasonably compatible with scenery. Exceptions occur, for instance along rugged coastline or in gorges, where the structures can detract from the natural grandeur of the scenery. In such conditions topography is rarely suited to high speed alignment. So new railways are seldom likely to be proposed through outstanding landscape.

Meanwhile, new high speed lines *are* being constructed or planned across land which is comparatively well populated yet still contains attractive countryside or even areas of designated beauty.

The scenic attraction is usually a pattern of mature semi-natural features interspersed by limited numbers of locally traditional structures. Such scenery can be more badly affected by developments other than railways, particularly large housing estates, major industrial or commercial enterprises or dual carriageway roads. Railways are relatively narrow features but new ones may damage the *pattern* of scenery because of their need for straight alignments.

Scenic damage can be avoided firstly by sensitive route selection, in which railway planners and engineers take the leading role; and then by mitigation measures achieved through architecture, planting or earthworks for which landscape specialists have their own techniques.

The most appropriate time to assess scenic impact in detail is after the first route selection and resolution of the most obvious transport and environmental issues, but before determination of the finer details of vertical and horizontal alignment, structures or environmental mitigation measures. Thus visually acceptable design can fit into an iterative process of route selection, design and cost estimation, and economic and environmental evaluation.

15.8.2 Route Selection

The alignment of a new railway has to connect the appropriate places, satisfy train operating requirements, avoid excessive cost and prove sensitive to landscape and other environmental requirements.

At the stage of examining broad route options, preference should be given to routes avoiding those types of scenery which earlier roads or railways have shown to be most sensitive to new development. In scenic terms this will include both officially designated scenery, such as National Parks or Areas of Outstanding Natural Beauty (AONBs), and other landscapes which reconnaissance identifies as sensitive to change. However, the way in which a railway line will affect a particular view or scenic feature is more significant than whether it crosses just inside or outside the boundary of a designated zone.

For any comparison of *alternative* routes, for instance along adjacent valleys or on either side of the same valley, a full comparison is required of the scenic features and the nature and magnitude of impacts upon them, classified and quantified as far as is practicable. These features are often intimately related to nature conservation or heritage land and water resources.

In more precise route definition within a chosen corridor, there have to be compromises between the needs of scenic protection and those of track alignment and cost. High speed routes cannot accommodate the tight curves that are ideal in narrow valleys; but the capability of modern trains to climb steeper gradients may allow more latitude than before in adjustments to reduce scenic intrusion in gently rising topography.

Rail alignment across undulating country inevitably involves sections on raised embankments. These can obscure views from points on one side of the embankment of what may be attractive scenery on the other. This obstruction is then an adverse impact on a local view.

An extreme measure of visual mitigation would be to lower a railway's vertical alignment so that it is hidden in cutting or cut-and-cover tunnel, even across valleys, where some parties consider it would otherwise intrude on the landscape. To test the acceptability in practice of such action there is no need to look further than the precedent set by Europe's many motorways; only rarely have they been sunk so as to avoid visual disturbance and even then only for short distances. For railways, vertical alignment considerations make short changes in level impracticable; besides, railway formation is so much narrower and traffic so much more intermittent that visual impacts are not comparable with those of major roads.

There is therefore no justification for *hiding* a railway—whether in continuous excavated or 'false' (contrived) cuttings or through shallow, unnecessarily long tunnels—purely so that it cannot be seen. A railway is not secret nor is it shameful. Trains may be unwelcome if they are noisy but to hear what you cannot see may aggravate the annoyance.

15.8.3 Sensitive Structural Design and Architecture

A bridge or viaduct makes only a partial obstruction but can constitute a disparate feature in local landscape. The design of any structure must therefore be visually attractive and sympathetic to the rest of the landscape, topographic and geological features and to other man-made structures. Particularly relevant are a façade of suitable texture and materials and, for bridges, acceptable proportions of height and span. Popular perception of existing structures and the advice of landscape and architectural bodies are relevant.

Earthwork such as an embankment may be stark in appearance when it is new but may mellow in due course, particularly if it has been planted.

15.8.4 Soft Landscaping—Planting

Preece (1991: 222) has written:

> In their contribution to visual effects and to shelter, trees represent in most situations the best value for money of any of the components of landscape design.

Planting of trees or lower vegetation can be undertaken in a pattern which, when mature, will blend with that of adjoining woodland, fields or hedgerows. A main objective is to soften the harsh straight line of track and earthworks; this may require some planting *beyond* fenced railway land. A mature tree belt requires up to 20 m width. Lineside tree planting has proved an attractive feature; but planting should not be so dense as to emphasise the linearity of the railway or to obscure vision totally from the trains; nor should it be so close to the line as to cause operational problems, such as excessive leaf drop in autumn.

Apart from blending with the other land cover in the vicinity, any plan for planting must take the following factors into account:

- The *suitability of species*—the visual and ecological qualities of trees, shrubs and ground cover; the types of vegetation already in place; soil characteristics, chemistry and depth; drainage and microclimate; habitat and nature conservation preferences.
- The *planting pattern and density*—allocation and combinations of tree types, alignment of boundary lines and density of planting (if trees are not to be harvested commercially the density planted need not exceed 5 to 10 per cent of that where timber quality is the ultimate requirement (Preece 1991: 232); for 'parkland' or open 'agro-forestry' densities are

even lower); avoidance of damage which might be caused by tree roots—to structures or even to archaeological features.

- The *time element*—how preservation of existing trees, planting of new ones and subsequent planting and management can provide a landscape of mixed maturity for centuries ahead.

Alternative planting schemes can be drafted using a range of varieties and boundary shapes; that most acceptable and practicable can be adopted. Acceptability can be judged in terms of how the planted vegetation will blend with the rest of the landscape and with the railway itself 10 or 20 years later. Practicability concerns implementation of planting schemes during construction and effective means of ensuring responsible management thereafter. Preece (1991: 223–262) has described how trees can be planted and maintained as an amenity and for landscaping at new developments.

Money spent on soft landscaping and subsequent growth management is an extra to the investment absolutely necessary to build a railway. But it makes a very visible contribution to the surrounding environment, as desirable an element in visual terms as attractively coloured rolling stock or bright, spacious stations.

15.8.4 Hard Landscaping—Modifying Earth Shapes

Much more contentious is physical adaptation of the shape of railway earthwork to make it blend with the neighbouring topography. It has been suggested, for instance, that the shape of embankments should be adjusted to make them look less obviously artificial against the adjacent slopes of a valley, e.g. by filling in any abrupt angles or by varying the cross-section of the embankment. This could be achieved by undertaking additional earthwork beyond that required for the structure itself. 'Hard' landscaping of this sort is contentious because there is little, if any, evidence of its use or necessity on existing railways. Engineering cost, topsoil stockpiling, archaeology and agricultural restoration considerations will mostly count against any measure which will decrease the steepness of stable side slopes and increase the area of land to be covered. Cost savings in disposal of any excess spoil *can* be a relevant benefit.

A first option in initial planning of a railway or road through undulating country is to balance cut against fill, i.e. to excavate an amount from cuttings equal to that which is placed in embankments. Occasionally, material excavated from cuttings may be unsuitable for use in load carrying embankments. More often, vertical alignment is such that, for certain long lengths, there is such an excess of cut or of fill that transfer of soil along the line is less practicable than disposing of soil elsewhere in the vicinity or of bringing extra fill from an acceptable nearby off-line source. The costs and environmental impacts of transporting soil, mining extra material or disposing of surplus spoil are key factors. Occasionally, more excavated soil may be available at an embankment site than is needed for its basic shape. Any surplus could then be used to adapt the appearance of slopes.

Besides the logistics of balancing cut and fill, the main engineering principles in earthworks design relate to the stability of slopes in cuttings or embankments in different types of soils. For any given vertical alignment there may be a number of alternative engineering solutions, for instance incorporating different soils in different cross-sectional designs and construction sequences or by partial use of retaining walls or bridge structures instead of earthwork. A range of options should be examined and a choice then made based on both cost and environmental factors. The latter include impacts on areas selected for borrowing extra fill or disposal of surplus cut; these impacts may be on amenity or conservation land in the corridor alongside the railway; or may concern the effect of embankments, spoil heaps and borrow areas on local scenery.

Therefore the final shape of an embankment can be influenced by *environmental* factors in making a choice among feasible *engineering* solutions. This may be a more practicable approach than one in which a single final earthwork design is considered and found to be in need of specific landscaping measures. Opportunity can be given to test engineering, professional landscape and popular response to alternative concepts in justifying the best approach.

Landscape architects have developed techniques in the design and construction of hard and soft landscaping for trunk roads and motorways. If it is decided that hard landscaping *is* necessary the planning for new railways can draw on these techniques; they should be modified to suit the different requirements for railways, i.e. narrower but higher embankments and cuttings. Earth shaping can be applied either formally, as in terracing, or to provide flowing curves to emulate natural land-forms. Railway cuttings through rock or embankments on hard material are essentially formal and utilitarian; it is more likely that their further sculpturing can be justified to suit adjacent structures or architecture than to modify hard land-forms. Approaches to naturalising the form of cuttings to reduce visual impacts on motorways (Preece 1991: 194) include curves to avoid cutting the skyline and S-shaped side slopes to resemble better the lesser gradients at the top and bottom of some natural hillsides. The first is not usually possible with long radius railway curves which, in any case, rarely break high level skylines; nor is expensive shaping of slopes so necessary for relatively narrow railway cuttings.

The relationships between visual benefits of hard landscaping, use of different materials in earthworks, and the economics and land use consequences of earthwork construction all warrant further research. Railway construction in undulating topography makes disturbance of ground contours inevitable; but the maximum possible preservation of natural land-forms which future generations may value should be a paramount consideration.

16 Environmental Evaluation of Land Resources

16.1 ISSUES IN ENVIRONMENTAL EVALUATION

Valuation is numerical estimation of a thing's worth—often the monetary price at which it can be traded.

Evaluation in project appraisal is commonly applied to *comparing* the value of alternative courses of action or, in the case of new railways, of alternative alignments. Comparison can be made in terms of

- financial expenditure and revenue or
- economic costs and benefits or
- non-monetary data and subjective preferences.

Alternative approaches to *environmental evaluation* were introduced in Chapter 2. One approach is to consider environmental issues separately; another is to express environmental costs and benefits in economic terms.

Agencies like the World Bank—which are expected to appraise projects with an eye to human welfare beyond immediate financial profit—require that economic evaluation should ensure a *desirable use of actual resources*. Desirable use is that which leaves sufficient for subsequent generations, i.e. is sustainable.

One aspect in which conventional evaluation of transport infrastructure projects fails to ensure sustainability is in the valuation and use of land resources. Undeveloped land is given little or no market value even if it represents a fragile and diminishing natural resource; but land that is built on, already 'lost' as such a resource, is allotted a high value equivalent to its current commercial price; as a result the economic solution for a road or railway is to route it through low-valued, possibly irreplaceable land rather than through areas where its construction would do little *long-term* damage that future generations could regret.

A new railway takes a relatively small proportion of land in any area which it crosses, say two hectares per kilometre; but the comparative straightness of its alignment makes it difficult to avoid all obstacles; and the barrier which it creates can affect adjacent land use. Any optimal strategy for removal or avoidance of obstacles or for changes in land use is difficult to devise without indicators of true resource values.

An objection to monetary valuation is that it might be insufficient to preserve something on the projected route that is unique, rare or irreplaceable; there could be wide disagreement on what price level is appropriate for an asset which, to some interests, is priceless.

One observation (Whitelegg 1993: 129) is

> that building these costs into decision making . . . may even make things worse as it assumes that no other kind of state action is needed to produce sustainability other than actions related to prices, taxes and charges.

Some new economic approaches have produced higher, more equitable resource values although conservationists find the means of valuation unscientific. Others have commented on the absurdity of sophisticated methods of pricing the unpriceable; so Coker (1992b: 74) suggests that an 'alternative approach is for conservationists to accept that the ends may justify the means, even if they disapprove of those means'.

This chapter reviews the unsolved problems of environmental land valuation as they affect railway route selection. Commercial pricing and alternatives for resource valuation are introduced in Sections 16.2 and 16.3. The issues in valuation and tentative ideas about how they might be tackled follow for conservation land (Section 16.4), heritage and amenity (Section 16.5) and landscape (Section 16.6).

It must be emphasised that experts are not yet agreed on what valuation methods are appropriate. Nevertheless, a broad strategy is proposed in Section 16.7. It prescribes strong protection against development, rather than monetary valuation, for certain areas of recognised exceptional long-term resource value; in other areas an upward adjustment of low market price for land of high natural resource value is proposed. Application of these adjusted values can be implemented first in initial route selection (Section 16.8). Subsequent inclusion of such values in formal project economic evaluation (Section 16.9) may have to be considered a longer-term goal; it may take decades before politicians and planners, or even economists and environmental specialists, can agree and accept specific valuation principles. Meanwhile, ideas are needed to improve economic-cum-environmental decision-making processes for transport projects; these might enable railways to take an appropriate place in twenty-first-century transport.

16.2 PROPERTY AND LAND PRICES

Property valuation is comparatively straightforward for developed or scheduled building land, for which there is an established market. The price directly reflects current demand for the residential amenity or economic production unit offered. Market prices also exist for agricultural and industrial land and can no doubt be calculated for land where mineral extraction is permitted.

Valuation is much more difficult and contentious for:

- historic buildings or ancient man-made features which no longer serve an economic role; any aesthetic attraction may be offset by a costly obligation to maintain the structure;
- land whose commercial value is low because it is reserved for agriculture or amenity use, such as in Green Belt zones where planning restrictions prevent more profitable exploitation;
- land representing a long-term natural resource such as heath, woodland and wetland (ecological habitat of low commercial value without 'improvement'), or rock, gravel, sand and clay (with potential use as building materials but where extraction is not currently permitted).

These types of land *can* occasionally be traded but the market is much less dynamic than that for residential property. Demand and prices are constrained by *environmental* limitations on land use.

In England and Wales land to be acquired for public works is assessed by the District Valuer (who also represents the taxation authority). He applies commercial valuation principles in fixing compensation for loss of property or access. In the case of exceptional historic buildings he may assess the 'reasonable cost of reinstatement' but is not likely to apply any such valuation or to assign enhanced property rights to undeveloped land.

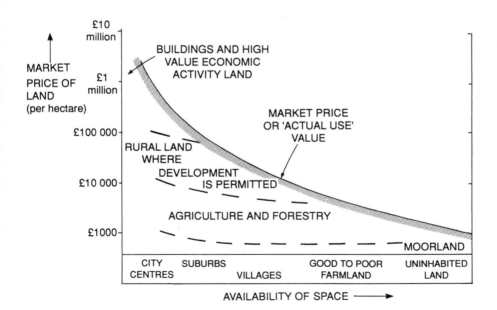

Figure 16.1. Typical land price levels

Figure 16.1 shows the wide range of market values. Land that is built on or available for building may be priced 100 or more times higher than any type of open space in the same locality. Valuation based on market conditions is not therefore a fair measure of the worth of natural land resources to future generations.

16.3 LAND RESOURCE VALUES

Various terms can be used to describe comprehensive values that reflect all the worth—capital and natural, direct and indirect, current and future—which is associated with an asset. Pearce et al. (1989: 60–62) discuss *total economic value* (TEV) which is explained in the equation

TEV = Actual use value + Option value + Existence value.

Actual use value is for now and the foreseeable future. Freehold property market prices theoretically represent actual use values; leasehold prices discount any future value after ownership ceases.

Option value is the value of retaining the option to use the resource in the future. It is thus a non-use value.

Existence value is what people might pay for the preservation of some resource when there is no expectation of directly using or experiencing it, such as a remote wildlife habitat. Economists measure values in terms of people's preferences; existence value may be thought of as arising out of the concepts of intrinsic (i.e. non-monetary) values discussed by ecologists. It is a non-use value.

Actual use, option and existence values are sometimes difficult to differentiate; in these circumstances and to avoid double counting, it may be easier to attempt to estimate total economic values as a whole.

There have been numerous attempts to determine the total economic value of land resources. Three methods used by economists comprise the following:

1. *Hedonic price methods*—these attempt to quantify willingness to pay for improved environmental quality by seeing how the price of a commodity varies with its environmental attributes. A common application is analysis of variations in house prices according to whether they possess certain welcome or adverse features, e.g. a scenic view or frequent train noise. Such methods require very comprehensive data collection and, if related to house prices, are subject to a great many variables concerning location, type of property and market circumstances.
2. *Travel cost method* is a means of determining use values by taking travel costs as a proxy market, for instance the costs of taking up opportunities to visit a stately home or national park.
3. *Contingent valuation methods (CVM)*—assessment of existence values

based on responses to survey questions about unmarketed assets in hypothetical markets. Essentially this is establishing how much people are willing to pay for pleasure or assurance, for themselves or their descendants, to ensure that something will continue to exist. One difficulty is to get people to express their true preference in monetary terms.

CVM methods have recently been the subject of rigorous examination (NOAA 1993: 4602–4614). This concluded that CVM can offer useful value indications if:

– the surveys are expertly designed, i.e. respondents are both asked the right questions and made aware of other ways of spending limited funds;
– comprehensive information is provided to enable the respondents to formulate well-considered replies.

Other research in the USA quoted by Pearce established that people *said* they were prepared to pay for the preservation of visibility of the Grand Canyon a charge 66 times as high as that actually to visit it (Brookshire et al. 1985); this crudely indicates that the existence value of the canyon is 66 times its use value.

Pearce et al. (1989: 70) find that CVM is frequently the *only* technique for benefit estimation and that it is applicable to most contexts of environmental policy. It is the only method currently available for estimating existence values.

Recommendations of economists concerning CVM may find favour with government policy makers and project planners. The fact that CVM sometimes indicates relatively high non-use values for natural land and water resources may interest conservationists, if only on the grounds that the means justify the end. As a result it seems possible that methods similar to CVM could be sanctioned or even made statutory on a wide basis in the USA. Nevertheless, it is a matter of some concern to specialists in conservation fields that decisions should be based on popular valuations rather than on expert assessment. Partly to allay this concern, and for whatever type of preference surveys are undertaken, information should be given or valuation criteria provided which reflect the *priorities of experts*. The expertise lies in judgement of the particular fragility or scarcity of land resources (environmental impact assessment), in practicalities of construction affecting these resources (engineering) and in experience of what has happened with similar developments in similar circumstances before (conservation).

A problem in drawing conclusions from CVM survey data is estimation of the numbers in 'user groups', now or in the future, who would be willing to pay the money estimated by the sample respondents. Which people do you ask and how far afield do you enquire? This may be more of an issue for sites whose significance is recognised only locally than for those of established national importance.

In actual railway route selection situations there are certain other approaches to asset evaluation which might be useful in estimating total economic values.

Alternative use valuation of land is estimation of its market value if it were applied to another use. Thus land where building has been banned by an environmental or planning restriction would soon be assessed at a much higher value if planning permission were granted. The value of land should be no less if a railway station is located on it than if houses are built.

Preventive expenditure value is the extra cost of rerouteing a railway, building a retaining wall, etc., so as to avoid, for instance, disturbance of a particular amenity or conservation area. This fudges the issue by assuming the answer the conservationist wants; but as a comparative figure this value has relevance in choosing priorities among environmental action options.

Replacement value is the cost of physically moving an historic structure or creating a new set for badgers. Where replacement of sensitive habitats is concerned there may be high risks involved; weighting factors enhancing the replacement value of the threatened resource could be applied in proportion to the measured failure rates of similar attempts in the past. Replacement value is similar to preventive expenditure in its implications. Both are important in planning mitigation.

16.4 THE VALUE OF CONSERVATION LAND—ASSIGNMENT OF INVIOLABLE CATEGORIES

16.4.1 Valuable Land or Irreplaceable Land?

Conservation groups or agencies are generally unwilling to put a value on any scarce or fragile resource. In their April 1990 submission to SACTRA, the former Nature Conservancy Council (1990) took the view that costing of compensating provision for threatened areas of 'general wildlife value' might be encouraged but that SSSIs are irreplaceable and of such importance that they must be avoided at all cost. The whole purpose of procuring a protective designation is to ensure that the parcel of land concerned will remain inviolate. Loss of a precious nature reserve is deemed absolute not relative. Whether a piece of land is valued at £1 million or £10 million, a developer may still think the price worth paying if it removes an obstacle to a project's commercial objective.

16.4.2 Valuation of Land of General Wildlife Value

Nevertheless, the need for an indicative value arises in the *comparative* assessment of conservation areas which are recognised only locally or are not in the highest national categories. Their protected status may be less stringent or unofficial whilst their commercial land values are usually much too low to be relevant. Consideration of conservation areas as resources—with non-use existence benefits—may reveal the total economic value to be more than that of land development.

For example, land is a scarce resource for both city offices and aquatic wildlife. Where these two are near each other, their value should therefore be closer than market prices indicate. The existence value of a small wetland area which might border the route into the Union Railway's London terminal could probably be rated at least as highly as the actual use value of nearby buildings. The existence of land crossed by the same line at the edge of the SSSI on Rainham Marshes is not necessarily less important than the utility of nearby housing or the Ford motor factory. In remoter regions the abundance of wildlife habitat reduces its existence value per hectare; but nor is there then much actual use value for office, factory or housing space.

In each case, judgement on the quality or comparative value of conservation land should be based on the ecologist's assessment of the valuable characteristics of the resource, its fragility or sustainability and its ranking for preservation.

16.4.3 What Land should be Inviolable?

The highest category of protection in Britain is afforded to any National Trust Land which has been declared 'inalienable'. Inviolable, immune, sacrosanct or 'No Go' are valid synonyms for land which is irreplaceable and where the seriousness of its loss is recognised. Even National Trust inalienable land could be built on if Parliament so decided but the justification would have to be a very vital public need.

English Nature or its other national equivalents seek similar immunity for SSSIs but the number of sites affected by recent road schemes indicates that they do not yet have it; meanwhile, sites designated under EC directives, such as Special Protection Areas (SPAs) which correspond to *some* of Britain's SSSIs, apparently have a firmer legal status.

At an early stage in route corridor planning a check should confirm whether there are any sites that have inviolable status or are recognised areas of *exceptional* conservation potential. Experts should also have an opportunity to apply for a site to be accorded inviolable status if it is recognised as exceptional. For instance, land cannot have National Trust status if it does not belong to the nation; whilst sometimes the true value of some sites, not yet declared SSSIs or SPAs, is not evident until they are threatened. Expert views are also needed as a basis for assigning values to less exceptional but still valuable conservation land.

If designation of levels of protected status is to be effective, then

1. any serious intrusion on inviolable land could only be approved at the highest level and for exceptional reasons;
2. the implications of any minor intrusions on to such land would have to be explored with the agencies seeking protected status in the first place;
3. the significance in *comparative value* terms of intrusion on other land could be realistically determined for different route options and development scenarios.

Items 1 and 2 represent a tighter interpretation of current protective planning; item 3 is seen as a less forceful but clear and commonly needed aid to ecologically-sensitive route selection.

16.5 THE VALUE OF HERITAGE AND AMENITY— PRACTICABLE PRESERVATION

Environmental valuation fell into disrepute for 20 years after it was attempted for the Roskill Third London Airport Inquiry in the 1970s (Adams 1992: 66). Evidently a Norman church was valued at £50 000, a reasonable insurance value at the time; but certainly it was not acceptable to assess the true value of such a building as that of a modern equivalent (which is all an insurance policy would cover).

Nor would any price valuation be acceptable to the guardians of an irreplaceable Ancient Monument if it was not enough to prevent it being lost; and agreement to any valuation would set an unacceptable precedent as to whether *anything* could be regarded as priceless.

On the other hand, less exceptional, more common heritage structures such as Grade II listed buildings or locally designated conservation areas, *may* exist in such quantity as to be comparable in value with commercially priced property. The commercial price might nevertheless be weighted upwards to reflect the scarcity of features which are uncommon or currently irreplaceable.

16.5.1 Historic Buildings and Monuments

Chapter 14 described the possibilities for avoiding, protecting or physically moving historic structures. It also pointed out the difficulties of reproducing the originals.

In the same way as declaring gems of natural habitat sacrosanct, it is appropriate to recognise as inviolable the more massive heritage structures or those exhibiting particularly exquisite craftsmanship.

For lesser structural treasures, including buildings of only local reputation, values could be adjusted by weighting indices, assessed by experts or through official classifications to allow for characteristics and rarity not reflected in market values. Note that market prices may be depressed for some historic buildings because they are expensive to maintain and their protected status inhibits their commercial attraction. In London nearly 1000 listed buildings (19 in Grade I) are threatened by decay and vandalism (Young 1991). Their poor actual use value and high maintenance costs discourage purchasers. Their total economic or heritage resource value might be assessed by adding a weighting factor to the market value of a similar-sized, well-maintained structure to reflect its exceptional features. This weighting could be based on the cost of complete protection *in situ* or removal to another site (preventive expenditure), or on construction and maintenance of a full size replica (replacement cost).

16.5.2 Private 'Heritage' Land

Commercial valuation data may be available for large country estates and smaller 'stately' homes; the total economic value of the historic parks or ecologically fragile land elements will have to be estimated in the same way as for other landscape or nature features.

16.5.3 Town Parks and Sports Grounds

For non-heritage recreational amenities relocation is simpler. However, total economic value of the land should be assessed for replacement space which, especially in an urban area, may be that of permitted building land. The full cost of satisfactory protection or replacement is fairly representative of the land resource value, whether or not such protection or replacement is actually undertaken.

In England many playing fields were provided to local authorities in the 1920s and 1930s for recreational use 'in perpetuity'. After 70 years those authorities' successors may be thinking that they are no longer bound by such guarantees (Shaw 1992); such land should be valued at the full cost of replacing it nearby.

Recreational land on urban fringes or near country villages is less likely to be built on even if planning permission is available. It can be valued accordingly—probably at levels between those of farmland and urban sports grounds but equivalent to an equal replacement locally.

16.6 THE VALUE OF LANDSCAPE—ALTERNATIVE USE OR CONTINGENT VALUATION

Landscape quality relates to areas available *for enjoyment*. It might be valued in terms of scenic resource scarcity: or for its protection or preference status in popular opinion; or by more specific analysis of the severity of road or railway intrusion in comparison with existing situations elsewhere.

'Valued landscapes' was the subject of several letters and a leader in *The*

Times early in 1991. The correspondence was directed at the evaluation of road projects, with particular reference to the crossing of Twyford Down by the M3 near Winchester; but the issues could be relevant also to new railways, for instance near the North Downs in Kent.

The Times leader of 4 January 1991 mentioned calculations 'which may seem almost farcical' of the 'green' price of a hillside likely to be destroyed by a new motorway. The article reported a ministerial decision that the landscape at Twyford Down was not worth saving. It deduced that the value of the landscape to be spoilt was something less than the £90 million (preventive expenditure) cost of an alternative tunnel solution. Presumably no actual valuation of the land as a scenic resource was made. Such valuation is in an area of speculation where most road planners fear to tread. Let us, therefore, rush in and conjecture just how contentious it could be.

Two approaches could be taken—one valuing the land for an alternative use (building houses), the other by a contingent valuation method (providing information for a popular survey on scenic land values).

16.6.1 Alternative Use Valuation

First assume that the true economic value of the land is that with planning permission for house building, say £1 million per hectare (£10 000 m^2). This value represents the opportunity cost of development—more than 100 times the modest agricultural land value. Then, if the width at the top of the road cutting is 100 metres and the total length of visible intrusion along the cutting and beyond is 2000 metres, the economic value of the strip to be taken is £20 million. This is less than the £90 million needed to prevent cutting the hillside; therefore the tunnel is not justified.

People who appreciate scenery will claim that the area *visually affected* is much more than the excavated strip and is actually 1000 metres wide, that the total value is therefore £200 million and that anyway the rarity of the landscape justifies a higher value than that of even the best development land. In response, the road lobby will protest that the land had been grossly overvalued (who would would pay even £20 million to preserve a barren hillside?) and that it is not first class scenery anyway.

Thus the problems lie in agreeing how much landscape is damaged and in valuation of the scenery. Only some of it is scarce grass downland; much of it is bare ploughed fields; the topography of all of it has scenic potential.

16.6.2 Contingent Valuation Method

Professor Pearce's (1991) rejoinder to *The Times* leader, printed two weeks later, protested that economists do not 'price the environment' itself but measure people's preference for it. As we have seen, CVM attempts to obtain people's preference in monetary terms.

Suppose that a sample of people are provided with comprehensive data about the scenery on the proposed route, how it will be affected by the proposed construction and how their wealth or taxes might be used to protect this scenery instead of for other social purposes. Suppose further that the following responses are obtained:

- From among the 100 000 people living in the district, half would be willing to pay an average of £500 each to preserve the scenery of the Down and that the other half would pay nothing—either because they did not care about landscape or because they knew what was behind the question and preferred that the (road or railway) should be built (contingent value £25 million).
- From among 10 million people living in the rest of south-east England, half were willing to pay £20 to preserve a type of scenery with which they had some slight familiarity (contingent value £100 million).
- From the rest of Britain, 10 per cent (say 4 million people) were willing to pay the same £20 to preserve any similar piece of national heritage (contingent value £80 million).

These figures are of course purely hypothetical. However, they illustrate the large number of variables introduced and the care which is required in planning the sampling survey, providing the data and calculating the total contingent value.

16.6.3 Conclusions

It is difficult to anticipate general agreement on either method. However, the principles involved might still be meaningfully employed for *comparative* valuation of *alternative route* options in scenic landscape.

The method of measurement and of comparison of intrusion into landscape should be specified in the scope of environmental assessment. The means of evaluation should then be determined by environmental and transport economists. Whether specialist economists or landscape architects make this evaluation or whether it is left to popular survey, the people concerned must be kept well informed of the technical issues.

This discussion has centred on the impact of new roads on the landscape, citing the example of Twyford Down, where a connection was sought between two separate sections of motorway. The principles are relevant also to railways. Railways have to tunnel through steep hills (and are therefore hidden) because trains cannot climb over them. But across deep valleys, railways need high and very visible viaducts. There is substantial scenic impact to be evaluated.

The main lesson to be learnt about planning through scenic countryside is that route options should be planned in their entirety and not on a piecemeal basis. The controversial choice between an economic and an environmental solution for a fragmented motorway has been seen. The lesser flexibility of railway lines, in terms of curvature and gradients, makes it all the more essential that they be planned *ab initio* over the full length; this planning should take into account the comparative value of land resources along the whole of each route option. In this respect any use of acceptable valuation methods for scenic resources could be complementary to environmental impact assessment.

16.7 A TWO-PART SOLUTION

The foregoing analyses indicate two conclusions for application in railway planning:

1. That monetary valuation should not be applied to the limited number of heritage structures and areas of conservation land which are recognised as inviolable; legislative instruments protecting them should be surmountable only in exceptional cases where it can be proved that routeing railway or other developments through them will achieve a benefit even greater than their loss; the onus for such 'proof' will no doubt fall on politicians who will face very strong arguments.
2. That less sacrosanct land features should be valued in ways better reflecting their resource or total economic value and making them less of a 'cheap' route option. For most open space the resource value is likely to be greater than current market prices and in scarce habitat, historic features or attractive landscape it could be as high or higher than any built environment.

These conclusions suggest the two parts of a solution: assignment of inviolable features and environmental valuation of land resources.

16.7.1 Assignment of Inviolable Features

Land whose intrinsic worth is to be taken as priceless should be identified in the earlier stages of project conception. All recognised conservation features crossed by possible route corridors should be considered and their status should be determined:

- by identifying places which already have virtually inalienable status;
- by assessing, rejecting or accepting and fully justifying the nomination of other features after close consultation with their appropriate guardians.

If any project should subsequently threaten an 'inviolable' feature, this should be clearly stated in any document likely to be used for assessment. If no value for the feature has been assigned in economic evaluation, this fact must be drawn to the attention of decision-makers.

16.7.2 Environmental Valuation of Land Resources

The market price of land represents one end of a range of possible valuations; it is the value usually adopted in transport project evaluation. The other end of the range could be represented by a practical suggestion by von Weizäcker, mentioned by Whitelegg (1993:143). This proposes application of taxes on all environmentally damaging activities. As well as levies on tonnes of waste, gigajoules of energy or cubic metres of polluted water, extra taxes are proposed of 100 ECUs (1988 values) per square metre of ground *newly covered.*

The latter approach would change the economics of land conversion in favour of conservation of existing resources; but by applying a flat rate tax it is sweeping in its assumptions about the similarity of all uncovered space. Certainly there must be a case for setting a minimum value on abundant *unexceptional* land so as to simplify assessment of different but not critically important land resources. However, for land that justifies any special consideration, whether as an opportunity for commercial development or because of its ecological rarity, higher, more appropriate values must be estimated.

Figure 16.2 is a suggestion (as a modification of Figure 16.1) of the range of total economic values that might result from adding option and existence values to actual use (or market) values. In effect there are three categories of land for valuation purposes:

1. Unexceptional land, valued at a minimum rate appropriate for the geographical region concerned and related to the future opportunity of using space for any development, use or planned non-use.
2. Special resource or opportunity land for which the total economic valuation falls above that minimum level and requires informed estimation.
3. Inalienable land which is priceless and therefore excluded from monetary valuation.

Value estimates may be derived using economic approaches such as the contingent valuation method (CVM). If people's estimates are the only way to derive total economic values, then one must ensure that people are given comprehensive information in a manner appropriate to the judgements that have to be made.

Whatever work may be done in the validation of CVM or other approaches they are likely to remain controversial to many decision-makers and their critics; it will always be difficult to agree valuation in terms that are representative of all environmental and commercial interests. However, *com-*

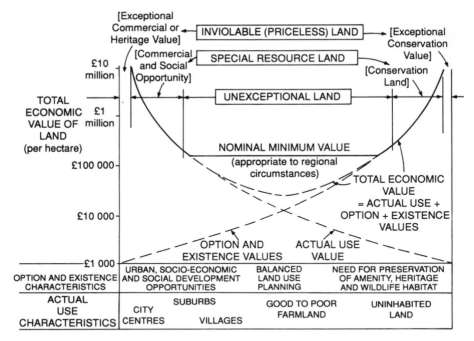

Figure 16.2. Land resource values

parative values—some in environmental rather than economic terms—can be used in planning exercises such as railway route selection. For example, particular features of one threatened recreation ground or the soil class of one field may be demonstrably superior to others in the vicinity. Comparison between different types of conservation land or property or between similar types at different locations along the route are more difficult. However, in either case an attempt at order-of-magnitude valuation may help to determine the most environmentally-sensitive solution.

A methodology for environmental evaluation has to be devised in each case to suit the railway project, the environmental features it will cross and the methods by which decision-makers will evaluate it. Issues to be taken into account will be:

- weighting factors, shadow pricing techniques or data resulting from work already undertaken on comparative land resource values;
- the type of earth science or ecological data which is available, particularly on maps or in geographical information systems;
- possible long-term changes in land use patterns; costs and benefits associated with particular activities or maintenance of special natural or semi-natural conditions;

- lower values which might be applied to land in existing transport corridors or on derelict land—to encourage preservation of virgin land and unspoilt resources elsewhere.

However, there are virtues as well as risks in simplicity. Current methods for assessing total economic values of land involve a formidable amount of

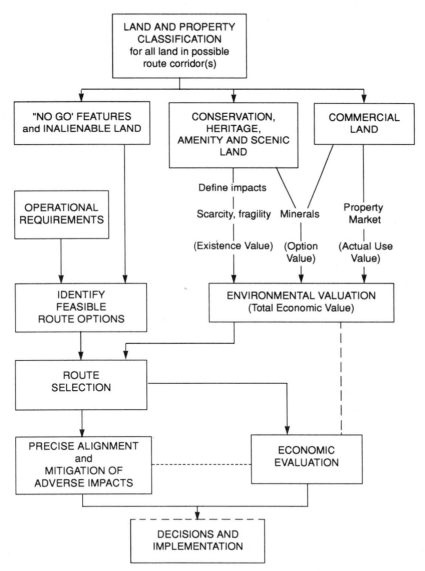

Figure 16.3. Land resource valuation and its application in railway planning

data and data-processing, although the techniques involved may be improving rapidly. Too much data in too many situations becomes unmanageably complex. The use of adjustments or shadow prices based on generalised data may give *comparative* rather than *absolute* measures adequate for evaluation and for decision-making.

Figure 16.3 summarises the steps in environmental land resource valuation and in its possible application for railway planning.

16.8 APPLICATION IN ROUTE SELECTION

The two steps proposed can be applied in initial railway route selection as follows:

1. Avoid inviolable land—unique or rare situations expertly identified for special protection against development. No route which takes such land or seriously damages it should be considered until all the implications of alternative routes have been determined. These implications could be those on train performance, commercial benefit and other environmental issues.
2. Avoid other sensitive land as far as is practicable; but, where this cannot be done, assign environmentally-aware total economic values in comparing the total costs of each route option. In heavily-inhabited country such evaluation could result, for instance, in knocking down more houses but preserving more natural resources; the latter would be more highly valued than by using market land values.

16.9 APPLICATION IN PROJECT EVALUATION AND IMPLEMENTATION

Inclusion of environmental values in formal evaluation of transport projects is contentious and not likely to be accepted quickly.

The purpose of project evaluation is to provide evidence for *decisions* on whether and how to implement railway or other development. Justification can be:

- economic, e.g. for the social benefit of mankind or to sustain resources;
- financial, to ensure that ventures are profitable.

16.9.1 Economic Evaluation

Traditional cost–benefit analysis provides for the early investment costs, such as those of initial construction, to be valued more highly than benefits and operating costs occurring later. This 'time preference' is applied through annual discounting. In economies linked to growth and long-term investment in which, over time, capital grows at a real rate of interest it is difficult to conceive any other approach.

But difficulties in 'discounting the future' arise if scarce land resources, whose value seems likely to *increase* in the future, are to be included in the equation.

One view is that discounting rates should be varied, reduced or abandoned in long-term evaluations. Another view is that proven discounting techniques should be preserved as agreement on alternatives may prove impractical; but that responsible and ingenious people should be able to value all future costs and benefits so as to respect the need for their long-term sustainability.

It is outside the scope of this chapter to conjecture how economic evaluation will develop in the future, which techniques will be used and at what discount rates. It seems likely, however, that modified procedures will have to be adopted to accommodate non-use resources. Economic evaluation does allow weighted or shadow prices to be introduced; these seek to reflect true social or environmental costs and benefits by allocating notional values, for instance to fragile land resources.

16.9.2 Financial Viability

Financial evaluations deal with expenditure and revenue, i.e. real money changing hands. Cash flow forecasts or balance sheets *cannot* include weighted or shadow prices; they *can* include taxes or subsidies applied by governments to correct environmental imbalances.

Whitelegg (1993: 142) proposes that

> If environmental problems are to be tackled by use of fiscal and economic instruments then they need to be embedded in a comprehensive strategy which tackles all sectors (waste, energy, water, agriculture, transport, etc.), do this on a European level and do it in a way which does not increase overall levels of taxation or cause distortions between different sectors of the economy.

Ultimately it is desirable to eliminate differences between financial evaluation, which requires precise expenditure and income calculations, and economic/environmental evaluation which introduces notional costs and benefits (like pollution damage or time-saving). Application of national or international environmental taxes, for instance on energy use or land development, could reduce these differences.

16.9.3 The Significance of Land Resource Valuation

The loss, in building a railway, of either currently valuable commercial assets or of natural resources or heritage must be balanced against the economic benefit, *while it lasts*, of an improved transport system. The total economic value of land may still remain a modest proportion of the total cost of new railway development. The construction cost alone of a modern railway similar to the Channel Tunnel Rail Link can be of the order of £3000 million for 100 km of double track length. If the average width of the land strip required is 20 metres then the total land-take is 200 hectares. Even at a high (development land) value of £1 million per hectare for *all* land, the total land value is still only £200 million, less than 10 per cent of the construction cost.

Railways, however, are not among the most rapacious users of land. In rural areas airports, roads and even some leisure facilities take up more space. The sort of evaluation we have discussed will only work if it is applied by the same methods and criteria to all forms of transport and development in the same region.

16.9.4 Decisions on Projects involving Land-Take

Whitelegg (1993) states that

> ... important issues are not those of monetary valuation but those of political decisions about priorities and tricky distributional issues about who gains and who loses ... based around the incorporation of environmental values into cost benefit analyses or investment appraisal decisions.

Sound decisions about transport development crossing a range of different land resources should be based on information resulting from the following steps:

1. *Allocation* of land areas suitable
   - for agricultural, industrial, commercial or residential development;
   - for road/rail land corridors, related to these development areas as well as to modern transport needs.
2. *Recognition of priorities* for preservation or enhancement of natural and heritage resources.
3. *Expert definition* of the extent and comparative merits of land categories.
4. *Equitable valuation* of these categories.

The requirements for Step 1 can be met by the application, if they exist, of national and local economic planning and environmental policies. Step 2 is dealt with by the first part of the two-part solution—assignment of inviolable or 'No Go' features. These should normally be avoided in route selection and need not then even appear in the final project evaluation process. Step 3 has already been discussed for various types of land in previous chapters and Step 4 has been the subject of this chapter.

Part IV

PLANNING FOR THE
TWENTY-FIRST CENTURY

17 Environmental Rail Transport Solutions

17.1 THE SIGNIFICANT IMPACTS OF RAILWAYS

The opening chapter of this book (Table 1.1) identified the known environmental impacts of railways. The remainder of Part I reviewed environmental appraisal and the operational and economic factors of rail services which affect the environment; Part II analysed impacts affecting people and Part III those on resources.

Any conclusions to be drawn should recognise first that railways are one of the modes of transport which have promoted global economic growth; but transport and other fuel burning activities are now polluting the air, using up the earth's precious resources and perhaps affecting our climate. The place of rail in these global impacts of transport is summarised in Section 17.2 below.

The most significant long-term impacts of new railways result from land acquisition. Land which is taken may comprise residential property whose loss is traumatic for a time but for which financial compensation can be offered; or it may be wildlife habitat where any loss is a permanent reduction of an often scarce natural resource. The effects on land resources are summarised in Section 17.3.

People living in the vicinity of proposed new railways are most concerned about disturbance by trains, particularly their noise, as well as by construction activity. Conclusions about impacts on people are drawn in Section 17.4.

Neither land acquisition nor noise impact is new. But some of the circumstances have changed in the century since most railways were built. Now land is much less plentiful and the population is both larger and, in Western Europe, wealthier. Planners and environmental groups are aware of the scarcity of wildland areas whilst modern home owners may resent the sound, smell and sometimes even sight of any new outside activity.

Section 17.5 summarises how environmental issues can be incorporated in evaluation of railway projects. Any rail development has to be justified as the most appropriate transport solution not only to meet the needs and conditions of our times but also to leave sustainable assets for future generations.

The great flexibility of road transport will ensure that rail never regains its nineteenth-century supremacy. However, a fair basis for economic comparison of the two modes, coupled to environmentally-protective policies concerning use of energy and land resources, could lead to a substantial shift back to rail in those sectors to which it remains suited. Currently these are high speed city-to-city passenger trains, urban rapid transit and long distance freight. Figure 17.1 relates these roles for rail traffic to the main environmental impacts. The last section of this book (Section 17.6) proposes how planning can enable railways to fulfil their economic and environmental potential.

17.2 THE GLOBAL IMPACTS OF TRANSPORT

The impacts of *transport operation* on the world's natural resources concern the following:

1. depletion of non-renewable fuel and possibly mineral resources;
2. air pollution and climatic change which may result from the combustion of those same fuels.

The impact of providing *new transport infrastructure* in hitherto undeveloped areas lies in:

3. opening up virgin land for development, providing economic opportunities but threatening natural resources previously protected by inaccessibility.

17.2.1 Depletion of Fuel and Mineral Resources

Industry, home heating and transport in the developed countries are fast using up the world reserves of petroleum—the fossil fuel most suitable for self-propelled vehicles. An energy crisis, arising from fear of a world oil shortage, was a phenomenon of the 1970s. Since then it has been forgotten in rich countries which continue to depend upon cheap and still plentiful petroleum.

Current and potential world energy shortages lie respectively in two very different forms of fuel:

- wood, which is the main but seriously depleted cooking fuel of the poor, particularly in the remoter areas of the world;
- oil and natural gas, relatively easily extracted and, as long as they remain plentiful, often the cheapest fuels.

The poor badly need oil derivatives like kerosene but cannot afford them; for the rich oil is so cheap that they have little incentive to conserve it or to seek alternatives. Meanwhile, the inhabitants of industrialised and transport conscious cities of the ex-Soviet bloc and the Third World make use of packed trains and buses; whilst they have barely started to acquire the proportion of private transport which has been regarded as analogous to prosperity in the West.

Coal fired the industrial revolution and is still relatively plentiful in many of the countries affected by that revolution. But cheap liquid and gaseous hydrocarbons in any free market economy make coal uncompetitive, even for generating electricity. Coal cannot directly fuel transport vehicles, except in heavy boiler-carrying steam locomotives. In less oil-dependent but still centrally planned economies of countries like China, steam traction is likely to remain viable for some years to come; market forces are causing its demise more quickly in Eastern Europe and South India.

Coal can be converted to liquid fuel; so can bituminous shales, which are plentiful where they occur. However, economic use of these processes awaits technological development and the higher energy prices which will enable them eventually to compete with petroleum products as fuels for transport. Some semi-liquid fuels like orimulsion are already cheap and plentiful; their acceptability depends on reduction at electricity generating stations of the emissions from their combustion.

Coal's major contribution to transport is also as fuel for generating electricity. Trains can be electric; most road vehicles cannot, although no doubt many will run off electric batteries later in the next century.

Thus railways can be powered by the relatively abundant fuels which can produce electricity. Where railways use diesel traction, they are joining in the depletion of petroleum resources; where they use electrical energy derived from coal or nuclear fuels there is less serious depletion of more abundant reserves. As ideal vehicles for electric power, railways offer a sustainable transport method; but only if that electricity is not itself generated from scarce energy resources.

Needs for minerals in railway construction are relatively modest. The use of metals in the manufacture of railway track and equipment compares favourably with that on roads, particularly because of the relatively long life and intensive use of rolling stock; less construction material is used in building railways because of their narrow width. Only in the use of hard crushed rock ballast do railways have a special requirement—between 1000 and 3000 tonnes per kilometre for new track and an annual requirement of about 100 tonnes per kilometre. Global resources are not significantly depleted provided materials are recycled and exploitation of limited local reserves is subject to suitable controls.

17.2.2 Air Pollution and Climatic Change

Coal combustion pollutes the atmosphere with particulates, SO_2 and NO_x; the first two can be substantially reduced, at a cost, in power station flue systems, the third less effectively during combustion. Fuel oil for electricity generation produces similar emissions of the two gases; orimulsion probably emits more SO_2, less NO_x and a different, finer particulate. Diesel in locomotives and lorries and petrol in cars are much more polluting in terms of CO and hydrocarbons. Natural gas is burnt without causing most of these pollutants but is a valuable, non-renewable resource, wasted in generating electricity. Combustion of all fossil fuels produce CO_2, the main greenhouse gas.

Eventually most electricity may be derived from non-fossil fuel energy sources (as is already the case in France). Therefore electricity is potentially the most sustainable and least polluting form of energy. Meanwhile, measures such as pre-combustion treatment or flue gas desulphurisation can be taken to reduce some of the harmful emissions of coal combustion.

In the late 1980s sudden concern arose about atmospheric pollution (particularly through 'acid rain') and then global warming ('greenhouse gases'). The possible effects cannot yet be accurately forecast but there is sufficient alarm that national governments are making undertakings about limiting power station sulphur emissions and reducing total carbon gases.

Figure 8.1 showed how various emissions from rail and other transport pollute the atmosphere and perhaps promote climatic change. Environmentally-sound improvements in the future can arise from a switch to electric traction (in trains and later light road vehicles), non-fossil fuels or cleaner combustion in power stations and phasing out of CFCs and other particularly damaging chemicals.

Polluting emissions from trains were much higher 40 years ago than they are today, mainly because of steam traction but partly because of the number of short trains, stops and shunting operations. In the 1950s more pollutants were emitted by trains per passenger-km or per tonne-km of freight than by petroleum-driven road vehicles. Today the reverse is broadly true. With the exception in each case of sulphur dioxide, pollution caused per passenger by diesel-hauled train travel is similar to that by buses but lower than that by cars. Diesel-hauled freight trains are in general much less pollutive than road lorries although there is evidence that the largest articulated lorries, when fully loaded, may approach the levels of performance of rail freight.

Electric trains are much less pollutive than all forms of road transport; again sulphur dioxide is an exception because of emissions from coal-fired power stations.

Diesel-powered railways add their emissions to other transport and industrial pollutants; but their relatively high efficiency makes rail an environmentally attractive option where good load factors are assured. Where passenger train services are impractical, for instance in rural districts, buses can carry passengers as efficiently and with no greater pollution. Where rail freight is not viable there is no choice but to accept the greater pollution per tonne-km of road lorries; the use of combined transport (intermodal road/rail) is a satisfactory compromise over long distances.

Where railways still use diesel (or steam) traction, it is usually because longer distances or lighter traffic volumes do not justify the capital cost of electrification. The density of emissions is therefore likely to be low. Nevertheless, standards are being set limiting the emissions of road vehicles and sometimes of diesel locomotives. Application of these standards requires that emissions be measured for compliance and predicted in environmental impact assessments for new lines or new traffic. The very measurement and prediction processes encourage measures to increase efficiency and to reduce emissions.

If electrification is affordable, safe and reliable, then electric trains are environmentally preferable. The responsibility for the method of producing electricity and the implications for acid rain and global warming lie with the generating companies. These companies are subject to national legislation or licences putting into effect international conventions agreeing to reduce CO_2, SO_2 and other emissions. Solutions are available which will both reduce emissions and conserve valuable liquid and gaseous fuel reserves.

17.2.3 Opening up Virgin Land

In well-populated countries the relationships between transport facilities and economic development are complex. Demand for transport infrastructure tends to be based on what suits particular industrial enterprises or on social preference rather than on any proven relationship between transport and economic growth.

However, there is no doubt about the importance of transport and communications in pioneering development on new land. Railways to the Pacific coast opened up the American West for human settlement and exploitation. More recently roads in the Amazon basin or railways in eastern Siberia have had the same effect. But the resources exploited are now scarcer and the need for environmental land use planning more urgent.

In these circumstances the first problem is to control man's exploitation of the resources. Rather than yielding to commercial expediency, long-term economic benefits of opening up virgin country must be balanced against the global environmental and socio-economic implications. If exploitation is justified, then the method of transportation will be one of the features which determine how to control that development in a sustainable way. In this respect railways are focuses for development only at terminals or stations, whereas roads offer opportunities at all points along their corridor.

In the already populated areas of the world, it is local impacts on specific parcels of land which are of critical concern.

17.3 EFFECTS ON LAND RESOURCES

17.3.1 Why is any More Land Needed for Railways?

In most well-populated regions a comprehensive network of railways exists, at least between major towns. Where the volume of total traffic now carried by main line railways is less than it was 50 years ago, these networks have spare capacity. It is therefore immediately questionable as to why any renaissance in rail traffic should require the acquisition of land for new routes.

Certainly across *flat land*, railways run generally in straight lines; they might therefore be upgraded without excessive difficulty to accommodate faster, heavier or wider trains. On straight and level alignments high speed trains require only adequate standards of permanent way and signalling. Heavier axle loads may require stronger under-bridge structures; larger loading gauge dimensions need higher and wider clearances at over-bridges or platforms and in tunnels. Electrification by overhead supply systems also requires high clearance space above the trains. Setting aside these structural considerations, however, higher speeds on straight track can be attained without any need for extra land acquisition; only an increasing density of traffic or a complex mix of fast and slower trains makes extra tracks necessary.

In more *undulating country*, we have seen that limitations to high speed on existing routes are much more evident. For a certain target average or maximum speed, the characteristics of curvature and gradients of a particular line may be so severe that upgrading becomes impracticable; the target can only be achieved by a completely new alignment. Justification of the latter depends on how the economic benefits of higher speed on a new line compare with its construction and environmental costs; foremost among the latter is the resource value of land taken.

Through *mixed topography*, such as across Kent, a high speed alignment necessarily includes:

- sections through hills where only a new alignment can accommodate very fast trains,
- sections across flatter country where straight and level routes already exist but existing track capacity is probably inadequate and a widening of the right-of-way is required to accommodate additional tracks.

On routes where *track capacity is still adequate*, such as much of the East and West Coast Main Lines or the Great Western routes outside the London area, many sections can be raised to 200 or 225 km/h standard without major realignment. On sections where operating restrictions remain, any justification for realigning or duplicating the track will depend upon how much real commercial advantage is gained by reducing journey times.

17.3.2 Alignment for Modern Trains

The average proportion of land needed for a railway is small—about 2 ha/km length or 20 m width. Significant impacts on land resources occur where valuable features are obstacles in the way of projected lines or where the new line cuts through a coherent land unit.

Alignment among obstacles is essentially a matter of choosing which features should be sacrificed in order to save others. Although some modern trains can climb steep gradients or even tilt on curves there remain only slight possibilities for adjustment to the straightness of high speed track. Therefore route planning has to consider all obstacles on as long a section as is defined between necessary fixed end points. The longer the route section and the wider the route corridor, the larger is the number of alignment options that can be considered. A possibility to hold in reserve for avoiding areas of very high preservation value is to reduce the radius of curves and hence the speed at which trains can negotiate them. However, too many such restrictions, requiring frequent braking and acceleration, will soon defeat the objective of high speed alignment.

Disturbance of communities and productive units cut in two by railway lines is something which can be overcome, at a cost and given time. Severance of a small nature conservation area, on the other hand, may threaten its viability as an ecosystem or habitat for valuable species. Railway lines sever wildlife corridors such as hedges; but they also create corridors along cuttings or embankments. Study of existing railways only reinforces the importance of considering them as part of the total land resources and of planning their alignment for the distant future rather than for instant convenience.

17.3.3 Classifying and Comparing Land-take Options

Where completely new railway alignments are required there are usually a number of route options with different implications for both environmental factors and construction cost. Even when only the widening of an existing route is required there is a choice as to which side of the formation should be extended.

There are two main types of *constraints* in acquiring land and converting it to carry roads or railways. The first, which is well understood, is the high cost of displacing active residential or commercial property. The second, which is much more intangible, is the need for conservation of land essential for non-commercial activity—wildlife habitat, amenity and recreation—or for deferred use such as for mineral extraction.

There must be close integration in long-term land use planning for economic, environmental or transport purposes if the most valuable resources are to be preserved for posterity. Railway route planning involves a complex set of choices among different categories of land and assessment of the relative values and sustainability of particular features in each category. The land taken by any particular route option has to be examined in terms of:

- *data* establishing the significant characteristics of each type of land concerned;
- an *inventory* of land resources of each type existing in an unspoilt state, both regionally and in any broad corridor within which routes are being considered.

Methods of classification of land and property should be devised to suit the nature of the particular territory crossed. However, the following categories are broadly suitable for consideration:

1. Urban and built-on land
   - Transport infrastructure: roads, railways and canals together with car parks, stations and airfields account for a wide pattern of covered ground. Railways have a strong preference for straight lines and valley bottoms. Use of existing transport corridors may reduce overall landtake and will prevent disturbance in previously unspoilt areas.
   - Industry: the basis of economic wealth, requiring transport access. Increasingly there is less need to locate factories near the sites of raw materials. Derelict heavy industrial sites offer land for railway as well as lighter industrial development.
   - Housing: usually restricted to areas defined by local government plans. The location of new estates should be related to transport facilities but also to levels of train noise or other disturbance. Existing housing areas are a major obstacle to new railways and to widening schemes.
   - Commerce, services: related to housing and industry. Like housing and most light industries, these services can be allocated to land of low or moderate conservation value.
2. Conservation land
   - Wildlife habitat: a wide range of ecosystems with different characteristics, variety, fragility and scarcity. Specialist public or private agencies keep inventories, varying in their coverage and numeracy. For new investigations seasonal factors may be important.
   - Heritage, recreation and scenic amenity: these include natural and man-made landscapes in biologically or geologically interesting areas. Man-made heritage includes industrial archaeology as well as historic structures. Visual effects may be relevant.
3. Other rural land
   - Areas containing minerals and construction materials: non-renewable but may be used for other purposes before or after exploitation. Eventually quarries and pits need rehabilitation and offer positive redevelop-

ment opportunities in a range from railway tunnel portals at quarry faces to conservation of wildlife in created habitat.

- Agriculture: generally has been scientifically graded by land class on maps. Currently land is plentiful in Britain but more food production may be required in the future. Paved areas or foundations are not easily convertible to agriculture, whilst agricultural land is not returnable to wilderness except over very long periods.
- Forestry: generally a form of long-term agriculture. However, broad-leaved woodland is often a valuable semi-natural heritage resource.

Each of these categories has many sub-categories by which an expert will wish to assess them.

Many inventories and descriptions already exist—for listed buildings. Areas of Outstanding Natural Beauty, Sites of Special Scientific Interest and agricultural land classes; others are more general or less well-defined and there may be gaps or discrepancies in official or less formal records. Often there is a necessity for further definition of standards or criteria whereby assessors can judge comparative land resource values. Actual investigations for a railway route may draw attention to unclassified features which require attention and consideration for protected status.

17.3.4 Planning to Alleviate Land-take Impacts

The following steps illustrate a typical sequence in railway route planning:

1. Identify opportunities to use existing railway capacity and land.
2. Identify features of sacrosanct quality or those which should be given high priority for avoidance by a new line.
3. Identify opportunities to use both existing and newly planned transport corridors effectively.
4. Determine locations for stations.
5. Identify relevant issues in land use planning for posterity; examine the implications of planning policies and guidelines; SACTRA (1992: 53, 10.07) has stressed that transport strategies should be set within wider planning strategies dealing with the *use and management of land* and the state of the environment, at each of the levels from national to local.
6. Selection of broad route corridors or route options in the light of 1 to 5.
7. Preliminary economic and environmental evaluation to confirm 6.
8. Detailed alignment studies involving:
   - identification and comparison of other natural or heritage resources to be avoided as far as practicable;
   - investigation of routes crossing land which has been built on, drained or 'improved' for agricultural use; whilst probably of relatively high current market price, this land may be of less value to posterity than more fragile resources crossed by alternative alignments;

– definition of measures for enhancement and management of land
resources which may be affected but not covered—within proposed
railway boundaries or on adjacent land.

17.4 IMPACTS ON PEOPLE

Social planning has to balance people's necessity or desire to travel against
any nuisance which transport may cause—often to different people.

Most people wish to travel. Many have to commute and some have to
make regular longer journeys; a majority need occasional transport from
home for shopping, recreation, education or social purposes and travel further afield for their holidays. All wish to travel in a pleasant environment
and one which is convenient and cost effective.

In general railway travel is pleasant, quick and convenient; sometimes it is
less than pleasant and quite often it is not cost-effective in comparison with
other available transport. Passenger environment in trains and stations as
well as costs (fares) are therefore considerations in establishing the viability
of railway services.

Many people notice transport operations and some are disturbed by them,
whether it is by road congestion or by the noise of aircraft, lorries or trains.
Occasionally there are claims that enjoyment of landscape, culture or recreation is affected by traffic operations; more often people complain about construction activities such as when new roads or railways are being built.

17.4.1 Environment of Railway Passengers

Safety is one direct indicator of a pleasant travelling environment. Rail
travel is the safest mode. Warner (1991), quoting Royal Society figures, says
that lives lost per billion passenger kilometres travelled are about 0.5 for rail
as against 8 for cars and 160 for motor-cyclists. TEST (1991: 43) quotes
1987 data of 1.7 for rail passengers, 0.4 for buses and 4.1 for cars. Clearly
these statistics use different data; they show cars being 16 or 2½ times as
fatally dangerous as trains; a third ratio (quoted by Howie 1992) is 8. In fact
there are numerous potentially dangerous situations on roads, some at specific locations. On railways, accidents are sufficiently rare and perhaps
random that when they do occur there is considerable publicity.

High degrees of comfort have been achieved on express trains by researched design of rolling stock as well as appropriate alignment of track. But
crowding on suburban and metro trains and stations is still common and,
like road drivers' frustration at traffic congestion, no totally satisfactory
solution is possible until people can work to a more flexible pattern. Maintenance and cleanliness in trains and stations make a marked impression on
people who travel widely. More fundamental on long journeys are the spaciousness and padding of seating and the quality of views from the train;

views are affected by the size and spacing of the windows and by whatever lineside and more distant features the design and alignment of the line allow. Facilities for the business person to work and for people to be able to obtain refreshments and walk about without restriction are very important; all contribute to relaxed arrival at the destination.

17.4.2 Environmental Impacts of Trains

The impacts of trains on people at the lineside are similar to those of machinery or other transport vehicles. The feature particular to trains is their intermittent passing compared with steady streams of road traffic.

Noise and related vibration are the only significant impacts of trains on the environment outside them. Noise can be measured and predicted and is generally louder for diesel than for electric traction. Diesel engine sound is superimposed on the basic wheel/rail sound of rolling stock whilst electric engines are quieter than even their ancillary equipment.

Fewer people live near railway lines than beside roads but most who do seldom notice the passing of trains; they may well prefer what they are familiar with to the uniform, lower pitched but continuous drone of road traffic. However, modern very fast trains appear so suddenly that an unfamiliar person or animal may be disturbed. Freight trains rumble past less dramatically but each incident may be more prolonged. At night, when freight trains can operate on otherwise more crowded mixed traffic lines, the level and duration of their sound may be sufficient to interrupt sleep; for this reason lower overall noise limits are sometimes proposed for night traffic.

The greatest potential for reducing railway noise is at source, i.e. in quieter trains. The noise level of certain services has been reduced by as much as 10 to 20 dB(A) when trains were replaced by more modern stock. It is probable that the technical limits of such improvements are being approached but there remains plenty of scope for applying the technology as stock is replaced. Other reductions can be made by appropriate design and maintenance standards of track and supporting structures.

Standards can be applied to limit the noise emitted by engines and rolling stock or to define the noise level at which insulation or protection should be provided in houses. The cost of providing noise barriers should be justified by a sufficient number of people benefiting and their design should be such as to preserve visual amenity. In heavily populated areas, restricting train speed may be a viable means of noise reduction.

17.4.3 Enjoyment of Landscape, Culture and Recreation

There is no convincing evidence that the *sight* of trains disturbs cultural, recreational or sporting activity.

The effects of railway infrastructure on historic buildings, amenities and landscape have been described in Chapters 14 and 15 in terms of the land resources involved. Scenic quality can only be appreciated by the beholder; but informed people's judgement of landscape is strongly influenced by their environmental awareness—for instance of the value of wildlife resources.

17.4.4 Impacts of Construction

Construction of a railway does not differ fundamentally from that on other extensive building sites. The narrow strip on which the line is built does call for some ingenuity in selection of access routes and plant areas and in construction of structures, particularly where crossings of conservation land are involved.

Serious inconvenience to nearby communities may be offset by temporary measures to protect property or provide access, by direct compensation and by a number of specific measures applied when certain thresholds for noise, traffic or discharge of wastes are likely to be exceeded.

17.5 PROJECT EVALUATION

Comparative evaluation is necessary before a decision can be made on whether a project should be implemented and in which particular form. In railway development the issues concern whether the transport requirements are met, how cost-effective will be the new facilities and how the environment is affected by trains and railway infrastructure.

Evaluation and hence decisions have first to take into account whether a project or each alternative satisfies relevant policies or other imposed constraints. These may place limitations, for instance, on the type of transport or on use of natural resources. The constraints may determine which options offer the best financial, economic or environmental performance.

In *preliminary* project evaluation, the alternatives examined should include a solution using another mode of transport (to prove that rail is the only, or an attractive, option). The case in which no action at all is taken should also be examined. The economic, social and environmental impacts on other transport systems and on natural or productive resources have to be assessed. *Final* project evaluation can then concentrate on those rail options which earlier screening has shown to be preferable and provisionally feasible.

17.5.1 Economic or Financial Evaluation

As we have seen in Chapter 2, monetary evaluation can be undertaken in economic (costs and benefits) or in financial (expenditure and revenue) terms. Even if they are subsidised, railways are usually operated on a commercial basis and can most easily be evaluated in financial terms. Except when tolls are imposed, roads are provided as a public service and are evaluated by economic cost–benefit analysis.

If rail and road alternatives are to be compared, then a common method of estimating the costs of supplying and maintaining infrastructure and the benefits of operating trains and road vehicles should be applied.

Fair comparison can be made by adopting either a road-type (economic) approach or a railway (financial) one, for instance, either by costing the railway transport benefits in economic terms, as are calculated by road transport planners, or by putting monetary values on the use of roads and vehicles, equivalent to railway operating costs or fares.

17.5.2 Environmental Evaluation

Estimates of environmental costs have in fact been quoted for some rail and road projects; but these are usually confined to the construction cost of mitigation measures such as noise barriers, landscaping or special deviations in horizontal or vertical alignment. The implications for less tangible resources which *cannot* be mitigated are seldom quantified.

Environmental issues can be taken into account:

- by rejecting any options which are deemed unacceptable in preliminary evaluation, e.g. that no route severing a certain SSSI is acceptable;
- by dealing with particular local issues as part of the engineering design, e.g. how least to damage the setting adjacent to an historic monument;
- by allotting *comparative* but quantified resource values to different obstacles in the route corridor, e.g. where it is impossible for any alignment to avoid all of several listed buildings and commercial properties.

A range of techniques for assigning values to environmental impacts is available. Monetary values have been assigned to such intangibles as people's leisure time or air pollution; there should be no insuperable difficulty, within defined limitations, in valuing conservation land, heritage features or amenities. However, the most satisfactory means of valuation arises if economic instruments, such as taxation, have already been applied to achieve environmental objectives. If taxes on diminishing fuel resources or land development can bring the comparative cost of running trains or cars or of building railways or roads to a more realistic level, then financial evaluation is immediately more equitable.

17.5.3 Project Evaluation beyond the year 2000

Whatever any planner's or economist's personal preference, it is likely that the actual method of evaluation will have to suit both the issues which are most significant and the decision-making process which will use the evaluation results.

The formal methods of project evaluation used by international banks, governments and transport planners have been established in recent decades to suit the pattern of international and public sector investment which has developed since about 1950. With the increased awareness of environmental issues and with novel methods of financing projects, it will be necessary to adapt these methods and to introduce new ones. It is not possible to predict what techniques will be contrived by economists, environmental planners and other professions not yet established; but it seems likely that changes will encompass:

- rules governing what issues have to be resolved before evaluation is even started;
- ways of differentiating economic and financial costs and of converting one to another;
- inclusion of environmental issues in economic costs and benefits;
- 'discounting the future'—should this continue to be done and at what discount rates?
- changes to suit new methods of financing and discriminatory use of economic instruments;
- the decision-making processes, which are the end result of these calculations and forecasts.

Meanwhile, in *environmental assessment*, the various impacts should be quantified, compared or described in terms best suited to their own characteristics and for a range of evaluation methods.

In *planning new railways or railway services*, features which are soon likely to be important in project evaluation include:

- the relationship between the cost of providing high speed routes and the incremental revenue which time-saving will then generate;
- the resource value of land which has to be crossed.

17.6 RAILWAYS IN FUTURE TRANSPORT PLANNING

17.6.1 Road and Rail Networks

The basic network of railways in Britain is well over 100 years old. The same was the case in much of Western Europe before the advent of very high speed routes in France, Spain, Italy and Germany. The modified and extended networks now being brought into use must be planned to meet the needs of the next 100 years. In Britain much of the old system can be adapted to achieve reasonably high speeds; an exception is the route from London to the Channel Tunnel where the existing lines have neither the capacity for the eventual traffic nor the alignment suitable for high speeds.

Everywhere where fast passenger services are needed a choice has to be made between the capital and environmental costs of a new line and use of a cheaper, probably slower and less revenue-earning service on an upgraded existing line.

The 1900 road network was even older than the railways but has undergone tremendous expansion, first with the introduction of motor vehicles and then with the building of motorways, bypasses and urban highways. The first half of the twentieth century saw two world wars but also establishment of motor transport as the successor to railways in rural areas and in distribution of freight. The second half saw, at least in the richer countries, unprecedented advances in standards of living, cheap petroleum, huge increases in car ownership and a transfer of freight transport to much more flexible and comprehensive road services. Any long journey uses roads which did not exist in the 1950s.

Against this background railway improvements have to be justified as the best long-term solution for a particular transport need. We have seen that the main impact of building railways is in taking the land the railway must cross. However, *new* railways need only be provided where no existing route is suitable. We have also seen that road transport is more wasteful in its use of land resources; it causes more pollution through the exhausts of millions of vehicles which, even if most of them eventually use electricity, are fundamentally less efficient than trains in carrying people and goods.

17.6.2 Railway Opportunities

Allied to the features of modern passenger and freight rail traffic, reviewed in Chapters 4 and 5, these environmental considerations point to certain fields of traffic for which railways are likely to remain well suited through most of the twenty-first century (see Figure 17.1). The traffic opportunities are as follows:

- High speed inter-city passenger traffic, over whatever distances rail can be competitive with or superior to roads and airways in terms of cost, time or convenience.
- Bulk freight, i.e. specific commodities between production areas and points for processing or transhipment, both being served by rail (e.g. coal from mines to power stations).
- Combined transport freight, usually with comparatively long rail sections between road/rail transfer stations.
- Commuter traffic from outer suburbs and metro (shared with buses) within cities.
- Leisure travel in particular circumstances.

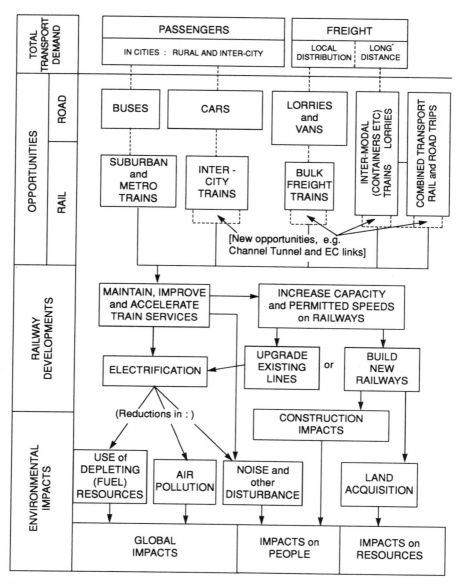

Figure 17.1. Rail opportunities and impacts

17.6.3 Thresholds and Priorities

All these rail opportunities are in services for which demand is shared with roads and in certain cases with air or water transport. There are threshold distances or volumes of traffic *above* which each type of rail traffic becomes more attractive than road. There are a number of ways in which these thresholds, and the preference of travellers or freight forwarders for one mode over another, can be influenced by economic, social or environmental action.

Economic steps can be taken by governments—to improve facilities or subsidise costs for one particular type of transport, or to reduce these facilities and impose charges to discourage certain traffic. Social and environmental goals, affecting use of particular transport modes, can be achieved through integrated transport planning; this may comprise:

- comparative analysis and evaluation, using common criteria, of alternative transport modes for meeting each and every transport need;
- allocation of duties and responsibilities for funding, for instance to provide public transport in towns where cars are restricted or banned; or inter-city rail services where a demand exists or can be created and where net environmental gains will result.

In cities transport is only part, although a most significant part, of integrated planning for all urban activities and land use. The European Community has already indicated the possibilities of cities without cars (Whitelegg 1993: 160). A decline in private car traffic in the centre of many cities seems likely within a decade or two; in such circumstances there is obviously scope for more trains and buses as well as cyclists and pedestrians. Meanwhile, the environmental and any other impacts of urban railways cannot be separated from city planning as a whole.

The problems for public transport are much more difficult in the modern megalopolis—vast, mainly urban, areas but with less densely inhabited, sprawling interstices. In these the abundance of cars and provision of multilane highways has created an environment in which it is necessary to travel comparatively long distances, generally by car, to do anything outside the home. The problems which result cannot be ascribed just to commuting; in Los Angeles 80 per cent of journeys are made for reasons *other* than going to work.

17.6.4 Policies, Taxes and Regulations

Apart from extensive and probably state-subsidised improvements in services by railway operators, only dramatic increases in fuel prices, toll charges or licensing costs of road vehicles can compel a reduction of car and lorry use and a transfer towards more fuel-efficient, less polluting public transport alternatives. If free market forces determine the matter, such increases may not occur for many years. Eventually diminishing fuel resources and environmental pollution could cause sudden and economically disastrous price rises.

Global and national policies must also be devised and implemented to enable transport to pay for using or abusing elements of the environment previously regarded as free. These elements, referred to by economists as 'environmental externalities', comprise land and materials used in construction as well as pollution caused by operation.

In environmental planning, each type of adverse impact can be analysed against the type of policy most likely to reduce it. Button (1990: 67) refers to 'three routes to solving environmental externalities' which have emerged in recent literature:

- assignment of property rights;
- pollution taxes;
- 'command and control' measures.

Assignment of property rights is an evaluation technique to give value to land where no property market exists. Chapter 16 shows how, in certain circumstances, this might be applied to land resources such as wildlife habitat or amenity land.

Pollution taxes can be applied to reduce directly visible (traffic congestion) or well-publicised (air pollution) impacts. Neither of these is particularly relevant to trains but both occur during railway construction. Carbon taxes are a proposed tax which could be levied on energy derived from fuels which pollute the atmosphere or interfere with the global climate. Development taxes could inhibit use of threatened land categories.

Subsidies have been given to railways to encourage their use in many countries. However, in Britain and elsewhere policies favouring market forces and privatisation make reliance on subsidies precarious. Much more satisfactory in the long term would be environmental (or 'ecological') taxation reform. In its application to railway and road transport such reform could affect:

- *trains and road vehicles* by taxes on fuel levied according to the polluting effect of the fuel and on the sustainability of the remaining reserves;
- *new railways and roads* by taxes on the development of land previously not built upon, thereby giving it 'shadow' property rights or scarcity value.

Command and control measures typically set legal standards, such as limitations on engine emissions. These are end-of-pipe measures, seldom of great significance in tackling major problems related to mass transport.

There was more direct interference through the *command* element in the policies of the former Soviet bloc where transport systems were selected partly to absorb the production of manufacturing industry; through *control*, fixing of usually low prices encouraged use of whatever public transport system was provided.

To be effective, *environmental taxes* may have to be high and therefore controversial, for example in compelling a change in popular, but polluting and wasteful, use of petroleum. However, such taxes would not be *extra*. Environmental taxation reform is seen as *replacement* for part of existing taxes on income, property, value added or sales.

17.6.5 Sequence for Planning New Railways or Railway Services

Practicable planning for new railway development should incorporate environmental assessment, total cost/environmental evaluation and project design, using any new techniques that may be developed. Probably the following steps will be involved. All stages of project planning are mentioned but with the intention of highlighting the environmental issues:

1. Analyse and quantify transport need; identify policies, regulations and development programmes that apply.
2. Formulate and assess preliminary solutions including:
   - alternative railway options;
   - other transport options;
   - 'Do nothing' option.
3. Preliminary environmental and economic assessment—for each option and to confirm whether rail is the best solution and which are the favourable options.
4. Feasibility study for railway development:
   - selection of preliminary route options in the light of operational, environmental and construction cost factors;
   - environmental assessment, defining detailed areas for attention;
   - operational and engineering design, including design of environmental mitigation factors;
   - costs estimates; revenue predictions, cf. speed, locations of stations.
5. Evaluation of railway options (economic/financial/environmental)—using accepted techniques.
6. Design of railway and detailed route alignment including design of environmental mitigation and protection and enhancement measures; also plan for railway maintenance including management of related environmental features.
7. Finance/funding plan for project implementation—including arrangements for capital and operating funds and to ensure continuing maintenance of environmental features.

17.6.6 Railways in the Modern Environment

There are three environments in which railways will have to operate in the twenty-first century.

The urban environment is that in which nearly half the world's population lives. A century ago a minority lived in towns; within a few decades most of humanity will do so. There will be urgent needs and challenging opportunities for providing high capacity rail transit services for those cities which are in their infancy or where transport systems are inadequate. The key must be found in city planning.

Well-inhabited but esteemed *countryside* is the jealously guarded heritage of many countries in Europe and elsewhere. The privilege of living or working in or visiting the countryside has resulted in enormous growth in road traffic. Roads are likely to remain the only practicable means of transport for most short distance country journeys or those to the smaller towns and villages. However, for inter-city passenger and freight transport the prospect of continuing trunk road expansion and wider motorways is no longer environmentally acceptable. Most basic elements of the 1900 railway network remain in place; often there is spare capacity although new alignments are sometimes necessary to achieve very high speeds. The next century should see profitable use of that capacity allied to development of new railway lines where this is economically necessary. The key is environmental evaluation.

Spacious land of opportunity is the third environment. Most of the empty areas of Siberia or Canada are climatically inferior and less fertile than more temperate alluvial river basins. Nevertheless, the population density of many countries of East and South Asia is far in excess of that in no less potentially productive regions elsewhere. Immigration and commercial expansion are bound to take place in at least the better favoured of the unfilled spaces. Railways have always been the main form of transport in these regions; indeed their construction has never entirely ceased. The great advantages of the discipline which operation on rail tracks imposes are those of high capacity and control. Control in this case includes determination of where terminals or stations shall be built and thus where development centres should be located, all to the benefit of the overall environment. The key lies in long-term planning of land use and natural resources.

The environmental advantages of railways can only be fully exploited if planning ability and political will can be harnessed in their support. Planning is vital in maintaining and upgrading existing railway infrastructure as well as in operational and environmental design of new routes. Political support is needed both to ensure funding of railway operation and development and to promote policies that will protect future quality of life and sustainability of resources.

References

Adams, J. (1992) 'Horse and rabbit stew', in: Coker, A. and Richards, C. (eds), *Valuing the Environment*, Belhaven, London.

Allaby, M. (1986) *Ecology Facts*, Hamlyn, London.

Anderson, M.A. (1989) 'Opportunities for habitat enhancement in commercial forestry practice', in: Buckley, G.P. (ed.), *Biological Habitat Reconstruction*, Belhaven, London.

Ardill, J. (1987) 'The environmental impact', in: Jones, B. (ed.), *The Tunnel: The Channel and Beyond*, Ellis Horwood, Chichester.

Baedeker, K. (1897) *Belgium and Holland*, Baedeker, Leipzig.

Bennett, A (1991) BR Railfreight contribution to Channel Tunnel Breakthrough Conference—Freight Traffic to Europe, London, February 1991.

Bevilacqua, O.M. (1978) *Transportation Planning and Technology*, quoted by Turton (1992).

Berrin, G. (1992) 'High speed track can be cheap to maintain', *Railway Gazette International*, June.

Binney, M. (1992) 'Fast track into the 21st century', *The Times*, 28 September.

Binney, M. and Pearce, D. (eds) (1985) *Railway Architecture*, Bloomsbury Books, London.

Bovey, E.C. (1986) Paper at Conference on Railway Noise—Prediction, Environmental Effects and Control, London. Report in *The Railway Engineer*.

Bray, J. (1993) *Avoiding the Bottleneck—The Case for London's Rail Freight Bypass*, Movement Transport Consultancy with London Roadwatch, London.

British Rail Channel Tunnel Project (BR CTRL) (1988) *Channel Tunnel Freight Train Speeds*, BR Briefing Paper Freight 6, London.

British Rail Channel Tunnel Project (BR CTRL) (1989), *Noise* (4 pamphlets), BR CTRL, Croydon.

British Railways Board (BR Board) (1988) *Channel Tunnel Rail Services—BR Study Report on Long-term Route and Terminal Capacity*, BR Board, London.

British Railway Board (BR Board) (1989) *International Rail Services for the United Kingdom*, BR Board, London.

BR Railfreight Distribution (1990) Channel Tunnel Freight, Policy Statement, BR.

British Standard
  BS 4142 (1990) *Method for rating industrial noise affecting mixed residential and industrial areas.*
  BS 5228 Part 2 (1984), *Guide to noise control legislation for construction and demolition.*
  BS 5228 Part 4 (1992), *Code of practice for noise and vibration control applicable to piling operations.*
  BS 6472 (1992) *Guide to evaluation of human exposure to vibration in buildings (1 Hz to 80 Hz).*

Brookshire, D., Schulze, W. and Thayer, M. (1985) *Some Unusual Aspects of Valuing a Unique Natural Resource*, University of Wyoming. Quoted in Pearce, D., Markyanda, A. and Barbier, E.B. (1989) *Blueprint for a Green Economy*, Earthscan, London.

Buckley, G.P. (ed.) (1989) *Biological Habitat Reconstruction*, Belhaven, London.

Buckley, G.P. and Knight, D.G. (1989) 'The feasibility of woodland reconstruction', in: Buckley, G.P. (ed.) *Biological Habitat Reconstruction*, Belhaven, London.

Bunce, R.G.H. and Jenkins, N.R. (1989) 'Land potential for habitat reconstruction in Britain', in: Buckley, G.P. (ed.) *Biological Habitat Reconstruction*, Belhaven, London.

Burrows, R. (1971) *The Naturalist in Devon and Cornwall*, David & Charles, Newton Abbot.

Button, K (1990) 'Environmental externalities and transport policy', *Oxford Review of Economic Policy*, **6**(2).

Chambers Encyclopedia (1955), 'Road—middle ages', George Newnes, London, Vol. 11, p. 721.

Chicken, J.C. (1993) private communication.

Coker, A. (1992a) 'Evaluating the environmental impact of coastal protection schemes', Chapters 7 and 9 in: Coker and Richards (eds), *Valuing the Environment*, Belhaven, London.

Coker, A. (1992b) Commentary on Chapter 4 in: Coker and Richards (eds), *Valuing the Environment*, Belhaven, London.

Condry, W. (1987) *Snowdonia*, David & Charles, Newton Abbot.

Construction Industry Research and Information Association (CIRIA) (1984), *Exposure of Construction Workers to Noise*, TN115, CIRIA, London.

Coppin, N. and Richards, I. (1990) *Use of Vegetation in Civil Engineering*, Butterworths/CIRIA, London.

Cox, S.J. (1994) 'Railway noise and vibration—track modelling', paper presented at seminar on Technology and Railway Noise, Pandrol UK Limited.

Davies, H. (1982) *A Walk Along the Tracks*, Arrow Books, London.

Department of the Environment (1992) *Coastal Planning*, Planning Policy Guidance Note No. 20, London.

Department of the Environment (1993) *Planning and Noise*, draft Policy Planning Guidance Note, London.

Department of Transport (1983), *Manual of Environmental Appraisal*, Department of Transport, London.

Department of Transport (1989) *Transport Statistics for Great Britain*, Department of Transport, London.

Department of Transport (1991) *Railway Noise and the Insulation of Dwellings* (also known as 'the Mitchell report'), Department of Transport, London.

Department of Transport (1993a) Design Manual for Roads and Bridges: Volume II *Environmental Assessment*, Department of Transport, London

Department of Transport (1993b) *Calculation of Railway Noise*, draft for consultation, HMSO, London.

Dickens, C. (1848) *Dombey and Son*, Chapman and Hall, London.

DNV Technica (1990) *Railway Noise Standards—Let's Get Them Right*, DNV Technica for Kent County Council.

DIN 4150 Pt 3 (1986), *Structural Vibration in Buildings: Effects on Structures*.

Down, G.S. and Morton, A.J. (1989), 'A case study for whole woodland transplanting', in: Buckley, G.P. (ed.), *Biological Habitat Reconstruction*, Belhaven, London.

Dynes, M. (1990), 'British Rail gives go-ahead for first freight village', *The Times*, 8 December.

Edward-Collins, R.F. (1992) 'Preserving the past', letter to *The Times*, February.

English Nature (1992) *Coastal Zone Conservation*, English Nature, Peterborough.

European Communities (EC) (1985) *Council Directive on the assessment of certain public and private projects on the environment* (85/335/EC), EC, Brussels.

European Communities (EC) (1992) *The Impact of Transport on the Environment* (COM 95), EC, Brussels.

Eurotunnel (1988) *The Channel Tunnel and the Environment*, Eurotunnel, London.

Eurotunnel (1991a) *Can You Reach the Channel Tunnel in 1993?* Eurotunnel, London.

Eurotunnel (1991b) *A Green Perspective*, Eurotunnel, Folkestone.

Eurotunnel (1993) *Beyond a Tunnel*, Eurotunnel, Folkestone.

Farrington, J. (1992) 'Transport, energy and environment', in Hoyle, B.S. and Knowles, R.D. (eds), *Modern Transport Geography*, Belhaven, London, p. 51.

Fields, J.M. and Walker, J.G. (1982) 'The response to railway noise in residential areas in Great Britain', *Journal of Sound and Vibration*, **85**(2).

Financial Times (various dates), London.

Foster, H.D. (ed.) (1993) *Advances in Resource Management*, Belhaven, London.

Fox, B. (1992) 'BR signals disaster for recording studios', *New Scientist*, 14 March, 22.

Freeman Allen, G. (ed.) (1990) *Janes World Railways (1990–91)*, Janes Transport Data, Coulsdon.

Friell, G. (1990) 'Past imperfect'; report on interview in *New Civil Engineer*, 25 October.

Gill, K. (1992) 'BR seeks a saviour for giant of railway heritage', *The Times*, 21 September.

Glasgow Museum of Transport (1992) Scotrail Exhibit.

Green, B.M. (1991) 'Railway route planning', in: Heraty, M.J. (ed.) *Developing Land Transport*, Grosvenor Press International, London.

Guild, A. (1990) 'Putting nature on the right line', *The Times*, 29 December.

Hansford, T.J. (1990) Address to *Port 2000 Conference*, Hull

Hardy, A.E.J. and Jones, R.R.K. (1986) 'Acoustic modelling in the design of railway passenger coaches', paper at Conference on Railway Noise-Prediction, Environmental Effects and Control, London. Report in *The Railway Engineer*.

Hawkes, N. (1991) 'Noting the key to a good sound', *Times Saturday Review*, 23 March.

Hayward, D. (1990) 'Jack be nimble', *New Civil Engineer*, 8 November.

Hemphill, R.W. and Bramley, M.E. (1989) *Protection of River and Canal Banks*, Butterworths/CIRIA, London.

Hemsworth, B. (1986) 'Railway noise propagation in the open and the effect of barriers', Paper at Conference on Railway Noise—Prediction, Environmental Effects and Control, London. Report in *The Railway Engineer*.

Hemsworth, B., (1987) 'Prediction of train noise,' in: Nelson et al. (eds), *Transportation Noise Reference Book*, Butterworth, Sevenoaks.

Hill, D.O. (1848) *Edinburgh Old and New*, painting in the National Gallery of Scotland, Edinburgh.

Hollingsworth, B. (1977) 'Civil engineering', in: Nock, O.S. (ed.), *Encyclopedia of Railways*, Octopus Books, London.

Howie, W. (1992), 'Risk reaction', *New Civil Engineer*, 4 June.

HR (Hydraulics Research) Wallingford (1988), *Assessing the Hydraulic Performance of Environmentally Acceptable Channels*, Wallingford, HR.

Institute of Terrestrial Ecology (1993), *Comparison of Land Cover Definitions*, Natural Environment Research Council, Huntingdon.

Institution of Civil Engineers (ICE) (1990), *Pollution and its Containment*, Thomas Telford, London.

International Railway Journal (1991), 'French to build 4700 km TGV network', *IRJ*, July.

International Railway Journal (1994), 'Swiss Make Tough Investment Decisions', *IRJ*, January.

International Road Federation (1989), *World Road Statistics 1984–88*. Quoted by Turtin, B. in Hoyle, B.S. and Knowles, R.D. (eds) (1992), *Modern Transport Geography*, Belhaven, London.

ISO 2631 Pt 3 (1986) *Structural vibrations in buildings: effects on structures*, International Standards Organisation

ISO 2631 Pt 2 (1989) Continuous and shock-induced vibrations in buildings (1 to 80 Hz), International Standards Organisation

Kershaw, K.R. and McCulloch, A. (1993) 'Environmental and planning issues' in: The Channel Tunnel, Part 2 Terminals, Proceedings 97 Special Issue 2, The Institution of Civil Engineers, London.

Kilvington, (1990) 'Macro-economic view of Europe—Freight market', Rail Engineers Forum meeting, Institution of Civil Engineers, London.

Kusler, L.A., Mitsch, W.J. and Larson, J.S. (1994), 'Wetlands' in Scientific American (50), January.

Leary, J.F. (1990) 'America adopts worn wheel profiles' in *Railway Gazette International*, July.

Leleux, R. (1984) *A Regional History of the Railways of Great Britain, Volume 9— The East Midlands*, David & Charles, Newton Abbot.

London Midland and Scottish Railway (LMS) (1947) *The Track of the Royal Scot*, LMS Route Book No. 3, LMS.

Marrs, C.H. and Gough, M.W. (1989) 'Soil fertility—a potential problem for habitat restoration', in: Buckley, G.P. (ed.), *Biological Habitat Reconstruction*, Belhaven, London.

Mathieu, G. (1991) 'TGV masterplan—SNCF's high speed rail programme', *Rail Engineering International Edition*, 1.

Moore, B. (1990) 'M40 extension—the environmental approach', *Highways and Transportation*, December.

Morgan, N. (1971) *Civil Engineering: Railways*, Longman, London. Quoted in Turnock (1990: 64).

National Oceanic and Atmospheric Administration (NOAA) (1993), *Report of the NOAA Panel on Contingent Valuation, US Federal Register*, **58**(10), 15 January.

Nature Conservancy Council (1990) *The Treatment of Nature Conservation in the Appraisal of Trunk Roads*. Submission to Dept of Transport, Peterborough, NCC.

Nature Conservancy Council (1991), *Earth Science Conservation in Great Britain*, Peterborough, NCC.

Netherlands Government, Ministry of Housing, Physical Planning and the Environment (1988), Environmental Impact Assessment Decree, The Hague.

Netherlands Government (1989), Environmental Protection (General Provisions) Act.

Netherlands Government (1990), Environmental Impact Assessment: The Netherlands—fit for future life.

New Civil Engineer (various dates) References against authors where known.

New Scientist (various dates) References in text against authors where known.

New Scientist (1990) 'Court challenge looms over Thames development', *New Scientist*.

O'Riordan, T. and O'Riordan, J. (1993) 'On evaluating public examination of public projects', in: Foster, H.D. (ed.), *Advances in Resource Management*, Belhaven, London.

Palmer, J.R. (1993) personal communication to Jo Hughes.

Parker, D. (1993) 'Euro leaders back infrastructure spend' *New Civil Engineer*, 16 December, 6.

Pearce, D. (1991) 'Valued landscapes', letter to *The Times*.

Pearce, D., Markyanda, A. and Barbier, E.B. (1989), *Blueprint for a Green Economy*, Earthscan, London.

Pederson, L.D. (1986) Paper at Conference on Railway Noise—Prediction, Environmental Effects and Control, London. Report in *The Railway Engineer*.

Preece, R.A. (1991) *Designs on the Landscape*, Belhaven, London.

Prideaux, J.D.C.A. (1991) 'Dr Prideaux prescribes IC 250', *Railway Gazette International*, April.

Pyrgidis, C. (1993) 'High-speed rail systems and the environment' in *Rail Engineering International Edition*, 1993, No. 4.

Rackham, O. (1986) *The History of the Countryside*, Dent, London.

Railway Age (1993) 'How Amtrak high speed trains will pay off', *Railway Age*, December.

Railway Gazette International (various dates).

Railway Magazine (1966) 'Fifty Years Ago' *Railway Magazine*, September.

Reid, Sir B. (1992) Letter to *The Times*, 29 October.

Remington, P.J., Kurzwell, L.G. and Towers, D.A. (1987) 'Practical examples of train noise and vibration control', in: Nelson et al. (eds), *Transportation Noise Reference Book*, Butterworth, Sevenoaks.

Reynolds, F. (1990) *Environmentalist's Response to the Pearce Report*, CPRE, London.

Richards, G., Green, C.H. and Coker, A. (1992) 'Conclusions', in: Coker, A. and Richards, C. (eds), *Valuing the Environment*, Belhaven, London.

Roumeguere, P. (1991) 'Keeping TGV green', *International Railway Journal*, February.

Runge, W-R. (1989) 'The 'rolling road' as an alternative for heavily-used sections of motorway', *Rail Engineering International*, **2**.

Sargent, C. (1984) *Britain's Railway Vegetation*, Institute of Terrestrial Ecology, Cambridge.

Sarre, P. (1975) *All about BART*, D101MB, The Open University, Milton Keynes.

Semmens, P.W.B. (1990) 'Bend-swinging Italian style', *Railway Magazine*, November, 786.

Semmens, P.W.B. (1991) 'Rifkind signals a new rail approach', *Railway Magazine*, July, 445.

Semmens, P.W.B. (1992) 'Railfreight orders inter-modal wagons', *Railway Magazine*, September, 46.

Semmens, P.W.B. (1994) 'Tools of the Trade', *Railway Magazine*, January, 31.

Seymer, N. (1992) 'Rail inflexibility', letter to *The Times*, 3 January

Shannon, P. (1994) 'German rail freight today', *Railway Magazine*, May, 45.

Shaw, D. (1992) 'Recreational land in need of protection', letter to *The Times*, 27 November.

Smart, J. (1984) 'Common-sense approaches to the construction of species-rich vegetation in urban areas', in: Buckley, G.P. (ed.) *Biological Habitat Reconstruction*, Belhaven, London.

Smith, L. (1985) *Investigating Old Buildings*, Batsford, London.

Standing Advisory Committee on Trunk Road Assessment (SACTRA) (1986) *Urban Road Appraisal*, HMSO, London.

Standing Advisory Committee on Trunk Road Assessment (SACTRA) (1992) *Assessing the Environmental Impact of Road Schemes*, HMSO, London.

Stanworth, C. (1987) 'Sources of railway noise', in: Nelson et al. (eds), *Transportation Noise Reference Book*, Butterworth, Sevenoaks.

Steiniger, P. (1991) 'Safety overkill slays base tunnel' repeat in December 1993 from original article in *Railway Gazette International*, July.

Strahler, A.N. and Strahler, A.H. (1977) *Geography and Man's Environment*, John Wiley, New York.

Sullivan, M. (1989), *How Green is your Railway?*, Council for the Preservation of Rural England, London

TEST (Transport and Environment Studies) (1991) *Wrong Side of the Traeks? Impacts of Road and Rail Transport on the Environment: a Basis for Discussion*, Transport and Environment Studies, London.

The Times (various dates) References against authors where known, London.

Turnock, D. (1990) *Railways in the British Isles*, David & Charles, Newton Abbot.

Turton, B. (1992) 'Urban Transport Patterns' in: Hoyle, B.S. and Knowles, R.D. (eds) *Modern Transport Geography*, Belhaven, London.

Turton, B. and Knowles, R.D. (1992), 'Urban transport problems and solutions', in: Hoyle, B.S. and Knowles, R.D. (eds), *Modern Transport Geography*, Belhaven, London.

Union Railways (1993) *British Railways Board Report*, Union Railways Ltd, Croydon.

Walker, J.W. (1986) 'Community response to railway noise', paper at Conference on Railway Noise—Prediction, Environmental Effects and Control, London. Report in *The Railway Engineer*.

Walker, J.W. (1989) 'Technology tames the noise problem, *Railway Gazette International*, July, 479.

Ward Lock (1897) *North Wales*, Ward Lock, London.

Warner, F. (1991) 'Relative cost of safety', letter to *The Times*, 17 January.

Watkins, L.H. (1981) *Environmental Impact of Roads and Traffic, Essex*, Elsevier Science Publishers (formerly Applied Science Publishers).

Wayson, R.L. and Bowlby, W. (1989), 'Noise and air pollution of high speed rail systems', *Journal of Transportation Engineering*, American Society of Civil Engineers.

Wellner, D. (1991) 'Reflections on a future German railway system', *Railway Engineering International*, 1.

Welsh Office (1992) *Planning and Noise*, Consultation draft for Planning Policy Guidance, Cardiff, Welsh Office.

Westwood, J.N. (1977) 'History—Great Britain', in: Nock, O.S. (ed.), *Encyclopedia of Railways*, Octopus Books, London.

White, H.P. (1986) *Forgotten Railways*, David and Charles, Newton Abbot.

Whitelegg, J. (1993) *Transport for a Sustainable Future*, Belhaven, London

Williams, O.T. (1991) Letter to *New Civil Engineer*, 31 January.

Young, J. (1991) 'Vandalism or neglect stalks scores of listed buildings', *The Times*, 31 January.

Index

Aberdeen area, train services/rail services
    in 20th century 63, 66, 298
access 235, 241, 255, 286
accidents 44, 74
'accommodation works' (bridges/
    crossings) 245
accounting for the environment 24, *see
    also* evaluation: environmental
acid deposition/'rain' 166–170, 180, 347
actual use value *see* valuation: actual use
adhesion, track 41, 70
advanced light transit 81
aerial photograph interpretation (API)
    or remote sensing 220, 272
aerial satellite photography 190, 272
aerodynamic noise *see* noise:
    aerodynamic
aerodynamics and air resistance 41, 70–
    71, 76, 142, *see also* pressure pulses
aeroplanes and airlines 28, 55, 132, 150
Afon Mawddach, Wales 309
aggregates *see* concrete aggregates
Agricultural Land Classification (ALC)
    244, 245
agricultural restoration 321
agriculture 219, 244–246, 351, *see also*
    arable land and crops
agro-forestry 320–321
air 14, *see also* air pollution
air conditioning 73, 74
air emissions 166–179
    dispersion models 181
    global impacts 169
    local impacts 166–168, 181
    long distance 168
    synergistic effects 167–168
air pollution 101–102, 115, 165–182,
    210–211, 345–347
    episodes 168
airport stations *see* stations: airport
air pressures due to passing trains 38,
    73–74
aldehydes 176
alignment 74–75, 236, 296, 307, 314,
    319, 349, 351
    horizontal 33, 34, 231
    route *see* route alignment, corridors:
        route
    vertical 34, 197, 231, 319, 321
allocation of transport demand among
    transport modes *see* modal split
Alps, the 107, 291, 294
Alta Velocidad España (AVE), Spanish
    high speed train 41, 70, 75
alternative use valuation *see* valuation
amenity 270, 271
    land 285, 286, 331
amphibians 256, 263
ancient countryside 247
ancient woodlands 251–252
animals 8, 154, 186, 201, 250, 259
aquatic pollution *see* water: pollution
aquicludes, aquifers 208, *see also*
    ground water
arable land and crops 253, 260, 296
archaeology 271–274, 321
architecture, railway 186, 278–285, 320
Areas of Outstanding Natural Beauty
    (AONBs) 195, 319
Areas of Special Significance for
    Agriculture 245
artists *see* painters
asbestos 167, 210, 244
ashes (cinders) 180, 183, 258, 265
Ashford 243, 288
atmospheric effects, electrical 176–177
atmospheric pollution *see* air pollution
Australia 33, 74, 75, 100
Austrian policy on heavy lorries 93
axle loads 32, 40, 100, 101, 102, 138

badgers 250, 256
Bahn 2000 Swiss rail plan 44
balancing ponds 264, 266
ballast 32, 163, 207, 213, 258, 265, 287,
    345
    tamping 132, 142

barriers, noise *see* noise: barriers
railways as physical 30, 221, 267,
269, 286, *see also* land: severance
baseline conditions 17
Bay Area Rapid Transit (BART), San
Francisco 81, 82, 208
beaches 306, 307
Beattock Bank 75, 252, 294, 295
Belgium 74, 104
Bern or Berne 44
(or UIC) loading gauge 105
Bern–Olten proposed line 123
benzene 167
Big Water of Fleet viaduct 282
Biggins Wood, Kent 252
birds 154, 201, 250, 259, 260, 266, 267
biodiversity 247, 251, 266
Birmingham 157–158, 284
bituminous shales 345
blackthorn 257, 258
Blaenau Ffestiniog 294
blasting and explosions 206, 207, 211
blight, property 241–243
block trains *see* freight: block trains
boat train routes through Kent 297
bog 259, 260
bogies 70, *see also* suspension of rolling
stock
borrow pits 205
boundary features, countryside 251–
253, 255, 256, 301, 302
Boxley, Kent 256
brakes 42, 44, 138, 139
breakeven or threshold distance for rail
freight transport *see* freight: road/
rail breakeven distance
bridges 36, 39, 44, 139, 262, 264, 320,
see also viaducts (long or high
multispan bridges across valleys)
construction of 209
Brighton 284, 288
Bristol (Temple Meads) 284
Parkway station 90, 285
Britain, image/perception of railways
117–119
Britain *see also references under headings
of original railway companies (e.g.
Great Central Railway) or routes
(e.g. Settle and Carlisle Line)*
British parliamentary procedures for new
railway projects 20, 21
British railways and railway network 4,
9, 63, 286

environmental characteristics 9
freight wagon plans for European
services 105
passenger services *see* passenger
environment: train services
traffic statistics *see* traffic and
transport statistics
broad-leaved woodland 251, 252
Brunel, I.K. 227, 228
buildings and building land 219, 238,
325, *see also* historic buildings, listed
buildings
bulk freight *see* freight: bulk
buses 47, 59, 78, 173, 346
business travel 58, 62
butterflies 257, 267

cable haulage 226, 227
Caledonian Railway 210
Cambrian Railway 310
Camley Street, London 263
canals 34, 184, 226, 227
cant or superelevation 33, 34, 35, 72
cant deficiency 33, 40, 72
capability of railways to provide
services 46
carbon dioxide (CO_2) 169–174, 176–
180, 346, 347
carbon fuels 95, 169
carbon monoxide (CO) 167, 171–176,
178–181, 182, 211, 346
carbon taxes 107, 360
carcinogens 167, 175
carrs (wet woodland) 248
car parks 77, 80, 90, 91, 127, 274, 285,
288
cars 78, 79, 173, 174, 359
Castle Hill, Kent 313, 314
cement 95, 211, 213, 298
centrally planned economies 98
centrifugal force 33, 72
chair lifts 291
chalk downland 253, 296, 311
Channel Tunnel 4, 14, 19, 40, 45, 51,
67, 82, 97, 99, 118, 122, 213, 228,
268, 304–305, 309, 312, *see also*
Cheriton terminal
Channel Tunnel Act 13, 18
Channel Tunnel traffic 67, 68
Channel Tunnel Rail Link (CTRL) 9,
12, 13, 57, 76, 82, 90, 117, 118, 119,
123, 124, 163, 213, 230, 233, 236,

242, 243, 248, 256, 274, 280–281, 288, 310, 312, 315, 340, *see also* Union Railway *for post-1993 routes*
Chat Moss 260
chemical waste 165, 210, 211
Cheriton (Folkestone) Channel Tunnel terminal 218, 242, 252, 268, 272, 277, 313, 314
Chicago 80
Chiltern Hills 312
China 84, 95, 98, 174
china clay 166
Chirk viaduct 282
chlorofluorocarbons (CFCs) 169, 346
cliffs 290, 292
    sea 267, 302–305, 309, 310
climate and climatic change 14, 169, 180, 346–347, *see also* micro-climate
civil engineering 34–39
coal, as a fuel 165, 166, 345
    combustion 166, 168, 174–175, 180, 345–346
    mining and land pollution 166
    production and transport 4, 94, 95, 97, 100, 101
coastal defence/protection 222–223, 267, 305
coastal scenery and railways 302–310
coastal zone management 222
collisions 44, 102
combined transport (CT) (intermodal) 47, 57, 94, 100, 103, 106, 108, 357, *see also* containers, demountable wagon bodies, 'piggy-back' transport, 'rolling road', swap-bodies
comfort of passengers 73, 74, 352–353
command-and-control 79, 360–361
commerce 239, 350
commercial development 55, 288
    at stations 88
commercial performance 31
commercial pricing or valuation *see* market prices
common carrier services 95
Common European Market 4
common land status 225
Community of European Railways 67, 74
commuter services, commuting 51, 59, 65, 78, 117, 126, 127, 357, *see also* urban railways

compaction in construction of earthworks 205, 209
    dynamic compaction in foundation preparation 208
compensation 241, 243
competition
    with airlines *see* aeroplanes and airlines
    with roads *see* road: competition with railways
compressed air lines 211
compulsory purchase 241
concrete aggregates 95, 201, 213
concrete in construction 202, 209, 211, 213
concrete 'slab' track 32, 155, 207
conifers, coniferous trees/plantations 219, 246, 252
conservation land 219, 247, 328–330, 350
conservation, nature 247, 248, 295
    on railway land 257–258, 265–266
construction of railways 199–214, 264, 354
    contracts 202
    in urban environments 212, 213
    materials for railway construction 201
    planning 201–204
    plant and machinery 210–212
    transport of construction materials and equipment 97, 213–214
    underground and in tunnels 83, 207, 208
    working space, work sites 209, 212, 213, 245
consultation, public 19, 121–125
containers 93, 96, 98, 100, 101, 106
contaminated ground 210, 244, *see also* land pollution
contingent valuation methods (CVM) 326–328, 332–333, 335
'controls' on power station emissions 176
convenience of rail travel 73
Conway river crossings 208, 283
    Castle and city walls 271, 275
conveyor belts 206, 211
Coquelles Channel Tunnel terminal, France 218–219
corridors, route 243, 349, 351
    transport 80, 105, 230, 232, 233, 301, 315, 316, 340, 350, 351
    wildlife 249, 256, 257, 349

cost-benefit analysis 21, 22, 338–339, 355
Countryside Commission 303
Countryside Survey 220
country transport 58, 62–65
coves 303, 306
Cromarty Firth 267
Cromford and High Peak Railway, Derbyshire 226
crop marks 272
cross-country train services 62, 63
crossing loops 43
Crossrail, London 82, 229
crush in commuter trains 47, 73
crushing rock 211
cultural heritage/activity, culture 14, 117, 270, 271
culverts 221, 263, see also drainage across railways
curvature, curves 33, 40, 75, 103, 138, 231, 236
cut-and-cover construction 38, 78, 81, 82, 123, 208
cut and fill, balanced 36, 206, 221, 321, 322
cuttings 36, 186, 265, 296, 314
    cuttings versus tunnels 37, 38
    false 319
cycleways 287

Darent river valley 280, 297, 298
Dawlish 306, 307, 309
Dawlish Warren 267
dealing with water during construction 209
decibels (dB) 129
decision-making 17, 19, 22, 23, 324, 336, 338, 340, 356
deer 250, 256
defensive expenditure 159
democratic planning processes 18–21
demolition of historic buildings 278
demountable wagon bodies 96
Denmark 51, 150, 154
Department of the Environment, UK 222, 261
Department of Transport, UK 17, 49, 131, 146, 152, 181, 241
    Manual of Environmental Appraisal (1973) 11 (superseded in 1993 by the Design Manual for Roads and Bridges Volume 10 (Environmental Design) and

Volume 11 (Environmental Assessment))
derailment 44, 45, 102
derelict land see disused or derelict land
developing (or Third World) countries 56, 77, 84, 116, 345
development land tax 360
diesel engine emissions 174, 175, 176, 180, 346
    fuel 345, 346
    locomotives 175, 182, 346
    multiple units (dmus) 39, 42, 132, 133, 134
direttissima Italian high speed lines 51, 74, 103
disbenefits 12
discounting in cost-benefit analysis 338, 339, 356
disposal of excess excavated material 206, 207, 321
    from Channel Tunnel 304, 309
District Valuer 241, 325
disused or derelict land 254, 350
    railway land 258, 265, 286–288
diversity see biodiversity
Dollands Moor sidings, Folkestone 218, 313
Dover, railways to 303, 311, 312
    White Cliffs of 303, 305, 309
Dovey estuary 309, 310
downland see chalk downland
drainage across railways 205, 221, 234, 262–264
    and derelict land 166
    characteristics 244, 245
    for agricultural/other purposes 219, 222, 254, 259, 260
    structures 36, 262
dump trucks 206, 211
dust 168
dust during construction 205, 206, 207, 209, 211, 214
    from freight train cargo 101, 166

earthshaping see landscaping: hard
earth science conservation see geological strata
earthwork 36, 37, 186, 198, 203, 205–207, 213, 273, 320, 321, 322
East Coast Main Line (ECML) 68, 70, 71, 75–76, 90, 303, 317, see also Selby diversion

East Grinstead, bypass road in old
    railway cutting 287
East Thames Corridor 9, 116, 261, 281
Ebbsfleet Marshes, Kent 261
ecology 5, 247
ecological assessment 248, 249
    of terrestrial habitat 254, 255
ecological reclamation 219
economic development 55–57
economic evaluation see evaluation:
    economic and financial
economic freight haulage distance see
    freight: road/rail break-even
ecosystems (biological communities of
    interacting organisms) 247, 248,
    249, 251, 255, 350
Edinburgh 80, 276, 277
    railway in city, scenery of 316, 317
Edinburgh and Glasgow Railway 317
effects 5
Eiger, Swiss Alps 291
electricity, electrification, electric power
    supply 29, 30, 41, 42, 199, 346, 347
    clearance/headroom 207, 236
    overhead line equipment (OLE) 35,
        39, 187, 283
    substations 142
    third rail 256–257
electricity generation 166, 174, 176, 182,
    183, 345–347
electricity transmission lines see power
    transmission lines
electric locomotives 41
    multiple units (emus) 39, 41, 134,
        136, 138
    trains/traction 29–30, 41, 73, 176,
        177, 180
elevated railways 38, 78, 82, 197, 233
embankments across estuaries 302, 306,
    308
    across valleys 186, 296
    and drainage on low ground 262
    construction of 205
    dual purpose 222
    versus viaducts 37, 186, 256, 263,
        307–308
energy 217
energy sources 29, 30
energy use 41, 48, 61, 71, 170, 173, 175,
    217
engine sheds 210
engineering see civil engineering
English Heritage 275, 280, 284

English Nature 247, 329, see also
    Nature Conservancy Council
enjoyment 270, 331, 353–354
    of train travel 58, 74
environment 5, 247
environmental appraisal 11, 12, 13–14
    timing of 25–27
environmental assessment or
    environmental impact assessment
    (EIA) 14, 17, 18, 19, 25, 26
    audit 13
    externalities 360
    evaluation see evaluation:
        environmental
Environmental Health Officers 202,
    211
environmental impact of freight trains
    101–103
    of new railways 59, 74, 343
    of railways (summaries) 6, 358
    of trains 353
    of urban railways 83
environmental policies 12–13
    pricing see valuation: monetary
    statements (ES) or environmental
        impact statements (EIS) 12, 17,
        19, 20, 21
    taxation reform 360, 361
Environmentally Sensitive Areas (ESAs)
    246
episodes (of emissions in air pollution)
    168
equal perception value (EPV) for
    vibration 162
erosion 236, 264, 267, 292, 296, 303
escarpments 312–315
estuaries 222, 223, 262, 267, 308–309,
    310
Eton College 228
Europe, Central or Eastern 54, 95, 98,
    165, 166, 174
    Continental or Western 54, 57, 95,
        96–98, 101, 105, 120, 174, 318
European Bank for Reconstruction and
    Development 14
European Community/Commission/
    Union 13, 248, 359
    Directive 85/335/EC 9, 14, 15, 16, 18,
        20
European high speed rail network 67,
    310
European (UIC or 'Berne') loading
    gauge 71, 101, 105

European Market, Common 4
Eurostar trains 70, 72, 138, 148, 150,
    197
Eurotrucks 93
Eurotunnel shuttle services 40, 54, 67,
    101
evaluation 323
    economic and financial 21, 337, 338–
        339, 354–356
    of noise impacts 159–160
    environmental 21–25, 323, 355
    of land resources 323–340
    project 22, 338, 354–356
excavation, soft 203, 205, 273, *see also*
    disposal of excess excavated
    material
    archaeological 272
    in rock 206
Exe estuary, Devon 267, 310
existence value *see* valuation: existence
existing c.f. new railway line
    performance 68, 69, 76
explosions *see* blasting and explosions
    in hazardous ground 210
Eynsford, Kent 190, 298

farms and farming *see* agriculture
fauna 14, 154, 201, 247, 250, 259, 263
fenced/enclosed railway land 295
fencing 44, 77, 286
fens, fenland 259, 260, 268
ferries, cross-Channel 54, 67
    for railway wagons 97
fertilisers 102, 253
Ffestiniog hydro-electric scheme 293
Ffestiniog Railway 293, 294, 309
fill (earthwork) 205
financial or commercial performance
    21, 48, 75
financial evaluation *see* evaluation:
    economic and financial
fire 102, 210
fish 259
flat land, railways across 234, 300, 301,
    348
    coastal 305–306
floods and flood protection 36, 45, 222,
    262, 267, 305
flora/plants 14, 201, 247, 249, 259, 263,
    265
flue gas desulphurisation (FGD) 176

Folkestone Channel Tunnel terminal *see*
    Cheriton
Folkestone to Ettinghill Escarpment
    SSSI 268
Folkestone Warren SSSI 38, 267, 268,
    304
footpaths 235, 239, 286, 287, 300
forecasting rail traffic 48–51
forestry 219, 246, 252, 254, 351
formation beneath permanent way 34,
    35
Forth bridges 283
fossil fuels 166, 168, 169, 174, 217,
    344–345
foundations of structures 208–209
    in wetland *see* bridges
four-track lines 43, 104, 233
France 13, 67, 101, 107, 129, 299, 302,
    310 *see also* French Railways,
    TGVs
    planning process in 19, 123
freight 92–109 *see also* combined
    transport
    block trains 95, 99, 100, 106
    bulk 94, 95, 357
    depots/terminals 106
    disadvantages of rail freight 107
freight facility grants 107
freight, intermodal *see* combined
    transport
    long distance 98
    motive power 99–100
    new railways for 93, 104–105
        through Switzerland 53
    proportion sent by rail 96–98
    road/rail breakeven distance 94, 96,
        98, 107
    shared or separate tracks for 104
    through the Channel Tunnel 51, 97,
        99, 102
    track capacity 47
    train brakes 139
    train noise *see* noise: freight trains
    trains 39, 99–101
        environmental impact of 101–103,
            107, 108
    wagon load 95, 96, 101, 106
French Railways (SNCF) *see also* TGVs
    freight 93, 99, 101, 102, 105
    pre-TGV routes 51, 68, 299
    TGV lines (*lignes à grand vitesse*) 36,
        40, 67, 74, 75, 76, 90, 103, 116,
        118, 123, 197, 231, 233

'Frontier Tradition' 299
fuel oil 180, 346
fuel on construction sites 211
fuels, pre-treatment of 176
  prices of 48, 59, 107, 217
fuel storage tanks 182
funicular railways 290
Furka–Oberalp Railway 294

gabions 209, 263
gantries, power line, appearance of 187
gas, natural 176, 180, 344, 346
gaseous emissions from trains 174–179
gauge see loading gauge, track gauge
geographical information systems (GIS)
  220, 336
geological strata and earth science
  conservation 265, 267, 268
geology 38, 83, 250, 272, 295, 296, 303
generated development 56
Geneva 89
Germany 13, 43, 53, 67, 74, 84, 122,
  123, 197, 232, 234, 310
  rail freight in 94, 96, 97, 99, 106, 107
Glasgow 82, 91, 210, 227
global sustainability 8, 48
global warming 169, 177, 347, see also
  'greenhouse effect'
golf courses 219, 271
'goods' 92
goods yards, disused 91, 287
gorges 295, 299, 300, 317
Gornergrat Railway, Switzerland 291
Gotthard Base Tunnel 53
gradients and alignment 34, see also
  alignment: vertical
  for freight trains 103
  for passenger trains 71, 75, 76
  on early railways 226, 227
Grand Union Canal 226
grassland 253, 267
gravel extraction 244
Gravesend West branch 237
grazing 253, 254
Great Britain see Britain
Great Central Railway 37, 191, 228,
  229, 280, 286
Great Eastern Railway 213, 287
Great North of Scotland Railway 63,
  66, 277, 298
Great Western Railway 226, 227, 228,
  282

Green Belt policies 311, 325
'greenhouse effect' 166, 169, 170, 346
ground level, railways hidden below see
  hiding railways
ground water 38, 208, 221, 222, 258,
  259, 262, 277
grout 211

habitat, wildlife 8, 246, 247, 248–249,
  250, 325, 350
  aquatic 209, 259, 262, 263, 264, 266
  creation and reconstruction 249, 257,
    263, 264, 266
  terrestrial 250–258, 287
Haddon Hall, Derbyshire 228
Harlech, North Wales 305, 306
haul roads for construction 213–214
hazardous freight 102
hazardous ground see contaminated
  ground
headlands 303, 307, 310
'headway' 43
health hazards 130, 167–168, 180, 181
heath 254
'heavy' metals 167
hedges and hedgerows 214, 245, 252–
  253, 257, 287, 302, 310, 315, 320
hedonic pricing 326
herbicides/weedkillers 253, 258, 264
heritage 270, 330–331, 350
  centres 117
Heritage Coastline 303, 304
heritage monuments 275, 330
Hertz (Hz) 130
hiding railways in tunnels, cuttings or
  cut-and-cover to meet objections
  38, 123, 124, 221, 228, 234, 261,
  319, 320
Highland Railway 225, 228, 294
high stack emissions 168
high speed railways 74–77
high speed trains and their
  environmental consequences 60,
  69–74, see also Alta Velocidad
  España (AVE), Eurostar, InterCity
  HST (UK), Inter City Express (ICE)
  (Germany), Shinkansen, Talgo trains
hill walking 292
historic buildings 241, 242, 274–278,
  325, 330–331, see also listed
  buildings and structures
  vibration in 161, 242

historic landscapes 251, 289
hoardings around construction sites 213
Holborough Marshes SSSI, Kent 315
Holycombe SSSI, Kent 268
hooters, horns 132, 142
horsepower (hp) *see* traction: power capacity
horses 77, 154, 186
houses and residential property 116, 151–152, 239–243
house prices 242–243, 326, *see also* market prices and noise levels 159
human beings 14, 15 *see also* people: impacts on
hydrocarbons (HC) 167, 171–176, 178–180, 346
hydroelectric power generation 176, 183, 290, 293

impacts 5, 14
impressionists *see* painters
'improved' land 245, 254, 351
inaccessible land 254
inalienable land *see* inviolable land
India 33, 84, 98, 106
industrial archaeology 278, 293, 294 development 108 property 243, 244
industrial revolution 55, 78, 165, 169, 300
industry 243–244, 350
information, environmental 15–17
inhabited countryside 310
inland waterways 54
Inner Thames Marshes SSSI 261
Insh Marshes, Scotland 267
integrated land use planning (including transport) 233, 257
integrated transport planning (including railways) 57, 80, 81, 359
interchange facilities for passengers 80
Inter City Express (ICE) German high speed trains 41, 70, 140, 141, 154
InterCity (IC) High Speed Trains (HSTs) in Britain
HST(IC) 125 40, 41, 70, 135, 136, 138, 140, 151, 153, 307
IC 225 70, 71, 138
IC 250 75
inter-city traffic/services 55, 59, 62, 67–69, 357

intermodal freight transport *see* combined transport
International Nature Reserves (INRs) 259
investigation, archaeological 272
inviolable (or inalienable, irreplaceable or sacrosanct) land 248, 328, 329–330, 334–335, 336, 337, 338, 340, 351
iron and steel production 95
Italy 51, 74, 302

Japan 9, 61, 151 *see also Shinkansen* trains image of railways and national policy 120, 126–128
job creation 116
Jungfraujoch Railway 291
justification, conceptual 12–13

Kazakhstan 79, 108
Kennet river, Kennet and Avon Canal by Great Western Railway 226, 227
Kent 51, 119, 197, 230, 233, 248, 251, 257, 274, 302, 303
Kenya 55
kinematic envelope for tilting trains 71
Korea 74

labour, construction 200
lagoons 266, 307
land acquisition 238
Land Compensation Act (1973) 241
land pollution/dereliction 165, 166, *see also* contaminated ground prices/values 324, 325 reclamation, landfill 206, 309 resources 8, 15, 92, 218–220, 270, 348–352 resource values 245, 326–328, 336, 337, 339–340 severance 54, 238, 241, 245 space clearance in cities 77 space occupied by railways 218, 257–258, *see also* width of railway land trapped or cut off 233, 236 use and land use mapping and land cover 220 use and transport 55, 80 virgin, development on 347

land-forms, active 222, 303, 322
landscape 14, *see also* scenic landscape,
    railways in
Landscape Advisory Committee 11, 283
landscape architects/specialists 318, 322
    valuation 331–334
landscaping, general 192, 193
    hard (modifying earth shapes) 205,
        206, 321, 322
    soft *see* planting
landslips 236, 268
land-take 74, 83, 238–239, 340, *see also*
    land resources
lead 167
leaves on track 42, 320
leisure travel 58, 357
Lelant Water SSSI, Cornwall 267
length of trains 40, 70, 100
level crossings 77, 235
Liberia 100
light rapid transit 81
Lille 67, 68, 116
limestone 295, 299
Lincoln area train services 64, 66
linearity of railway lines 290, 300, 301,
    302, 305, 315, 318, 320
line capacity 43, 44, 47, 99
    metro 81, 83
line equipment, appearance of 187
lignes à grand vitesse see French
    Railways: TGV lines
lineside disturbance 60 *see also* noise:
    annoyance and disturbance
lineside land and vegetation 257, 258,
    265, 295, 320
liquid effluents/wastes 165, 182
listed buildings and structures 275, 280,
    283, 330
littoral drift and offshore currents 222,
    267, 305, 307
Liverpool and Manchester Railway 3,
    227, 260
Llangollen Canal 226, 282
loading gauge 39, 40, 57, 71, 82, 101,
    105, 236
loads (weights) tare and pay 40
Local Nature Reserves 248
locomotives *see* traction, diesel, steam
    locomotives
London 168, 170
    routes to or through 224, 229, 312
London and Birmingham Railway 226,
    227

London and North Western Railway
    229
London airport stations 89
London Docklands Light Railway 151,
    208, 316
London terminus stations 88, 218
    Euston 227, 284
    Kings Cross 88, 263, 288
    Liverpool Street 284
    Marylebone 229
    Paddington 81, 88, 284
    St Pancras 65, 88, 184, 232, 263, 284,
        288
    Victoria 284
    Waterloo 88, 284
London, Tilbury and Southend Railway
    233
London to Greenwich Railway 37
London Transport and London
    Underground 82, 91
Los Angeles 168, 359
Lotschberg tunnel duplication and 'base'
    tunnel, Switzerland 44, 53
Luxembourg 317
Lyon city and station redevelopment
    91, 284

machinery, construction *see* construction
    plant and machinery
Madrid 88
maglev (magnetic levitation) trains 73,
    75, 139, 233
managed retreat 222
management for nature conservation
    249, 250, 253, 254, 255, 258
Manchester Central station 284
Manchester Metrolink 83
Manchester viaducts 283
Mansion House, London 207–208
Marazion, Cornwall 266, 310
market prices/values 127, 235, 239, 241,
    242, 323, 324–326, 330, 331
marshalling yards 95, 106, 184, 218,
    240, 287
marshes 259, 261, 268
mass transit, mass rapid transit 80, 81
material assets 7, 14, 15
material resources in permanent way 32
matrices in environmental evaluation 23
meadows 249, 253 *see also* water:
    meadows
Mediterranean (Riviera) 302, 303, 305

Medway, river   261, 283, 297, 298, 315
methane gas (CH₄)   206, 210, 244
metro railways   81 *see also* urban
      railways
   for cities of one million population
      79
Metropolitan Railway   81, 229
Meuse valley freight route   104
micro-climate   37, 245, 249, 262, 265
Midland Railway   196, 226, 228, 229,
      299
Miller's Dale, Derbyshire   300
Milton Keynes   90
mine railways   104
mineral resources and raw materials
      345
mining   166, 184, 219, 291, 293, 294
Ministry of Agriculture, Fisheries and
      Food (MAFF), UK   244, 246
mini-topographyy   314
Misbourne viaduct   191, 280
Mitchell, Dr CGB—'s committee   152–
      153
mitigation of environmental impacts   18
mixed traffic (passenger and freight)
      operations   9, 103
modal split/allocation of traffic   50
modes of transport   28, 29, 31
Monaco   305
monitoring environmental impact   18
Monsal Dale, Derbyshire   300
moorland   254
Moscow Metro   82, 91
motive power *see* steam locomotives,
      traction
motorway characteristics compared with
      railways   199, 242, 314
   corridors shared with railways   232,
      233
   routes
      M1   11
      M2   233, 283, 315
      M3   314, 332
      M20   233, 298, 315
      M23   314
      M25   90, 191, 280
      M40   124, 256, 257, 277
mountain scenery and mountain
      railways   290–294
mud   205, 206, 214
mudflats   267, 307
multiple unit trains   39, 70
Musée d'Orsay, Paris   284

museums   117
mutagens   167

Narex, Narita airport expresses, Japan
      128
narrow gauge railways   33
National Nature Reserves (NNRs)   248
National Parks   319
National Rivers Authority   260
National Trust   248, 275, 329
Nature Conservancy Council (NCC)
      247, 328, *see also* English Nature
      which, with Scottish and Welsh
      counterparts, replaced NCC
nature conservation   247–269, 295
nature reserves   265–269, 288
navigation of boats   264
'navigators' (navvies)   200
neo-industrial development   56, 57
Netherlands, The   120, 240, 249, 259,
      301
   EIA procedures in   16, 17, 19, 20, 26,
      27, 123
neubaustrecke lines, Germany   44, 75,
      157, 232
Newton Dale, Yorkshire   300
new towns   90, 91
night working during construction   211,
      213
nitrogen oxides (NOₓ)   167–169, 171–
      174, 176, 178–182, 211, 345–346
noise   8, 129–160
   action levels   212
   aerodynamic   73, 139
   air absorption of   145
   ambient levels   142–143
      change in   150, 153
   annoyance and disturbance by   114,
      130, 149, 150
   attenuation (ground, air)   144, 145
   barriers and screens   127, 145, 156–
      157, 187, 198, 283, 353
   construction   206, 207, 211, 212
   dangerous levels of   130, 148
   decibels in the A range of frequencies
      (dB(A))   130
   evaluation of impacts and necessary
      mitigation   158–160
   façade level   145
   freight trains   102, 103, 135, 137, 148
   frequency   130
   geometric speed of   144, 145

habituation to 148
inside stations 84, 157
inside trains 154, 155
insulation in buildings 157–158
level/loudness 129, 130
    equivalent single event level (SEL) 131
    maximum L_{max} 131
    over a period (L_{Aeq}, L_{10}, L_{90}) 131
    typical for trains 140, 141
    typical for transport 132
locomotive or engine 102, 130, 132, 137, 142
measurement 142, 143
mitigation of 155–158
night time 102, 152, 153, 353
passenger trains 72, 73
people's response to 146, 148–150
pantograph (overhead current collection) or 'arcing' noise 137, 157
prediction 143–147
propagation 144
receptor site conditions 145
re-radiated 160, 162
road traffic 132, 150
of shunting 106
standards 150–155
structural 139, 156
tone 130
track joints and condition, effects on 134–135, 138, 139, 155
track maintenance factors 142
underground trains 82, 84
wheel–rail contact/interaction 73, 130, 132, 138, 139, 154, 155
North Downs 297, 298, 311, 312, 313, 314
Northern Viaduct Trust 283
North London Railway 232
North Pole Channel Tunnel train depot 213
North Yorkshire Moors Railway 300
Nozomi trains see Shinkansen
nuclear energy 176, 180, 183
nuclear waste transport 102
nuisance 129, 177
Nullabor Plain, Western Australia 33, 301

objective data 18
offices 153, 243

oil see petroleum
open countryside 253, 254
open space 223, 224
operation of trains 42–45
opportunities for railways in the future 47, 56, 357, 358
option value see valuation: option
orimulsion 345, 346
outdoor activities 271
overhead line equipment (OLE) see electrification
overhead railways see elevated railways
Oxford Canal 226
Oxleas Wood, London 251
Oxted line, Surrey 312, 313
Ozone (O_3)
    atmospheric 168
    stratospheric ('ozone layer') 169

painters, landscape 5, 185, 186, 301, 317
Pakistan 33, 106
pantographs 137
Paris stations
    Charles de Gaulle airport 89, 284
    Gare de Lyon 88
    Gare du Nord 88
    Gare Montparnasse 88, 231
    Quai d'Orsay see Musée d'Orsay
parkland (open grassland with widely spaced trees) 257, 320
parkway and park-and-ride stations 90, 267, 274, 285
parliamentary bills 20
particulate matter 167, 168, 174–180, 182, 211, 345
passenger environment 58, 73–74, 83, 84, 352–353
    traffic see traffic and transport statistics
    train services 61–69
pasture 253
    flood 249
Paxton, Joseph 284
Pendolino Italian tilting trains 72, 73
people, impacts on 5, 8, 113–117, 214
perception see public perception of railways
peripherality 57
permanent way see track
petrol engines 176
petroleum and oil 344–345

photographs 189, 196, 197 *see also*
    aerial photography
photo-montage 190
photosynthesis forming smog 168
'piggy-back' transport 53, 93, 94, 101,
    105, 107
pile driving 208, 211
pioneering development 55–56
pipelines 95, 235
planning for railways 46–51, 361
planning permission, policies,
    restrictions 328, 332, 351, 354,
    359–360, 362
planting trees and vegetation 256, 257,
    320, 321
plants *see* flora
plateways 294
platforms and platform lengths 40, 71,
    83, 85, 87–89, 91
ploughing 253, 254, 272
policies for railway development 12, 13
'polluter pays' principle 214
pollution 165, 166 *see also* air pollution,
    land pollution, water pollution,
    wastes
ponds 263, 266
Pontcysyllte 226
Pont du Gard 283
Portmadoc 'Cob' 308, 309
Portugal 33, 57, 75
post-industrial development 56, 57
power requirements *see* traction: power
    capacity
power stations and pollution *see*
    electricity generation
power transmission lines 42, 187, 235,
    293, 302
    through disused railway tunnel 287
'Preservation of Wilderness' philosophy
    299
preservation of countryside 224
    of historic buildings 273, 330–331
    of historic railway structures 278, 279
pressure pulses 73–74
    of trains crossing in tunnels 76
preventive expenditure valuation *see*
    valuation: preventive expenditure
privacy, visual intrusion into 187
property 238–241, 255 *see also* blight,
    houses
    loss or damage 15, 115, 241, 316
    prices *see* market prices

protection during construction 200
    rights 360
    severance *see* land: severance
protected status 247–248, 275, 328, 351
protest groups 119
public consultation *see* consultation
    inquiry 21, 122
    participation *see* consultation
    perception of railways 117–121

quality assurance in construction 202
quality of life 5, 114, 115
quarries 206, 219, 244, 293, 296, 298,
    314, 315, 350–351
quietness of rural areas 224

rack-and-pinion railways 34, 290
rails 29, 30, 32
rail wear 33, 103, 138
rail/wheel contact *see* noise
Rainham Marshes 261, 329
Ramsar wetland sites 248, 259, 261, 267
Rannoch Moor 254, 260, 295
rapid transit 81, 82
realignment 236
reclamation of land 206, 222, 309
recreation 117, 219, 270, 288, 331
recycling materials 218
regional railway services 62–65
regular interval train services 44
reliability 45
relocation of historic buildings 277, 278
    of recreational amenities 331
    of river channels 263
    of woodland 252, 257
remoteness 223, 224, 285, 295, 302
remote sensing *see* aerial photograph
    interpretation
reopening or revival of closed railways
    59
replacement value *see* valuation:
    replacement
restoration of construction sites 212
retaining walls 37, 39, 233, 240, 277,
    322
Ribblehead Viaduct 184, 280
right-of-way *see* wayleave
river banks and riverine areas 221, 259,
    262, 264
    crossings 36, 221, *see also* bridges,
        watercourses

diversion/relocation 263
road assessment features relevant to
    railways 13, 17, 125, 195, 242
  bridge design 315
  competition with railways 48, 92–94,
    107–109, 357
  congestion and possible relief by
    railways 101, 105, 107, 108
  construction compared with that of
    railways 199
road/rail freight interchange 53, 288,
    see also combined transport
roads and railways in common route
    corridors see motorway corridors
roads, effects of railway development
    on 53, 54, 235
  for railway construction 202, 214
  in historical transport networks 225
  on old railway routes 287
rock excavation see excavation: in rock
'rolling road' 93, 94
rolling stock 39, 72
roll-on/roll-off operations 101
roofs of stations and railway buildings
    284
Roskill Third London Airport Inquiry
    330
Rotterdam Metro 280
route alignment 74, 192, 231–236
route improvement 236–237, 256
route selection 255, 315
  and scenery 319–320
  'broad band' 229–231
  environmental objectives in 230
Royal Fine Arts Commission 196
Royal Gorge, Colorado, USA 299
Ruskin, John 300
Russia 79
  rail freight in 95, 98, 99, 106, 108

sacrosanct land see inviolable land
safety 44, 45, 77, 236, 352
  in construction 202
St Erth to St Ives branch line, Cornwall
    267
salt marsh 259, 267
sand dunes 254, 267, 307
  windblown 45
satellite photography see aerial
    photography
scenery 195, 289

scenic compatibility 318–319
scenic landscape, railways in 184, 188,
    289–322
Scotland, Highlands of 58, 225, 293,
    294, 295, 301
  peripheral situation of 57
  sea coast of 310
  train services in north-east 63, 66
scour around bridge foundations 262
scrapers 205, 206
scrub 254, 268
seaborne traffic 54
seawalls 267, 268, 303, 307
Selby diversion of East Coast Main
    Line 118, 224, 234, 236
Settle and Carlisle line 280
settlement over tunnels 83, 207, 208
severance see land severance
  of roads 53, 54
Seville station 284
shadow prices 230, 339
shafts for access to underground railway
    works 83
Shakespeare Cliff, Dover 304
Shap, Cumbria 71, 75, 294, 295
Sheffield Supertram 83
Sheppey, Isle of 261
Shinkansen, Japanese high speed trains
    70, 74, 121, 128, 137, 155, 156,
    207
  Nozomi version 70, 75
Siberia 55, 104–105, 108, 246, 347
sidings 102, 106, 108, 240, 244, 266
signalling 44, 45, 76
Singapore 80, 82
Single European Act 13
Single European Market 118
single event level (SEL) see noise
single line working 45
single track operations 43
Sites of Nature Conservation Interest
    (SNCI) 248, 261
Sites of Special Scientific Interest
    (SSSIs) 247, 248, 259, 266–268,
    328, 329
slate industry 294
sleepers 32
slope stability 37, 38, 322
Smardale Gill viaduct 283
smog 167, 168
smoke 167, 170, 174, 175
smooth-riding of rolling stock 32, 72

Snowdon Mountain Railway 292–293
Snowdonia National Park 293, 294, 300
snow sheds 198
social impacts 114–115, 200
socio-economic impacts 115–117
soil 14, 37, 244, 250, 320
    marks 272
solar power arrays/farms 301
solitude 293, 295, 301 see also
    remoteness
soot 167
sound (c.f. noise) 129
South African coal traffic 100, 104
South Darenth, Kent 117
    viaduct 280, 281, 298
South Downs 311
south-east England 74–75, 220, 244,
    251, 261, 310–316
South Eastern Railway 228
Soviet Bloc, former 79, 84, 345, 361
Soviet Union, former (USSR) 33, 79,
    84, 87, 98, 100, 108
space requirements see land space
    occupied by railways
Spain 33, 69, 74
Spalding 301–302
Special Areas for Conservation (SACs)
    248
Special Protection Areas (SPAs) 248,
    259, 329
speed of trains 59, 63, 75, 76, 99, 104,
    348
    restrictions 235
spiral tunnelling 227
sport venues beside railways 271, 285
Sri Lanka 302
stability of slopes see slope stability
stability of trains 33
Stamford station 279
standards 18
Standing Advisory Committee on Trunk
    Road Assessment (SACTRA) 13,
    25
stations 84–91, 242, 316
    airport 89–90
    city 87–89, 284, 285
    country 285
    environmental factors 85
    hauptbahnhof 88
    in Japan 127
    roofs 89, 284
    parkway see parkway stations

underground 91
steam engines, stationary 226
steam locomotives 3, 29, 41, 99, 174,
    175, 293
Stephenson, Robert 227
Stevenage 90
stockpiles 206, 209, 211, 212, 321
Stockport 316
Stockton and Darlington Railway 239
Stoke bank, Lincolnshire 75–76
Stokesay Castle, Shropshire 271
Stratford, East London 86, 213, 287,
    288
streams see watercourses
strip banks 252
structure in rivers or wetland 36, 221,
    261, 262 see also bridges
subgrade 32
subsidence over tunnels see settlement
    risks at Selby coalfield 236
subsidies 48, 339, 360
suburban train services 65
suburbs 78
Sudan 45, 100
sulphurous emissions and sulphur
    dioxide (SO₂) 167–169, 171–174,
    176–180, 182
sumps for collection of polluted liquids
    or drainage water 209–210, 264
superelevation see cant
surplus excavated material see disposal
    of excess material
suspension of rolling stock 42, 73, 102,
    103, 138, 139, 163
sustainability 8, 217, 218, 323
swamps 259
swap bodies 96
Sweden 51, 93
Swedish X-2000 tilting trains 72
switch-back reversal 227
Swiss Railways (SBB) Bahn 2000 plans
    44
Switzerland 13, 19, 53, 74, 75, 93, 122,
    123, 291

taiga forest, Siberia 105
Taiwan 74
Talgos 70, 72, 73, 138
Tamar bridges 283
Tan-Zam Railway 56
taxes 48, 218, 335, 339, 355, 359–361

Tay Bridge 44
téléfériques 291
telegraph poles 187
TGVs *see trains à grand vitesse* (TGVs)
Thames river and estuary 37, 261, 297
Thameslink 65, 229
thermal efficiency 175
Third World *see* developing countries
time pollution 59, 60
tilting trains 71, 72, 73
topsoil 205, 212, 258, 321
total economic value (TEV) *see*
    valuation: total economic
tourism 290, 293
track/permanent way 30–34
    capacity *see* line capacity, *see also*
        four-track lines
track-laying 199, 207, 213
track gauge 33, 56
    maintenance noise 142
    slab *see* concrete 'slab' track
    spacing 76, 236
    wear on curves 33, 43
traction 41, 42, 217
    power capacity 3, 41, 71, 99, 100,
        234
traffic and transport statistics 53, 58,
    59, 61, 96, 97
traffic forecasts 48–51
training walls for watercourses 262
train paths 43, 99, 103
trains 28, 39, 40
trains à grand vitesse (TGVs) 40, 41,
    68–72, 75, 88, 137, 138, 140, 141,
    148, 151, 154–155
trains, visual appearance of 185, 256,
    301
trams 47, 78, 81, 318
trans-boundary air pollution 168
transitions to horizontal and vertical
    curves 35
transport characteristics 31
transport corridors *see* corridors
    total planning for cities 79, 80
    total requirements 48, 49
Trans-Siberian Railway 54, 98
travel cost method of valuation 326
trees affected by construction 201, 205
trees, general; *see* woodland
Trent and Mersey Canal 226
'tube' railways 82
tunnels, Alpine 53, 294

as an environmental solution 38, *see
    also* hiding railways
compared with cuttings 37, 38
construction of 207, 208
cut-and-cover *see* cut-and-cover
    construction
design issues 38
diameter of bores 82
speed in 74, 76
'turbo trains' (Class 165/166 dmus) 73,
    133, 135, 153
twenty-foot equivalent (container) units
    (TEUs) 96
two-track operation 43
two-directional line operation 43
Twyford Down, Hampshire 34, 332–
    333
Tyneside Metro/Tyne and Wear PTE
    151, 157

Uganda 55
ultra-violet radiation 169
uncertainty 243
underground construction 207, 208
underground railways 82
undulating country and landscape 298,
    299, 315, 348
unexceptional land 335, 336
Union Railway 9, 51, 88, 116, 121, 148,
    229, 232, 233, 251, 256, 261, 263,
    277, 283, 284, 313, 314, 315, 329,
    see also Channel Tunnel Rail Link
    for pre-1993 routes
United Kingdom (UK) *see* Britain
United Nations Economic Commission
    for Europe (UNECE) 168
United States of America (USA) 61, 74,
    84, 104, 109, 150, 233
    environmental requirements/
        evaluation 15, 16, 24, 260, 327
    rail freight 98, 99, 106, 108
unpaved land 219
'unspoilt' countryside 223, 224, 232
upgrading railways for higher speeds
    76, 236
uplands, main line railways across 294,
    295, 301
urban environment 77, 362
urban railways 38, 78, 81–84
    in landscape 316–318
Utrecht 88, 284

valleys, routes along 296–298
valley scenery 295, 296
valuation, value 323 see also
    contingent valuation methods,
    hedonic pricing, market prices/
    values, travel cost method
  actual use 325, 326, 331, 335, 336
  alternative use 328, 332
  existence 326, 335, 336
  heritage resource 331
  insurance 330
  intrinsic (non-monetary) 326
  monetary 24, 25, 324, 334
  option 326, 336
  preventive expenditure 328, 331
  property 324–326
  replacement 328, 331
  scenic resource 187
  total economic (TEV) 326, 331, 335,
    336, 337, 338
vandalism 242, 331
ventilation of tunnels 38, 82, 207
viaducts 36, 37, 186, 280–283, 298, 315,
    316, 317 see also embankments
    versus viaducts
vibrating structures 139
vibration 83, 129, 130, 160–164, 206,
    207
Vibration Dose Value (VDV) 161, 162
Victoria, Queen 228
view from the train 196–198, 352
  into private property 187
views, description and illustration of
    189–190
village green status 225
visual envelopes 189
  impact appraisal 187–193
  impacts 184–198
  influence, areas of intrusion 189
  intrusion, zones of 189
  obstruction 189, 191, 192, 193, 194
volatile organic compounds (VOCs)
    167, 171–174, 180

wagon load freight see freight: wagon
    load
wagon retarders 106
Wainwright, A. 300
walls, stone 293, 294
Walthamstow marshes 268
Warwick Gardens, South London 213
washouts 45

wastes from construction 206, 210, 211
  from power stations 180, 182, 183
  from trains and operations 182, 183
water 14 see also watercourses, wetland
  dealing with in construction 209, 210
  meadows 259
  pollution 165, 209, 210, 211, 264
  quality 210, 264
  quantity 221, 263
  table see ground water
waterborne transport 54, 93
watercourses 36, 221–222, 258, 259,
    262–264
Watford Tunnel 228
Watkin, Sir Edward 228, 229
wayleave or right-of-way 35, 36, 201
  of roads 225
weather and construction 205
weedkillers see herbicides
Welsh Highland Railway 300
Wennington Marshes, East London
    261
West Coast Main Line (WCML) 68,
    70, 71, 72, 75, 76, 138, 229, 252,
    294–295
West Highland Railway 260, 295
West London Extension Railway 229
Westminster, Statute of 225
West Thurrock Marshes 261
wetland 221, 222, 234, 247, 249, 258–
    264, 296, 307
  of international importance see
    Ramsar wetland sites
  verification (USA) 260
wheels see noise: wheel/track contact
  small diameter 42, 138, 139
widening railways 39
width of railway land 36, 199, 218, 233,
    236
wild flowers 205, 258, 265
wildland 219, 247, 254
wildlife 8, 247, 249–250 see also habitat
windows, train 74
Windsor branch 37, 228
wood fuel 344
Woodhead Tunnel, Old 287
woodland 249, 250–252, 293, 296, 310
  ancient 251–252
  management 219, 251, 254, 257
  secondary 258
World Bank 14, 247, 323
World Health Organisation (WHO)
    167

working hours for construction 202
works, disused locomotive, carriage or
    wagon 279, 287, 288

yards, freight 106
    disused 287–288

York 87, 284, 316

Zoetermeer, The Netherlands 91
zones in land use planning 80
Zurich 89